Safety Culture

Dedication

James dedicates this book to his loving wife, my friend, my lifelong partner of 40+ years, who has always been patient with me in my endeavors to enhance the safety profession and my love of social media. She has always given me the freedom to pursue my dreams.

Nathan dedicates this book to his wife, Bonnie, for the support in completing this endeavor given all the time and attention that had to be diverted from the mainstream of our days.

Safety Culture

An Innovative Leadership Approach

Nathan Crutchfield, CSP

James Roughton, CSP
Six-Sigma Black Belt

ELSEVIER

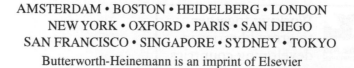

AMSTERDAM • BOSTON • HEIDELBERG • LONDON
NEW YORK • OXFORD • PARIS • SAN DIEGO
SAN FRANCISCO • SINGAPORE • SYDNEY • TOKYO
Butterworth-Heinemann is an imprint of Elsevier

Butterworth-Heinemann is an imprint of Elsevier
The Boulevard, Langford Lane, Kidlington, Oxford, OX5 1GB, UK
225 Wyman Street, Waltham, MA 02451, USA

First published 2014

British Library Cataloguing in Publication Data
A catalogue record for this book is available from the British Library

Library of Congress Cataloguing in Publication Data
A catalog record for this book is available from the Library of Congress

ISBN: 978-0-12-396496-0

For information on all Butterworth-Heinemann publications
visit our website at store.elsevier.com

Printed and bound in United States

13 14 15 16 10 9 8 7 6 5 4 3 2 1

Contents

Foreword

Developing Mission and Intent – Building on the Basics

We have found that while we followed different paths and are from different professional backgrounds (i.e., manufacturing and risk management), we have had similar experiences in a

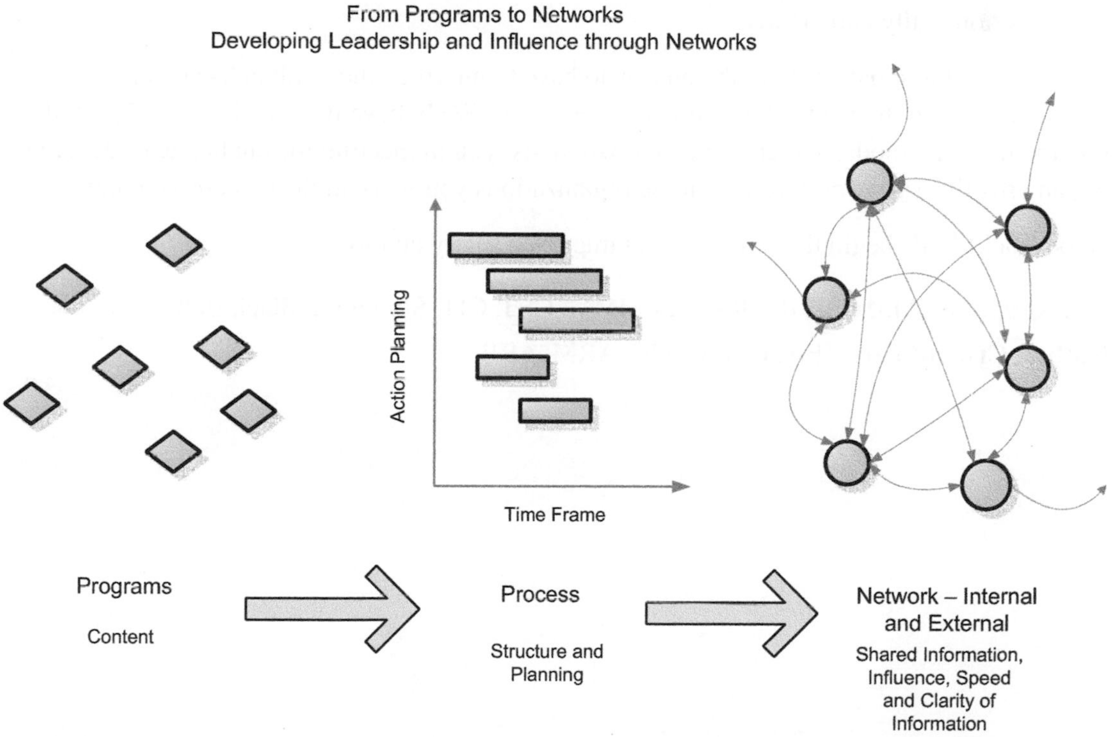

wide range of diverse industries. This difference in perspective has allowed us to create diverse points of view based on our personal histories. In this book, we will provide a hands-on, comprehensive approach to building and enhancing a safety culture. Both of us have lived the experience, from both the inside and outside of organizations. We have seen both the positive and the negative in developing safety management process and program design, administration, and leadership.

In our discussions about concepts for this book, we wanted it to be more than simply a retooling of traditional safety programs in the hopes that an effective safety culture can be achieved by further refinement of safety management systems. We believe a new perspective is needed to bring together a wider array of new tools and techniques used by safety professionals. These areas include a wider understanding of the models of human performance improvement, social networking, and safety management systems that bring the safety process to life.

As we will discuss, safety is the result of the dynamics of an ever changing organizational environment. It is a jungle "out there" and, like a jungle ecosystem, changes can occur overnight and conditions can quickly evolve. The key is that you must keep an open mind, constantly check your mental map, adjust, and be flexible in adapting to the needs of your organization. You must organize your thoughts carefully so that you can present and defend your positions fully and clearly.

As safety professionals, we are the ones who have to interpret and explain hazards and associated risk to the leadership team and employees. We believe that this book will provide you the tools, methods, and concepts that will assist you in meeting the challenge of developing an effective safety culture within the organizations you work in throughout your career.

Good Luck! Let's begin the journey to an improved safety culture.

Jim Roughton, MS, CSP, R-CRSP, R-CHMM, CET, CIT, Six Sigma Black Belt
Nathan Crutchfield, MBA, CSP, CPCU, ARM, ARP

Preface

Part 1 – Laying the Foundation

Chapter 1, The Perception of Safety

The development and sustainment of an organization's safety culture requires a multidisciplinary approach that entails understanding the work environment, perceptions about safety, compliance, basic safety management systems, human error performance, communications, etc. For a safety culture to be developed and sustained, the leadership team and employees must change their perception about safety itself and the management system.

Chapter 2, Analyzing the Organizational Culture

An organization, whether public or private, large or small, must be constantly adjusting to its environment in order to meet a wide range of internal and external pressures and demands. To understand an organization's safety culture, you need to define how you want the safety management system to work within this environment of constant change. To establish a level of stability, a system of desired behaviors and beliefs must be developed that provides consistency in problem solving, decision-making and general relationships, and resilience to loss-producing events.

Chapter 3, Analyzing and Using Your Network

The importance of organizational networking in the development of an effective safety culture cannot be overstated. We are all part of a complex web of personal and professional networks. These networks establish how we send and receive information about ourselves, our environment, its issues, problems, and concerns as well as our successes and values. We use this information to establish how we function, operate, and become a part of organizations, families, communities, and social or professional groups.

Chapter 4, Setting the Direction for the Safety Culture

When trying to improve, develop, or ensure that the safety management system can be sustained, you have to develop a planning process that will move the organization forward.

This planning process ranges from daily, weekly, and monthly activities to annual and multi-year planning. The overall big picture is to ensure that the safety culture is consistently maintained and that your safety management system is effective and efficient in the identification and control of hazards and associated risks.

Part 2, Safety Management Systems Defined

Chapter 5, Overview of Basic Safety Management Systems

When you are in the beginning stage of developing and implementing a safety management system for your organization, communication is the key to success. To get all employees involved at all levels of the organization, the quality and depth of your network is crucial to ensuring your messages get transmitted. Both the leadership team and employers will gain from a shared collaborative effort and the system will be better as a result of everyone's involvement.

Chapter 6, Selecting Your Process

Based on experience, a safety culture requires support and commitment from the entire organization, especially from leadership. While commitment starts with leadership, it is necessary to get employees' involvement to make the safety management system work. Effective leadership is an indicator of how well the organization is working to create a safe work environment.

Chapter 7, Leadership and the Effective Safety Culture

When the word "leader" is heard, it immediately conveys an image of someone who is able to step to the front of a group, clearly communicate what he or she desires the group to do. Leaders inspire the group to do more than what is normally expected, and get the group to achieve what others did not think was possible. Real leaders are rare and leadership is something you know when you see it.

Chapter 8, Getting Your Employees Involved in the Safety Management System

For an effective safety culture, one of the core elements is employee involvement in the safety management system. As part of the leadership team, you do not have to solve all of your safety culture issues alone. Your organization already has great "built-in" resources available that may not be fully utilized as effectively as they should. These resources are your employees! If fully engaged, your employees can be your best problem solvers as they are closest to the real and/or potential hazards and associated risks.

Part 3, How to Handle the Perception of Risk

Chapter 9, Risk Perception—Defining How to Identify Personal Responsibility

An understanding of risk and its control must permeate your organization if an in-depth safety culture is to be sustained. A safety management system using only inspections and observations to identify hazards will not provide a full appreciation of the potential for injury and damage without linking the results to the potential risk.

As safety is an emergent property of all aspects of an organization, without constant focus on the potential changes in the organizations, the potential for loss-producing events may have a way of slipping out of control. Too often, the true scope of hazards and associated risk are only identified after a loss-producing event has occurred.

Chapter 10, Risk Management Principles

Risk management is an essential element of a strong safety culture. Safety management systems such as ANSI Z10-2012 have criteria for a risk assessment to be completed as part of the overall analysis of an organization. The concepts of risk management should be considered an essential part of the leadership team's decision-making. All employees need a basic understanding of the terms risk, risk control, and risk management if the organization's safety culture is to be sustained.

Chapter 11, Developing an Activity-Based Safety System

In the ideal safety culture, the leadership team is constantly communicating and emphasizing its vision, conveying the goals and objectives it believes are required for the organization to be successful. A safety process is best implemented using a systematic approach that focuses efforts on key essential activities to drive improvement of communication, rapid feedback on issues and concerns, and improved coordination of safety activities.

Chapter 12, Developing the Job Hazard Analysis

The job hazard analysis (JHA) is the foundation for any successful safety management system. A safety culture can only exist when a full understanding of ongoing jobs, steps, and tasks are defined and the various hazards and associated risk are managed and controlled. As such, the JHA is an essential element in assessing the depth and scope of risk within the organization.

Part 4, Tools to Enhance Your Safety Management System

Chapter 13, Education and Training—Assessing Safety Training Needs

The safety culture and safety management system are supported by effective comprehensive education and training. The culture of an organization has powerful influence over employee norms, habits, and behaviors as they complete their daily assignments and tasks. The strength of the underlying unconscious perceptions, thoughts, beliefs, etc., of the organization can override the best safety-related training designed to influence behavior.

Chapter 14, Assessing Your Safety Management System

A safety management system assessment provides a comprehensive assessment of the current state of the safety culture. The effectiveness of the safety management system should reflect the real values and beliefs of the organization. A structured and detailed assessment should provide specific details about potential gaps and opportunities for further improvement, not just in the safety management system but in your organization as well.

Chapter 15, Becoming a Curator for the Safety Management System

Throughout this book, a central theme has been the emphasis on the communication and the flow of critical hazard and associated risk information through an organization. The safety culture is dependent on leadership and employees receiving and understanding the importance of maintaining open lines of communication, working toward reducing barriers and resistance to the message being sent.

We hope that you enjoy what we have presented. We wish you all the best in your endeavors for the improvement of your safety culture.

James Roughton
Nathan Crutchfield

About the Authors

James Roughton - MS, CSP, R-CRSP, R-CHMM, CET, CIT, Six Sigma Black Belt.

Mr. Roughton is an accomplished author and manages his own web sites, www.safetycultureplus.com and www.jamesroughton.com. He has received awards for his efforts in safety and was named Project Safe Georgia Safety Professional for 2008 and the Georgia ASSE Chapter Safety Professional of the Year (SPY) 1998–1999. Mr. Roughton is an active member of the Safety Advisory Board of the Departments of Labor/Insurance of Georgia and has been an adjunct instructor for several universities.

Mr. Roughton received his Bachelor of Science degree in Business Administration from Christopher Newport College and his Master in Safety Science degree from Indiana University of Pennsylvania (IUP).

Mr. Roughton has been very active in developing expertise in social media productivity and its use in communication of safety culture and safety management system concepts and information.

Youtube - MrJamesRoughton - http://bit.ly/PTXU0o

http://twitter.com/jamesroughton

http://www.linkedin.com/in/jamesroughtoncsp

Google+ at http://bit.ly/19gSpU1

Nathan Crutchfield - CSP, CPCU, ARM, ARP; Crutchfield Consulting, LLC.

www.myjobhazardanalysis.com; www.crutchfieldconsulting.com

Nathan has a professional history that encompasses a full range of risk control program design, development, implementation, and evaluation. He has provided consulting expertise to a broad array of clients that include public entities, associations, media, newspaper, cable, and general industry. His current interest is the development of effective safety culture and safety management systems, job hazard analysis, and network analysis for safety communications.

He has a Master of Business Administration from Georgia State University and a Bachelor of Civil Engineering Technology from Southern Polytechnic State University. Nathan was awarded the National Safety Council's Distinguished Service to Safety Award in 2001. He serves on the Planning Board for the Georgia Department of Labor/Insurance Annual Safety, Health and Environmental Conference; James and Nathan coauthored, *Job Hazard Analysis: A Guide for Voluntary Compliance and Beyond*; Butterworth Heinemann, 2007

Acknowledgments

For assistance in researching, discussing the concepts, developing, and writing this book, we would like to acknowledge the essential people and organizations that made the endeavor possible:

Many thanks to Dr Mike Waite and William (Bill) Montante, CSP, for their friendship, insights, and discussions on the nature and scope of safety management systems. They have led the way in thoughts on the direction the safety profession should consider pursuing. Bill is a true scientist who has done extensive personal research that combines exciting new approaches that move the definition of safety into a dynamic model. Mike has provided insights from his experience as a safety manager, a consultant to major corporations, and as a professor, providing a multitude of viewpoints and perspectives about the profession of safety.

We would like to acknowledge the following individuals and organizations that were kind enough to provide insights drawn from their materials and permissions to reprint, modify, and/or adapt as necessary. Their input was invaluable.

Presenter Media for the use of their graphics used to emphasize concepts and thoughts throughout the book. We cannot thank them enough for the quality of their products and services. (Presenter Media, n.d.)

David F. Coble, MS, CSP, President of Coble, Taylor and Jones Safety Associates in Cary, North Carolina. (Coble, Taylors, & Jones, n.d.)

Colleen K. Eubanks, CIH, CSP, Palmetto EHS. (Colleen K. Eubanks, n.d.)

Steve Geigle, M.A., CET, CSHM, CET, Senior Safety Education Specialist, former online training program manager with the Oregon OSHA, and former Manager of Health, Safety, and Environmental (HSE) Training for Vestas Americas. (Geigle, n.d.)

Robert Lapidus, CSP, CSMS Safety Management Consultant Lapidus Safety Consulting.

Stephen Newell, Esq. Principal, Mercer ORC Networks. (Newell, 2001)

Alberta Occupational Health and Safety. (Building an Effective Health and Safety Management System, 1989)

Information and Research Services. (Safety Culture Snapshot User Guide, 2010)

Management Study Guide. (Delegation of Authority, n.d.)

tutorialspoint.com. (Six Sigma Measure Phase, Data Collection Plan and Data Collection, n.d.)

National Archives Health and Safety Executive. (Successful Health and Safety Management, hsg65, 1997)

Oregon OSHA. (Safety and Health Management Basics, Module One: Management Commitment, n.d.)

The University of Edinburgh. (Research Data Management Guidance, Permission to reprint, modify, and adapt for use, Creative Commons Attribution 2.5 UK: Scotland License, 2011)

References

Building an Effective Health and Safety Management System. (1989). Partnerships in Health and safety program, Partnerships in Injury Reduction (Partnerships), Reprinted/Modified and/or adapted with permission. Retrieved from http://bit.ly/WsHteC

Coble, D., Taylors, B., & Jones, J. (n.d.). Central safety and Health management System, Adpated with permission. CTJ Safety Associates, LLC, Adpated with permission. Retrieved from http://bit.ly/14Eu85r

Colleen K. Eubanks, C. CIH. (n.d.). How Does Your Company's Safety Culture Rate? Palmetto EHS. Retrieved from http://bit.ly/Xsy1Y2

Delegation of Authority. (n.d.). management Study Guide, Pay Your way to Success, permission to Reprint/Modify/Adapted for use. Retrieved from http://bit.ly/VQe5xg

Geigle, S. J. (n.d.). Introduction to OSH training, Train the Trainer Series., OSHAcademy (TM), Course 703 Study Guide, permission to reprint, modify, and/or adapt as necessary. Retrieved from http://bit.ly/VAZrxt

Newell, S. A. (2001). A new Paradigm for safety and Health Metrics: safety and Health Metrics: Framework, Tools, Applications, Framework, Tools, Applications, and Opportunities, Reprinted with permission. ORC. Presenter Media. (n.d.). Retrieved from http://bit.ly/XLpgGC

Research Data Management Guidance, Permission to reprint, modify, and adapt for use, Creative Commons Attribution 2.5 UK: Scotland License. (2011, August). Edinburgh University Data Library. Retrieved from http://www.ed.ac.uk/is/data-management.

Safety and Health Management Basics, Module One: management Commitment. (n.d.). Oregon Occupational Safety and Health Division (Oregon OSHA), Public Domain, Permission to Reprint, Modify, and/or Adapt as necessary. Retrieved from http://bit.ly/Yztx5p

Safety Culture Snapshot User Guide. (2010). Department of Labor, TE TARI MAHI, Reprinted with Permission.

Six Sigma Measure Phase, Data Collection Plan and Data Collection. (n.d.). tutorialspoint,, Permission to Reprint/Modify/Adapt for Use. Retrieved from http://www.tutorialspoint.com/six_sigma/index.htm

Successful Health and Safety Management, hsg65. (1997). Health and safety Executive, Crown Publishing, permission to Reprint. Retrieved from http://bit.ly/YKDzRk

Introduction

I came to see ... that culture isn't just one aspect of the game – it is the game.

—Lewis, V. Gerstner, Jr., CEO, IBM

Why Understanding Safety Management Systems and Safety Culture Matters

The vision of a total safety culture should be to ensure that the organization has the essential elements in place that make it resilient to hazard and associated risk. Not only must it be resilient to hazards inherent in its operations and routine activities, it must be resilient to the ongoing changes in the organizational business climate. It must maintain a safety process in spite of changes in leadership, management style, personnel, technology, and all aspects of what goes into an organization's survival and success.

A culture can be dynamic, static, evolving, or restrictive. Culture, which we will discuss, is the essence of an organization and determines how it makes decisions, maintains or loses direction and coherency, and meets its objectives. Unless the safety professional understands the safety management system and safety culture, probability for long-term success will be restricted.

One of the initial perceptions that must be overcome is the idea that safety programs are permanent. Their intent may be to incorporate fundamental principles, but the environment in which they are used is ever changing—people change, conditions change, budgets change, management direction changes, etc. At the same time, a culture changes slowly and adjust to these ever changing conditions. If the safety process does not take into account what culture is, safety efforts, while well intentioned, will be resisted just as a body resists a virus.

A multilevel approach must be used to sustain the safety culture.

First, accept the fact that you have been given the responsibility for managing the safety process for your organization or have been tasked with assisting in building a process. Your goal is to get others to buy into your vision of a strong safety culture. It is not asking for "management commitment" per se, it is your taking on a leadership role in this specialized area of the organization—the control of hazards and associated risks.

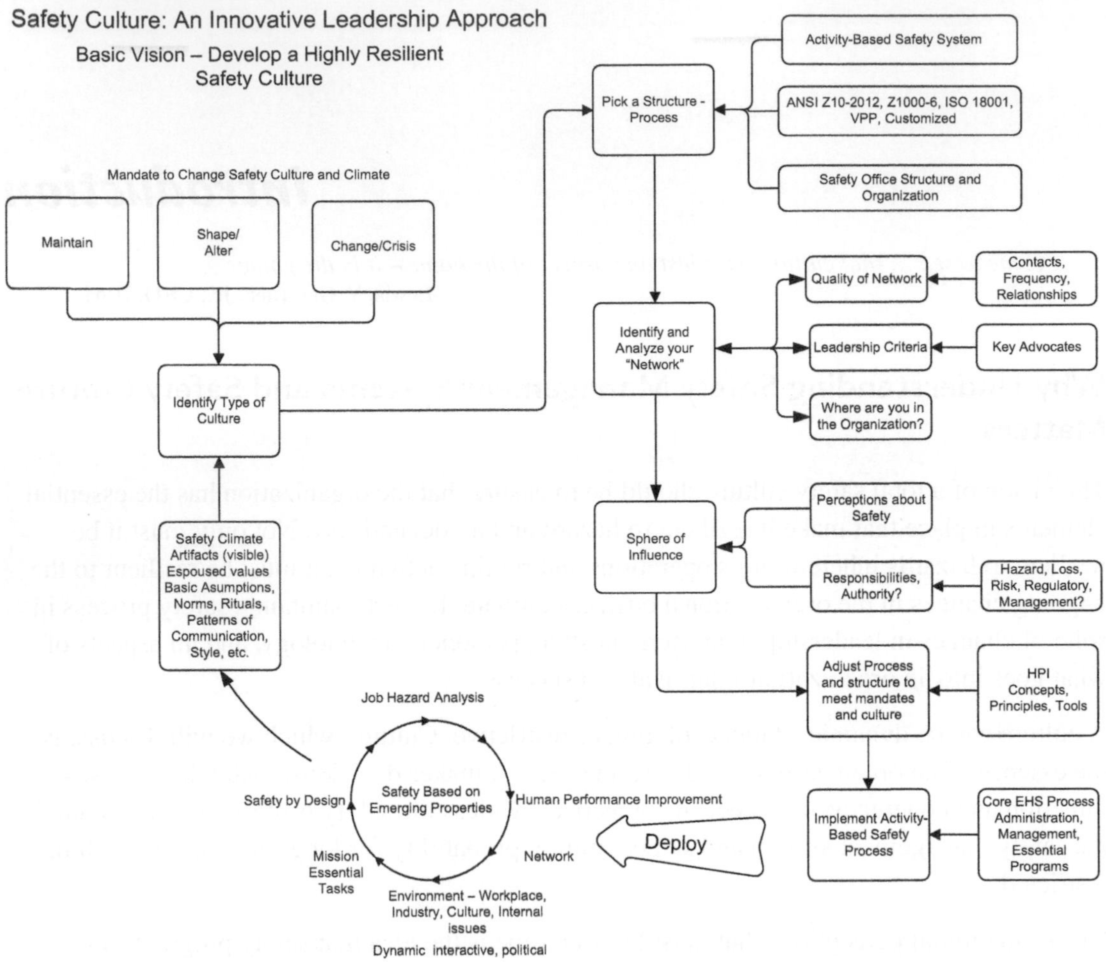

Second, it is building a network of support and communication that extends through the organization. Your network should allow you to quickly respond to issues and concerns. It should recognize the politics found in any human endeavor and work to increase your influence.

Third, you are the corporate memory of the organization. You gather hazard and associated risk data using your network. This data is analyzed and prioritized for corrective actions. Your library, contacts, and information system should place you in the position of being a "go to" professional for any type of issue. This is a continuous effort requiring open and rapid communications.

Fourth, you provide the safety management system structure for the organization. This structure must be complementary to the organization.

PART 1

Laying the Foundation

CHAPTER 1

The Perception of Safety

Change the way you look at things and the things you look at change.
—Wayne W. Dyer

Introduction

The development and sustaining of an organization's safety culture requires a multi-disciplinary approach that entails understanding the work environment, perceptions about safety, compliance, basic safety management systems, human error performance, communications, etc. For a safety culture to be developed and sustained, the leadership team and employees must change their perception about safety itself and the management system.

> Organizations are not static and are constantly in flux and change. Safety management desires stability and wants to build processes that are permanent and unchanging. This sets up an adversarial relationship.

Safety is not simple! It is complicated, as it is a complex network of many business skills and psychological and scientific interactions combined with internal and external resources. Safety management and a safety culture requires following a "Long, Hard, and Winding Road" (Pearse, Gallagher, & Bluff, 2001). Improving the safety culture is also dependent on your sphere of influence within the organization and how well networked you are into the organization.

The extent of safety-related information readily available from many resources (Internet, government, professional associations, media publications, etc.) has created an environment in which organizations cannot be excused for not knowing or finding content regarding safety-related issues. We have found through experience that we have shifted from being the sole primary information resource to expanding into researching, teaching, and mentoring how to deploy a safety management system and ensure safety information is utilized.

> *If you give a man a fish, you feed him for a day. If you teach a man to fish, you feed him for a lifetime.*
> **Ancient Chinese proverb**

Safety Culture.

The need to shift from a loss-based safety process that uses injury data to drive decision-making to a risk-based approach that decreases the probability of severe loss has been advocated in safety management systems such as ANSI Z10-2012.

We believe in the need for the safety professional to not only be proficient in hazard and risk control, but to understand the impact of communications, social networking (not just the Internet), and the perception of safety when presenting a case for the safety culture to the leadership team.

When this chapter is completed, you should be able to:

- Develop a personal working definition for "Safety".
- Discuss why perception is important for the safety culture.
- Identify ways you may be perceived in the organization.
- Identify perceptions about safety and shaping those perceptions.

Defining Safety

Let us start by covering an issue that impacts the safety culture, namely the definition of safety. The assumption is made that everyone understands the term "safety", but that is not necessarily the case. If a safety culture is to be sustained or developed, a mutually agreed upon definition is needed for the term "safety". While this may seem to be unnecessary, a second look shows that after all these years, a clear concise definition is still debated and discussed. If an organization is working with multiple definitions or vague concepts, then the potential for improving the safety culture is reduced.

Dictionary definitions for safety include:

- "Safety—(1) The quality of being safe; (2) freedom from danger or risk of injury".
- "*Safety*: the condition of being safe from undergoing or causing hurt, injury, or loss". (Safety, n.d.)

These definitions define safety in terms of itself and imply safety is something you know when you see it. They do nothing for determining latent hidden hazards that may be a high risk with potential that has not as yet been identified. A better definition is:

> *Relative freedom from danger, risk, or threat of harm, injury, or loss to personnel and/or property, whether caused deliberately or by accident.*
>
> **(Safety, n.d.)**

ANSI/ASSE Z590.3—2011 defines safety as "freedom from unacceptable risk" with risk defined as "An estimate of the probability of a hazard-related incident or exposure occurring". Hazard is defined as "The potential for harm" ("Prevention through Design Guidelines for Addressing Occupational Hazards and Risks in Design and Redesign Processes", 2012). These definitions from Z590 for hazard and risk allow for a relative level of risk acceptance as a practical matter. A higher risk can be acceptable as certain jobs or tasks retain an element of risk even after intense efforts are made to mitigate or control their risks. Examples range from firemen, police, astronauts, race car drivers, stuntmen/women, and so forth where the risk is considered acceptable by society to achieve a goal or necessary or desired activity. High risk must be evaluated and controlled to the degree possible to ensure all feasible protective devices and procedures are effective. We discuss the concepts of risk perception in Chapter 9, "Risk Perception—Defining How to Identify Personal Responsibility" and the concepts for risk management in Chapter 10, "Risk Management Principles".

Lesson Learned # 1

A colleague and risk control consultant, William Montante, has asked supervisors and managers over a number of years to define safety in supervisor training classes and presentations. He cites William W. Lowrance from "Of Acceptable Risk" (Lowrance, 1976) that "much of the widespread confusion about the nature of safety…would be dispelled if the meaning of the terms safety were clarified". Lowrance's definition for safety is "A thing is safe if its risks are judged to be acceptable" (Lowrance, 1976). "Safety is that state of being when risk and the hazards derived from it are judged acceptable or in control" (Montante, 2006). The issue of defining safety is not new!

Montante has been given dozens of definitions for safety. In an article for *Professional Safety* (Montante, 2006), Montante listed responses from 130 safety leaders within one organization. The definitions he received include:

- Preventing accidents or injuries;
- Freedom from harm or injury;
- Being safe;
- Being aware of your surroundings;
- Not getting hurt;
- It is number one;
- Following procedures and rules;
- It is a state of being;
- Looking out for each other;
- Complying with Occupational Safety and Health Administration (OSHA);
- Going home the same way you came to work.

Montante states, "simply put, safety is no more and no less than a condition or judgment of acceptable control over hazards and risk inherent to what one is doing at a point in time or chooses to do at some future point. That state of being can be personal or a reflection of the *business culture* (our italics)" (Montante, 2006).

To better align a definition of safety with an emphasis on hazard and associated risk, Montante suggests that safety be defined more in terms of hazard control: "Replace the traditional mantras of 'Safety first', 'Think safety', and 'Safety is your responsibility'...Strive for the personal and organizational mastery where each 'hazard control manager' can state with confidence and certainty that s/he intimately knows safety and how s/he and the company manage control" (Montante, 2006).

Note that many of the safety definitions as discussed are vague and more of an "essence", something not quite definable. "Safety" can become a crutch. It is a crutch in that it is used to cover over problems stemming from other hidden issues in the organization. These issues develop within a culture that allows poor planning, not understanding personnel issues, ineffective training, poor maintenance, poor quality, technology problems, etc. In evaluating safety-related requests, work orders, and/or operational issues, we have seen that many of these resulted in the lack of clear processes working together. Refer to Figure 1.1 for an overview of "The Organizational Puzzle".

In our ongoing discussions with William Montante and Dr Michael Waite, we believe that safety is a property that results from the effective meshing and intertwining of a multitude of interrelated processes and elements of an organization. The role of the safety professional is to work toward improved relationships between all the elements and aspects of the organization and an in-depth understanding of its culture. Knowing the culture requires considering how people react and work together, what they believe, and how they communicate.

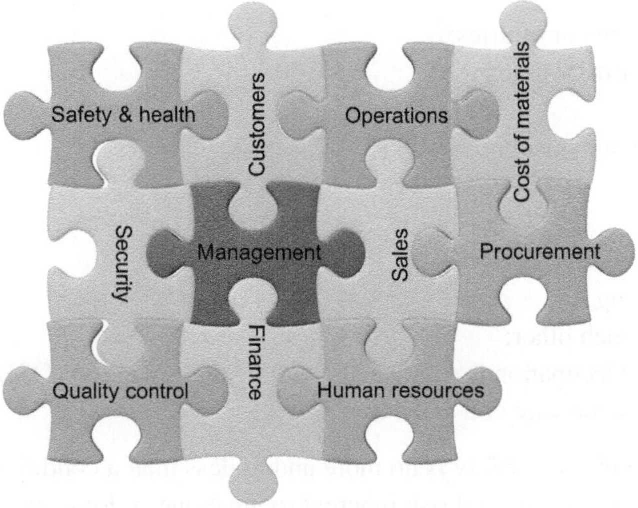

Figure 1.1
The Organizational Puzzle.

> *Emergent property: Any unique property that "emerges" when component objects are joined together in constraining relations to "construct" a higher-level aggregate object, a novel property that unpredictably comes from a combination of two simpler constituents.*
>
> **(Emergent Property, n.d.)**

For years we have used an organizational model by Wendell French and Cecil Bell from their book, *Organizational Development, Behavioral Science Interventions for Organizational Improvement* (French & Bell, 1984). This model consists of six interlocking areas—environment, technology, task, human-social, structure, and goals. Our approach is that when there is a misalignment of these areas with their inherent hazards and associated risks, the potential and probability for loss increases.

How does this emergent property model impact the safety culture? The role of the safety professional shifts from a narrow focus on developing programs to a more organizational development process that ensures that each of these areas have their own inherent hazards and associated risk in a state of acceptable control. The effort shifts from just managing programs to ensuring that, to the degree possible, all the elements of the organization are working together and that communications are in place to adjust the areas where hazards and risk are increasing or in need of better control.

> Acceptable control? What is acceptable and from what perspective? Internal? External? From each individual? Community? Political? Engineering? Safety Professional?

The Perception of Safety

An organization's leadership team (management and supervisors at all levels) must constantly make decisions that are believed to bring success to the organization. These leadership decisions are based on a continuous flow of information from both inside and outside of the organization. Decisions are made based on this information and feed into the perceptions about the state of the business and operational environment.

Leadership has only so much time and its finite attention is placed on actions and decisions that appear to best continue the movement toward specific vision/mission, business goals, and objectives with minimal interference. The leadership team bases its actions and decisions on past experience and expertise, preconceptions, and personal opinions developed from data streaming in from core functions such as finance, operations, engineering, human resources, legal, sales, and marketing, as well as general economic and industry conditions. Refer to Figure 1.2 for an overview of who has the "Leadership's Attention".

Figure 1.2
Who Has Management's Attention.

Within this swirl of data and decisions, your task is to keep the leadership team aware of the scope of the hazards and associated risk inherent in operations. Your main effort is to influence leadership decisions on recommended hazard and risk controls necessary to reduce injuries and loss-producing events. To do this, you must understand how communication works within the organization to ensure that correct information concerning hazards and risk flows quickly through the network without distortion.

> *Perception: The process by which people select, organize, interpret, retrieve, and respond to information from the world around them.*
>
> **(Hunt, Osborn, & Schermerhorn, 1997)**

If you are reading this book, you are searching for ways to improve how safety-related efforts are managed and implemented in your organization. The perception may exist that you alone are fully responsible for all "safety" in the organization as evidenced by your job description or direct mandates from leadership. In Appendix A there are several job descriptions and an outline from the *Dictionary of Occupational Titles* (Dictionary of Occupational Titles, n.d.) for review. Note the perception of management—the reality of the current status of safety—in the descriptions that show the safety manager as the primary individual that does everything related to safety, as well as the lengthy list of skills required! The descriptions also show an absence of defined authority and leadership role. For a detailed discussion on the value of developing and implementing written job descriptions, refer to Chapter 7, "Leadership and the Effective Safety Culture".

Changing the Perception

Perceptions are indicators about the safety culture of an organization. When you hear the words "Safety Manager" or "Safety Leader or Coordinator", what is the immediate image that comes to mind? Because of a long history of imagery that has developed about safety, most of us will see a person wearing a hard hat and carrying a clipboard, which have become the symbols of the safety profession. It is the image that has been fostered by various publications and traditions over time. Nothing is implicitly wrong with this perception. However, while it is necessary to advocate the use of personal protective equipment and taking notes, if the perception stops there, then leadership does not have a full understanding of what you should be providing to the organization in managing and improving the safety management system.

Most professions have a stereotype, for better or worse, that is associated with them. Think of what comes to mind when you hear titles such a "banker", "engineer", "teacher", "police officer", "coach", "politician", etc. These immediate images are shortcut associations that can bring out an established stereotype that we immediately apply to that person. This imagery can also imply what the expected behavior of the person in that role should be and reinforces the norms, habits, and behaviors about that role.

The image of safety has been a person who goes through a work environment looking for things that are wrong and quotes regulations or finds personal protective equipment issues. While this image portrays only one facet of the role we play, it remains the dominant image. This is the image that our experience has found that begins the process of placing the safety professional in an adversarial role with organizational leadership. Your image must be changed if you want to enhance your professional role and be seen as having the leadership skills necessary to shape the safety culture.

Lesson Learned # 2

Nathan once had a visit from a management consulting firm. The question was posed, "Do you have a safety algorithm that can easily be followed and implemented by a corporation?" The consultants went on to say why they were asking this question, "We have an issue with safety managers. We have safety mangers being promoted into senior management. However, they do not speak the same language as the other managers. We are being asked, What are these people talking about? We do not understand what they are saying!"

How Are You Perceived?

When you enter a room, hold a meeting, submit a report, etc., have you considered how you are perceived? Based on our experience, the image you project has a direct impact on the safety culture. What is the image you are projecting to your peers, employees, and the leadership team? The previous safety person may also have had an impact on your professional image. Your image is rounded out with the general public perception of the "safety guy", the OSHA compliance officer, movie/TV roles, and other cultural baggage.

Refer to Table 1.1 for examples of perception issues that provides insights on differences in core beliefs that can create perceptions.

Table 1.1: Example Perception Issues

Safety Objectives	Leadership Objectives
Prevents things from happening—the safety manager immediately brings up the various hazards and risk that may be present.	Ensure things happen—leadership may propose a new process or activity.
"*Risk*" reduction is desired—the safety manager searches for ways to reduce risk.	"*Risk*" taking is desired (different definition of risk!)—leadership wants to take entrepreneurial risk and sees risk taking as an essential part of business.
Requires expenses to fix problems—the safety manager identifies areas that need redesign, specific controls, and additional training and materials.	Keep expenses in check, be profitable—leadership must keep projects within budget.
Identifies management defects in system—the safety manager makes recommendations that imply leadership defects or gaps.	Seeks recognition and rewards—leadership has an ego, reputation, and personal authority image to maintain.
Relies on standards, regulations, rules—stay within rules—the safety manager must keep the organization in compliance with a multitude of regulations.	Bends rules to reach objectives ("get the job done") under pressure—leadership sees regulations as restrictions that reduce the potential success of the enterprise.
Seeks compliance and control—the safety manager seeks stability and control.	Seeks independence—leadership sees independence and thinking "outside the box".

> If you find yourself asking the questions, "Why don't they listen to me?" or "Why do they not see the importance of what we do?" it implies that the role of safety is perceived as being separate from the rest of the organization.

Refer to Appendix B to assess your perception of about how you are perceived in the organization.

Personal Branding

The safety culture may initially be dependent on how you are perceived. Reviewing your personal "brand" and its impact on the perception of safety within the organization's overall culture begins the process for change. Having a "brand" applies to you as you have a service that the rest of the organization "buys". If your brand brings no value, then your customers (leadership and employees) will not take your services seriously.

> *A brand is the set of expectations, memories, stories, and relationships that, taken together, account for a consumer's decision to choose one product or service over another. If the consumer (whether it is a business, buyer, voter, or donor) does not pay a premium, make a selection, or spread the word, then no brand value exists for that consumer.*
>
> **(Godin, 2009)**

What Is Your Mental Model?

Peter Senge in the *Fifth Discipline Handbook* defined mental models as the "images, assumptions, and stories which we carry in our minds of ourselves, other people, institutions, and every aspect of our World" (Senge, 2006). Critical to the understanding of the needs of a safety culture is determining how the organization views or perceives safety. Developing this understanding requires careful discussions with all levels of leadership and employees.

> *That is, the world is made up of billions of self-interested individuals who provide for one another by pursuing their own self interests. If an individual or business fails to provide for others, then it will fail in providing for itself.*
>
> **(Walling, 2008)**

A perception may exist based on the mental models held about safety and how your predecessor(s), if any, managed the process. What has been the management style used to deal with safety-related issues in the past and what methods are being used now?

- Is the concern only about regulatory compliance, rules, or regulations?
- Is there a minimal approach that just lets things happen with the intent of making change only when necessary?
- Is an established safety management approach mandated from leadership?
- Is there any indication that the current safety process is perceived as in conflict with the organization's vision or mission?

> *A mental model is a kind of internal symbol or representation of external reality, hypothesized to play a major role in cognition, reasoning, and decision-making.*
>
> **(Mental model, n.d.)**

The past and current style of safety management used, as well as the personalities involved, may have skewed the perceptions about safety-related criteria and must be understood in order to determine the impact on the safety culture and what steps are needed to change that perception.

Jay Wright Forrester defined general mental models as: *The image of the world around us, which we carry in our head, is just a model. Nobody in his head imagines all the world, government or country. He has only selected concepts, and relationships between them, and uses those to represent the real system*

(Mental model, n.d.)

Safety—A Multi-Disciplinary Profession

Depending on the size and scope of the organization, you may be tasked with a wide range of responsibilities and have to wear many hats. As evidenced by the job descriptions shown in Appendix A, including examples of duties and responsibilities, you must be knowledgeable in a wide range of topics and have the ability to work across all boundaries within an organization, as we will discuss in Chapter 2, "Analyzing the Organizational Culture". Refer to Figure 1.3 for "Safety—A Multi-Disciplinary Profession". The figure is not to be considered complete by any means.

These assigned roles and responsibilities can cause a drift away from the primary goals and objectives of the safety management system. As you move from responsibility to

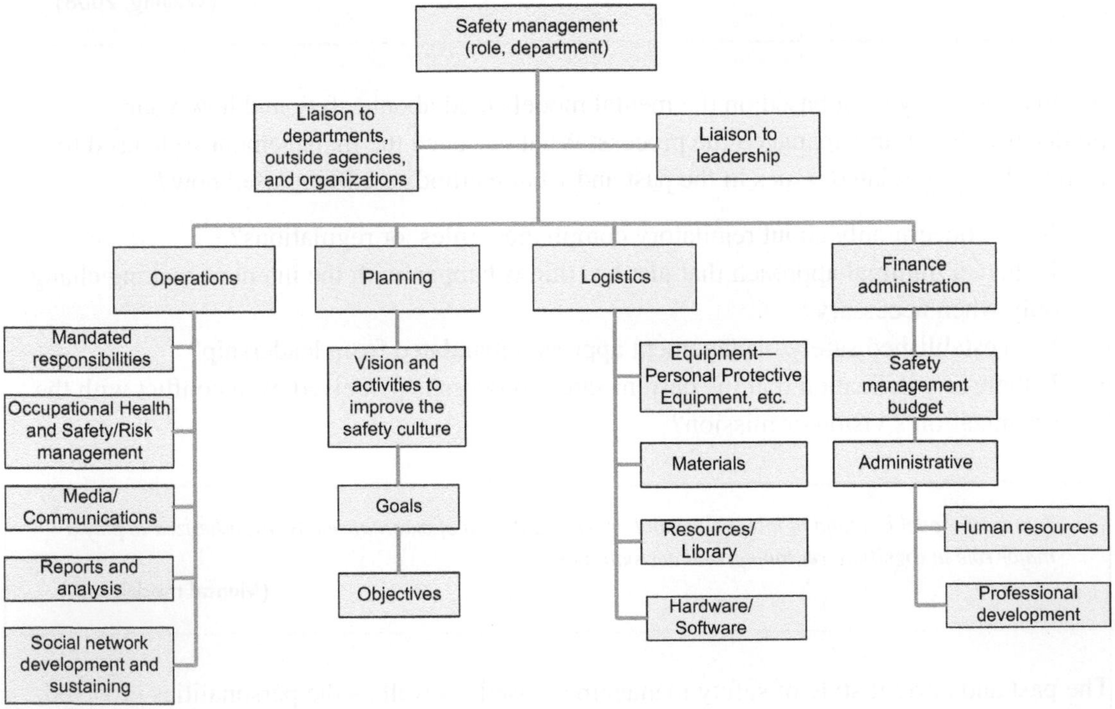

Figure 1.3

Safety–A Multi-Disciplinary Profession. *Source: Adapted from Incident Command Model* https://en.wikipedia.org/wiki/Incident_Command_System.

responsibility, you must ignore or back burner elements that may not be urgent yet can have a major impact on the long-term development of the safety culture. Steven Covey discusses this drain of attention and emphasized focus on the important and nonurgent activities for growth (Covey, 1990, p. 151). As an example, job hazard analyses and emergency planning are fundamentally important activities that are not urgent per se, yet they can have a major impact on the safety culture. Refer to Table 1.2 for an example time management matrix.

The Important/Not Urgent quadrant is where Covey says long term achievement is attained. The Important/Urgent Quadrant consumes immediate time and is not normally part of the planning process (Covey, 1990, p. 151). They take time away from the development of the safety culture.

Safety at a Crossroads

Safety is at the crossroads of all that happens within an organization. If an understanding of the needs of the specific organizational culture can be developed, your influence becomes essential to meeting those needs. Your major strength will be in the multi-disciplinary skills you can bring to the table. Your goal, at a minimum, is to become a "go to" person, the person who can:

- Address reported safety concerns from leadership and from any level or part of the organization.
- Curate and access information that supports your position and conclusions. We discuss this concept in more details in Chapter 15, "Becoming a Curator for the Safety Management System".

Table 1.2: Time Management Matrix

	Urgent	Not Urgent
Important	We have some important things to do that are urgent. These items must be done immediately. Post-loss actions, accident investigations, and claims management fall into this category.	In your roles and responsibility, you have things that are important, but they are not urgent. Important to us and need to be part of an overall plan but may not need immediate attention. However, they may not be considered important to the organization.
Not important	These items consume time as they take time away from important/not urgent items. These may be important to others and urgent to them, especially leadership, and become an important/urgent item that you had not planned for! (Oh, by the way projects from Leadership.)	E-mail responses, web searches not associated with the safety management system, hazards, and associated risk assessment control. These you think are important, but leadership thinks they are not.

Source: Covey, 1990, p. 151.

- Keep communication channels open and positive with all levels of the organization. Being able to "short circuit" the formal communications channels using your social network is crucial in communicating risk and hazard issues. Refer to Chapter 3, "Analyzing and Using Your Network".
- Analyze problems that are of concern about actual or perceived hazards and associated risk.
- Be influential enough to get action taken on safety management system issues.

Your role is to influence and shape the perception that the safety management system is a benefit and of value, not an impediment to operations, during crisis, new production or facility startups, operational expansion, changes in leadership, or business philosophy, etc.: The strength of this perception can reduce or increase the potential that core safety-related criteria will be bypassed or ignored. The bottom line is, the safety culture must be strong enough to withstand the pressures and test of time as the organization evolves and changes under different business conditions.

Safety Is an Espoused Value

Management must set priorities based on time and budget constraints. Safety is an espoused value in most organizations. By espoused, we mean the visible communications and expressions about safety. The "espoused" values of both safety management and the organization have to be aligned to ensure values are real and underlay the decisions that establish the priorities that drive the organization. We will discuss espoused values in more detail in Chapter 2, "Analyzing the Organizational Culture".

> *Espoused values are those values that are adopted and supported by a person or organization. Information about espoused values can be obtained by asking questions about the things that you observe or feel. Espoused values are those values that people say that they support. Values are preferred states about the way things should be.*
>
> **(Safety Culture in Nuclear Installations, 2002)**

While the needs of the safety management system are crucial from your perspective, you cannot change the safety culture without an understanding of the internal "politics" and interpersonal relationships that exist in all human undertakings. This understanding is essential for you to better position yourself as an essential part of the organization and influence leadership decisions.

As presented in Figure 1.1, each of the professional disciplines in an organization is competing for the attention of the decision makers who must allocate resources. The various departments, locations, specialties, and support units are all requesting or needing resources in order to maintain their own specific mandated or desired programs. They also believe that they should have a priority on the time and budgets established by leadership. These other departments and groups are competing and in possible conflict with you and each has a vested interest in how resources are allocated.

A further complication may be in the way problems are approached and how decisions are made by leadership and the way traditional safety-related issues are addressed. Many management decisions have to be made based on limited information and uncertainty. Leadership has to address the needs of multiple entities and resolve issues in the hope that the right decision or decisions are selected for the problems to be addressed. The decisions selected will have a potential for failure. As safety professionals, we have used a logical approach that suggests that optimal decisions can be reached based on quantifiable benefits—"If we implement policy X, then Y incident will not happen". In reality, we are also dealing with uncertainty and the end results may or may not be desired. The difference in how situations and decisions are approached may be one reason conflicts can develop between leadership and safety managers. Refer to Figure 1.4, "Differences in Approaching Decisions" (Radford, 1988).

If you are not aware of the politics involved in an organization, you may become very shortsighted and see the world from only your perspective and the needs of your specific process. While it is true that what you do is critical to the organization, there are only so many resources available in time, budget, and personnel.

> *Politics: Political affairs or business; especially competition between competing interest groups or individuals for power and leadership (as in a government)*
>
> **(Politics, n.d.)**

The leadership team generally understands that all elements of an organization are important. However, the leadership must determine how to best allocate finite resources. The traditional threat from safety management has been to raise the issue that "compliance agencies may fine us if I do not get my budget." This statement also sets up an adversarial relationship, does not add value to your position in the organization, and degrades the perception of the safety culture and you as a professional. This may be a valid comment at times, but it should not be the only reason for having a safety management process as it does not provide any visible path to improvement that leadership can factor into its overall decision process. A true safety culture does not need to use threats to attain its goals and objectives.

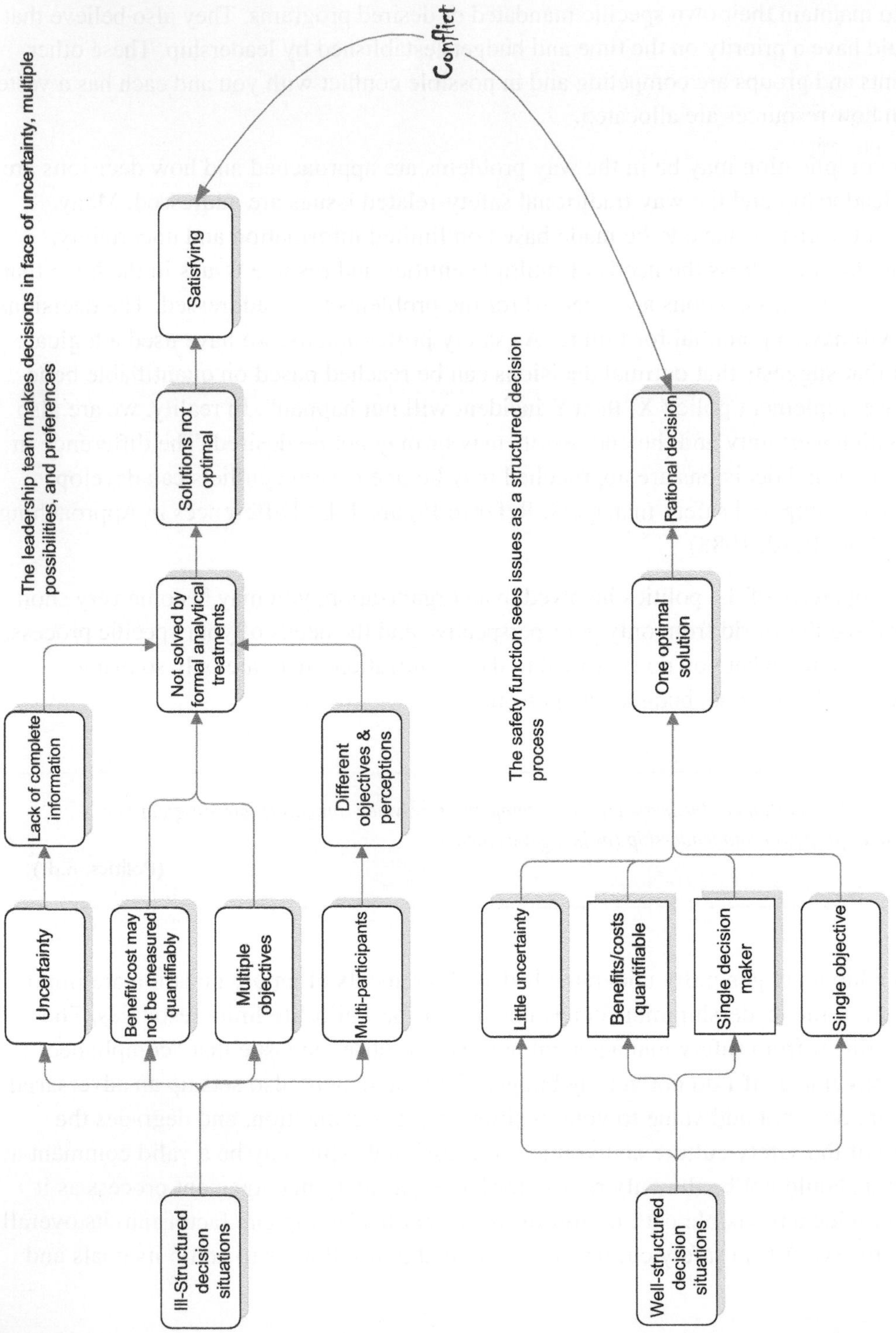

Figure 1.4

Differences in Approaching Decisions. *Source: Adapted from Radford, K. J. (1988). **Strategic and tactical decisions.** Springer-Verlag.*

Do You Speak the Same Language?

A key step is to determine if you are speaking the same business language that your organization speaks. You communicate through your daily actions, presentations, attitude, communication style, mannerisms, professional presence, and other characteristics that are constantly on view. That language determines how your communications are received by peers, leadership, and employees.

You have to sell your ideas and make your case constantly and consistently. Do not assume that just because safety may be considered a value that it will get the desired resources. The game becomes setting a strategy the leads you to be able to consistently show that your process and its programs are effective and efficient in relation to other specialties. What you must do is determine how to best present your case that hazards and associated risks are defined in the same language and terminology as used by the organization.

To determine what the current perception about safety is within the organization requires observation, discussions with other departments, and researching the overall business environment. Refer to Table 1.3 for a brief overview of information sources that offer insights on the perception of safety.

Table 1.3: Information Sources to Determine the Perception of Safety

Annual reports	Give an indication of the health of the organization, what has been ongoing, and its general finances.
Mission statements	Provide insight on the espoused values of the organization. Determining the gap between the mission and the real world is an indicator of the organizational culture.
Promotions	Who gets promoted and in what departments?
Presentations and meeting agendas	Types of presentations, who is involved, who speaks first and last; where are they held, what did they accomplish? Are discussions about safety and quality on the agenda? Are they allowed adequate time on the agenda? Are they limited to post-loss or injury related data?
Industry trade magazines and associations industry	How is this status affecting the safety culture? Give perspective on the organization and how it fits into the general industry. Are you in an organization that is held in high regard? One that is up and coming? One that is trending downward due to marketing global changes? What is the status of your organization and its industry?
Organizational newsletters and media	Provide insights—What is their source? Who contributes to them? Who edits them? How are they distributed and to whom? Are they considered of value by employees? Does the safety function have a byline and is it allowed to contribute?
Organizational charts	Who reports to whom and what are the levels of responsibility? Where is safety in the hierarchy? This will be discussed in Chapter 3, "Analyzing and Using Your Network".
Planning goals and objectives	How are they developed? How are they communicated and to what levels in the organization? Are safety-related criteria part of the established goals and objectives?

Table 1.4 provides examples of statements you should consider about yourself to better assess the perception of safety and yourself.

Table 1.4: Check Your Brand—Questions to Consider

• I am considered knowledgeable and a resource to the organization. • I have solid communication skills and a positive rapport across the organization. • I have a professional image and appearance. • I am approachable and have good interpersonal skills. • I understand the organization and its issues and problems. • I present information about hazards and associated risk in a style that is understandable at all levels of the organization. • I base my recommendations and suggested changes on in-depth assessment and current information. • I provide alternatives to my recommendations based on risk assessment.

From Startup to Status Quo

The potential for maintaining a strong safety culture increases as it is woven into the fabric of the organization and based on a positive perception. A strong, well-designed, flexible safety management system will enhance the safety culture and will be resilient and resist being shattered by such events as:

- Changes in leadership personalities and attitudes.
- Employee relations and communications.
- Changes in organizational risk and hazards.
- Any type of crisis that impacts normal routines (or example—workplace violence, fire, severe injury, nature disaster, etc.)
- Training demands.
- Operational changes.
- Mergers and acquisitions.
- Poor planning and decision-making.
- Technology change.
- Many other potential complications impacting the organization.

Refer to Part 2, "Safety Management Systems Defined" for a detailed discussion.

One reason for this ever present potential for failure is the dynamics of an organization are ever shifting. As new managers, supervisors, employees, technology, other projects, and priorities move into the operation from all directions, the safety effort will remain in a perpetual startup, reinventing the wheel. This can be naturally demoralizing.

However, if we accept that safety is an emerging property in an ever changing environment, we can increase the potential for maintaining a safe culture by accepting change as a given trait of all organizations. This is not change management per se, but a shift in perspective. Acceptance of change as ongoing reduces the frustration level. Understanding the importance of a strong network and culture of communicating the value of safety becomes a part of overall planning and discussion. Accepting a dynamic approach rather than a static approach shifts how a safety process is developed.

We will be discussing concepts that we believe will enhance the ability of the safety professional to build a network and communicate to the organization in a more effective manner.

Over the course of our experience, we have seen numerous well-written safety programs that after a brief time simply evaporated and could not even be found in an organization. The day to day pressures and priorities had allowed "safety" to be degraded to a secondary status.

To be effective, you must mentally accept that the safety management system can be flexible and adaptable to rapid change as organizational and business conditions evolve. You want a safety management system that can bend with the needs of the organization and not break under the demands of the overall organizational culture.

Summary

In this chapter, we discussed how various perceptions can affect the sustaining and continued improvement of the safety culture. This perception is based not only on the current opinions about the role of safety management held by leadership and employees but includes perceptions based on what has gone on in the past, how your predecessors managed themselves, and how safety-related criteria were communicated in style and message.

Organizations are not static and are constantly in flux and change. Therefore, safety is not something that is static. The safety management system requires stability yet must have processes that are flexible to meet the new demands of business environments. Situations can develop as simply a matter of course as conditions shift and change. For example:

- Turnover of employees brings inexperience into the workplace and potential different perspectives.
- Different levels of skills and education may require new training methods.
- New leadership may still be establishing how they will work as a team and their communication styles may not have been fully established.
- Hazards and associated risks may be evolving as new processes are being brought on line.

These changes may be increasing the probability of loss-producing situations.

Safety is an emergency property that is determined by an ever emerging set of organizational and environmental conditions. These conditions include the competition for resources from other areas of the organization.

A primary goal is to assist and advise the organization on effective methods and concepts that will sustain the safety culture. It is essential that you fully understand the environment you are in and how to approach and communicate with leadership and employees. To accomplish this goal, you must understand the current perception your organization has about the role of safety.

In Chapter 2, we will discuss how understanding the culture of your organization is essential to increasing the potential for the effective development and implementation of a safety management system. We will demonstrate how organizational cultures develop under different conditions and how their history creates a unique style and personality. In addition, we will provide details on how it is essential that you identify the unique characteristics of your organization given the different styles, needs, and structure that will require different approaches when deploying and sustaining the safety management system.

Chapter Review Questions

1. Why does developing and sustaining an organization's safety culture require a multi-disciplinary approach?
2. In your own words, how do you define a safety culture? Conduct additional research and provide at least three additional resources that define safety culture and explain the difference in the definitions.
3. What is ANSI/ASSE Z590.3 and how does it define a safety culture?
4. Define the term "emergent property" and explain how it is used in the safety culture.
5. How would you define the perception of safety?
6. Explain the value of developing and implementing written job descriptions. List seven benefits of a well-written job description.
7. Discuss the procedure to review the existing organization. Identify the five methods of assessing an organization to define the roles.
8. Discuss why perceptions are indicators about the safety culture of an organization.
9. Discuss the elements as outlined in Table 1.1, Example Perception Issues. Research using the Internet and provide at least three additional elements.
10. Discuss why you need to change the perception of a safety culture.
11. Explain the makeup of a mental model. Provide examples of a mental model.
12. Discuss the elements of a time management matrix. Explain the important of this matrix.
13. Discuss the problem of not speaking the "language" of the organization.
14. Discuss Table 1.3, and why this information is important.
15. List the eight reasons that you need to check your brand. Explain why this is important in a safety management system.

Bibliography

Covey, S. R. (1990). *The 7 habits of highly effective people: Powerful lessons in personal change*. New York: Simon and Schuster, Inc. Retrieved from http://amzn.to/WsCA5g.

Dictionary of Occupational Titles. (n.d.). Photius Coutsoukis and Information Technology Associates, Public Domain. Retrieved from http://bit.ly/XlUQcJ.

Emergent Property. (n.d.). Dictionary.com. Retrieved from http://bit.ly/YQRNKT

French, W., & Bell, C. (1984). *Organizational development: Behavioral science interventions for organizational movement*. Englewood Cliffs, NJ: Prentice-Hall.

Godin, S. (December 2009). *Define: Brand*. Retrieved from http://bit.ly/XJ6vnX.

Hunt, J., Osborn, R., & Schermerhorn, J. (1997). *Basic organizational behavior*. New York: John Wiley and Sons.

Lowrance, W. W. (1976). *Of acceptable risk: Science and the determination of safety*. William Kaufmann, Inc, Los Altos, CA.

Mental model. (n.d.). Wikipedia, the free encyclopedia. Retrieved from http://bit.ly/124eseh

Montante, W. M. (November 2006). The essence of safety – what's in your mental model? *Professional Safety*, 36–39.

Pearse, W., Gallagher, C., & Bluff, L. (Eds.). (2001). *Occupational health & safety management systems, proceedings of the first national conference*. Crown Content. WorkCover. Retrieved from http://bit.ly/WLdr5y.

Politics. (n.d.). Merriam Webster.com. Retrieved from http://bit.ly/15HqUzF

Prevention through Design Guidelines for Addressing Occupational Hazards and Risks in Design and Redesign Processes. (2012). ANSI/ASSE Z590.3-Secretariat, American Society of Safety Engineers.

Radford, K. J. (1988). *Strategic and tactical decisions*. New York: Springer-Verlag.

Safety. (n.d.). Farlex. Retrieved from http://www.thefreedictionary.com/safety

Safety Culture in Nuclear Installations. (2002). International Atomic Energy Agency (IAEA), Guidance for use in the enhancement of safety culture, IAEA-TECDOC-1329. Retrieved from http://bit.ly/V1KsKt.

Senge, P. M. (2006). *The fifth discipline: The art & practice of the learning organization. A currency book*. New York: Crown Publishing Group.

Walling, A. (August 2008). *Spontaneous order revealed*. Marketing-Based Management Institute. Retrieved from http://bit.ly/WrEsMa.

CHAPTER 2

Analyzing the Organizational Culture

Cultural legacies are powerful forces. They have deep roots and long lives. They persist, generation after generation, virtually intact, even as the economic and social and demographic conditions that spawned them have vanished, and they play such a role in directing attitudes and behavior that we cannot make sense of our world without them.

—Malcolm Gladwell (Gladwell, 2008)

Introduction

Understanding the culture of your organization is essential to increase the potential for the effective development and implementation of a safety management system. Organizational cultures develop under different conditions and their history creates a unique style and personality. It is essential that you identify the unique characteristics of your organization given the different styles, needs, and structures that will require different approaches when deploying and sustaining the safety management system.

An organization, whether public or private, large or small, must be constantly adjusting to its environment in order to meet a wide range of internal and external pressures and demands. To understand an organization's safety culture, you need to define how you want the safety management system to work within this environment of constant change. To establish a level of stability, a system of desired behaviors and beliefs must be developed that provides consistency in problem solving, decision-making and general relationships, and resilience to loss-producing events.

> *The last thing a fish would recognize is water.*
>
> **Anonymous**

Considerable research has been developed concerning the definitions and concepts of organizational culture. The old saying that "the last thing a fish would recognize is water" is true. Without an understanding of the culture of your organization, your efforts to shape the "safety culture" will be restricted by hidden restraints and limitations that are not visible. Starting with an understanding of the culture provides you with the advantage of knowing how to modify and fine-tune your safety management system.

Upon completion of this chapter you should be able to:

- Identify the elements of an organizational culture.
- Identify and discuss the three levels of an organizational culture.

- Define and discuss the elements of a safety culture.
- Identify the possible types of organizational cultures.
- Identify ways you can change a culture.
- Identify the nine warning flags that can defeat controls.

What Is Organizational Culture?

What is organizational culture? Why is it considered essential to the development of an effective safety management system? Let's begin with an overview of definitions of culture.

Dr Edgar Schein, perhaps the leading expert on the study of organizational culture, provides a definition (as cited in the US Department of Energy, Volume 1: Concepts and principles, human performance improvement handbook, 2009):

> *...a pattern of shared basic assumptions that the group learned as it evolved its problems of external adaptation and internal integration. Over time this pattern of shared assumptions has worked well enough to be considered valid and, therefore, to be taught to new members as the correct way you perceive, think, and feel in relation to those problems.*
>
> **(Volume 1: Concepts and principles, human performance improvement handbook, 2009)**

Refer to Table 2.1 for components of the organizational culture.

The US Department of Energy's action plan on the Challenger Disaster and Davis Besse Reactor Pressure-Vessel Head Corrosion Event provides the following:

> *Organizational culture refers to the basic values, norms, beliefs, and practices that characterize the functioning of a particular institution. At the most basic level, organizational culture defines the assumptions that employees make as they carry out their work. An organization's culture is a powerful force that persists through reorganizations and the departure of key personnel.*
>
> **(Columbia Space Shuttle Accident and Davis-Besse Reactor Pressure-Vessel Head Corrosion Event, 2005)**

> Assumption—Are employees listening to the system?
> In the Columbia accident: "The machine was talking to us, but nobody was listening." Deviations from requirements had become normal business for NASA. The Columbia Accident Investigation Board (CAIB) report referred to this as the "normalization of deviations" (Columbia Space Shuttle Accident and Davis-Besse Reactor Pressure-Vessel Head Corrosion Event, 2005).

Table 2.1: Components of Organizational Culture

Culture can be defined as holding basic assumptions on how things work, the perceived values of the organization, what employees believe about how they should act and behave, and work practices, all of which combine into: • Problem solving and coping with day to day duties and issues necessary for the completion of defined job requirements. • Behaviors and activities that are considered successful and are routinely used over time. • Education and training of new employees about these behaviors and activities that can be imparted to new members of the group. • Traditions and customs about the proper way to perceive, think, and feel about tasks and actions.

Source: Adapted from Schein as cited in Volume 1: Concepts and principles, human performance improvement handbook (2009).

These components can work in favor of sustaining a safety culture or can become a barrier to improving that culture. The key is that daily feedback strengthens these assumptions and ingrains them into the fabric of the organization. Refer to Chapter 11, "Developing an Activity-Based Safety System" for insights on how to strengthen feedback and Chapter 9, "Risk Perception—Defining How to Identify Personal Responsibility" for insights on identifying and changing assumptions and beliefs about risk. Refer to Table 2.1 for a discussion on the "components of organizational culture".

An organizational culture may be forced into attempting to change itself due to crisis and extreme external pressures. A severe loss such as that faced by NASA created the need for change. The organizations that cannot change and adapt to new conditions and changes in their existing environments may not survive. Refer to Figure 2.1 for "Determining Cultural Change".

The key is to develop attributes within the culture that keep hazards and associated risk visible. The strong organizational culture maintains a memory of the nature and scope of the hazards and associated risks. Refer to Appendix C, D, and E for further insights on culture traits, questions to ask, and potential impact.

The acceptance of known hazards and associated risk must not be allowed to become the norm! If it becomes the norm, it becomes part of the belief system.

When excellence in safety is measured by zero failures, a self-limiting organizational viewpoint and very dangerous employee belief is created: "If safety means no incidents, then anything that I do that doesn't result in an incident or get me hurt, must be safe." When this occurs, risk will be overlooked, complacency will set in, an important and healthy degree of vulnerability of risk will be lost, and organizations will be surprised by an incident that occurs out of nowhere. This will often sound like: "I can't believe that experienced, well-trained employee did that," or, "How did we miss that?".

(Galloway, 2012)

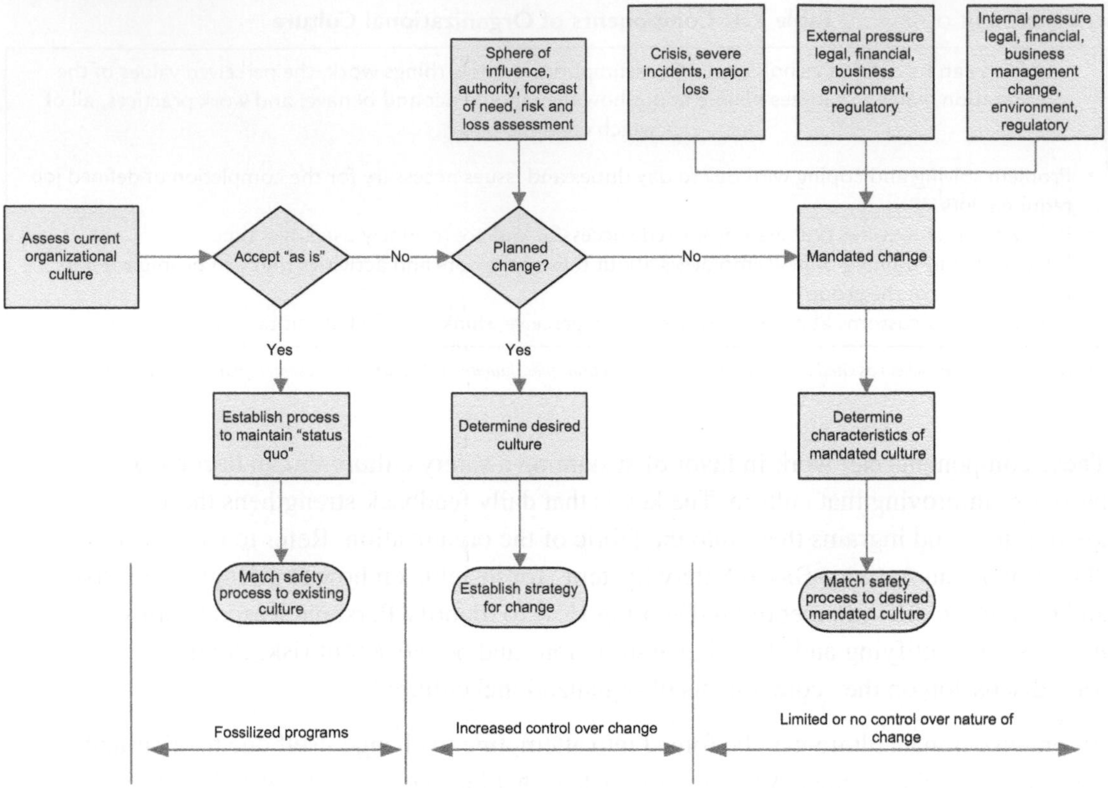

Figure 2.1
Determining Cultural Change.

> *"Git 'er done" is a saying that implies do anything necessary to get a job done regardless of the hazards or associated risk.*
>
> **(git r done, n.d.)**

Three Levels of Culture Defined

Edgar Schein provides a foundational approach to review a culture. He describes a culture as having three primary levels. His three levels consist of Artifacts, Espoused Values, and Underlying Assumptions (Schein, 2004). The following discussion provides insights on how these levels can be used to assess the depth and scope of the safety culture.

Artifacts

The first level consists of Artifacts that are the visible organizational structures and processes. These are the tangible things you can see as you move about within facilities or review the environment of the organization. For example, general housekeeping that includes appearance of equipment and its maintenance, the visible work environment that includes the type and design of facilities, warning signage, written safety materials, policies, training rooms, and dress codes indicate what the underlying assumptions and beliefs are in the organization.

Espoused Values

The second level of culture involves the Espoused Values that consist of the content of the various strategies, goals, and core philosophies that are used by the leadership to guide the organization. What does the organization say about itself? We see these espoused values in the form of safety slogans, safety mission statements, and various commitments made with regard to safety.

One of the techniques that we will be discussing is the use of the activity-based safety process, as described in Chapter 11, "Developing an Activity-Based Safety System". We will discuss the use of the preshift meeting to set the tone for each workday as well as other techniques designed to directly shift the perception of safety at the interface between the worker and the hazard.

Underlying Assumptions and Unspoken Rules

The third level of culture is the foundation that is invisible but must be addressed if a safety culture is to be fully ingrained into the organization. As we previously discussed, these are the underlying assumptions and unspoken rules followed by all employees of the organization. By reviewing the first and second levels, plus interviews with employees and leadership, it may be possible to determine what many of the underlying beliefs are that truly drive the organization. This is not an easy task. The organization may espouse "safety as a value", yet the underlying belief is that it is not. We discuss this in Chapter 14, "Assessing Your Safety Management System" (Schein, 2004; Schein, n.d.; Volume 1: Concepts and principles, human performance improvement handbook, 2009). Refer to Figure 2.2, "Three Levels of Culture".

Safety Culture Defined

A safety culture goes beyond simply managing a basic series of required programs for regulatory compliance. The structure and shape of the safety management system should take into consideration the variations in cultures. In our experience, the test of an organization's safety management system is in how it is sustained during times of ongoing change in leadership, new personnel, crisis, or other situations that can be positive and negative as discussed in Chapter 1. Basic safety programs or activities can quickly disappear unless the organization has fully internalized the concepts necessary to keep inherent operational hazards and associated risk within acceptable control.

While the fundamental concepts of a safety management system can be the same, we will discuss in Chapter 5, "Overview of Basic Safety Management Systems", how it is deployed and sustained will vary based on the organization's culture. Program elements, safety-related

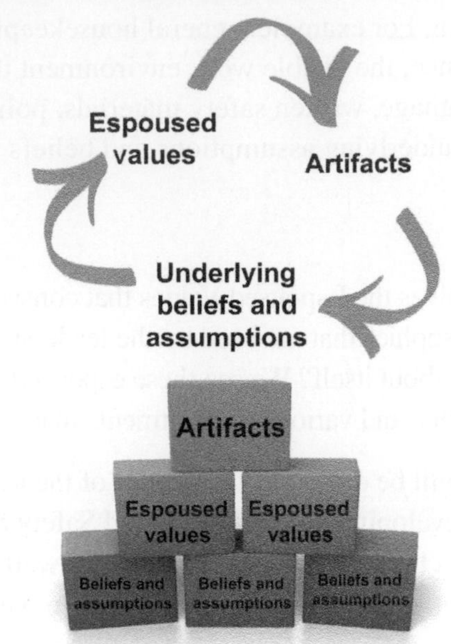

Figure 2.2

Three Levels of Culture. *Source: Adapted from Schein, E. H. (2010). Organizational culture and leadership. Jossey-Bass.*

activities, etc. that work in one operation, department, or even sister companies may or may not be effective elsewhere.

> Dr Michael Waite at California State University, Fresno, has an interesting analogy for safety culture. He equates safety culture to what is known in the wine industry as "terroir". It is the terroir that creates the characteristics of the wine. The correlation is that if we wish to develop a safety culture, we look at all of the conditions, work environment, personnel, industry characteristics, etc., and use these to determine what the nature of the culture is and to define actions that can improve the safety culture.

Assessing the Current Safety Culture

> *Safety culture is also an amalgamation of values, standards, morals and norms of acceptable behavior. These are aimed at maintaining a self-disciplined approach to the enhancement of safety beyond legislative and regulatory requirements. Therefore, safety culture has to be inherent in the thoughts and actions of all the individuals at every level in an organization. The leadership provided by top management is crucial.*
>
> **(Management of Safety, Safety Culture and Self Assessment, 2000)**

Nathan was involved in the insurance industry for more than 20 years. As part of underwriting studies for insurance placement, it was necessary to visit prospective companies to discuss their safety process and visit operations and facilities to determine their overall risk status. In retrospect, as previously discussed under organizational culture, what was being done was an attempt to determine what the safety culture was through observation of the various Artifacts and discussions about the Espoused Values as evidenced by mission statements, policies, procedures, etc.

The question the underwriters wanted an answer for was "Is this prospective client maintaining a safe work environment and is it serious about keeping hazards and associated risk under control?" If the observed Artifacts indicated that issues existed that increased the potential for loss and did not reflect the Espoused Values, then the perception left was that the Underlying Belief system was different from the Espoused Values. An example, if inspections were to be routinely done and deficiencies immediately corrected but observation noted maintenance and housekeeping problems, then the real belief is that inspections were not considered important.

> *An organization with a good safety culture relies on the close interdependence between technical safety and organizational processes. In practice, a high level of safety culture means the systematic organization and implementation of activities aimed at creating high quality technical, human, and organizational systems.*
>
> **(Management of Safety, Safety Culture and Self Assessment, 2000)**

Lesson Learned # 1

James, after 30 years of reviewing facilities and implementing safety management processes as a safety manager, developed skills that by combining a walk through of a facility, making visual observations, and talking with employees and supervisors he was able to identify and target gaps in the safety culture.

Habits as Part of the Culture

As culture involves patterns of behavior, habits are defined as "an acquired behavior pattern regularly followed until it has become almost involuntary: the habit of looking both ways before crossing the street" (Habit, n.d.). Unless old beliefs are eliminated that allow risk behavior to be acceptable, the unsafe activities become habits and will continue to exist.

Charles Duhigg suggests that an understanding of what he calls "keystone habits" is needed, as these are habits that have the "power to start a chain reaction, changing other habits as they move through an organization". He states that these keystone habits "start a process that, over time, transforms everything". Duhigg believes that "the habits that matter most are the ones that, when they start to shift, dislodge and remake other patterns" (Duhigg, 2012).

As an example of how changing a keystone habit impacts the culture and in turn changes an entire organization, Duhigg uses the example of how Alcoa began changing its overall culture in the late 1980s, when its new CEO, Paul O'Neill, used strong emphasis on Alcoa's safety program to change core habits that impacted the entire Corporation. O'Neill wanted to improve the way Alcoa did business and improve its quality and methods of production. Per Duhigg, Alcoa still has a strong safety process long after O'Neill has left the company, as safety became an ingrained habit and a strong underlying belief (Duhigg, 2012).

> *Cultures grow out of the Keystone habits in every organization, whether leaders are aware of them or not.... Keystone habits transform us by creating cultures that make clear the values that, in the heat of a difficult decision or a moment of uncertainty, we might otherwise forget.*
>
> **(Duhigg, 2012)**

In this case, the CEO went straight to the core of Alcoa's assumptions and beliefs. An assumption existed that safety was unrelated to other issues impacting the success of the organization. By realizing that safety touched on all areas of the organization (an emerging property, as we previously discussed), he was able to change underlying beliefs, creating a different series of key habits.

Possible Characteristics of a Culture

To determine the nature of the culture you work in, each aspect of the culture must be considered to determine the driving personalities and characteristics of the organization and how they are interconnected. With this understanding, your potential for shaping the safety culture is enhanced and the required time frame for change may be decreased. We will discuss in more detail the use of interconnections in Chapter 3, "Analyzing and Using Your Network".

Knowing the type of management types can provide insights on how to approach and work with leadership. Refer to Table 2.2 for an outline of the "types of cultures" that have been suggested by several authors. The types followed by each of these authors appear to have similar characteristics and follow an overall theme.

Once you understand the type of culture you work within, you can more effectively design and develop your communications to leadership.

A cautionary note is that these definitions represent a spectrum of types. There is no line you cross to move from one to the other. The types may change as leadership changes.

Table 2.2: Types of Cultures

Seidman (Seidman & Clinton, 2011)	Westrum (Westrum, 2004)	Craig (Craig, 1978)	Likert (Likert's Management Systems, n.d.)
Anarchy and lawless	Power oriented	Authoritarian/Exploitative	Exploitive/Authoritarian
Blind obedience	Rule oriented	Authoritarian/Benevolent	Benevolent/Authoritarian
Informed acquiescence	Performance oriented	Consultative	Consultative
Self-governance		Participative	Participative

Lesson Learned # 2

Nathan knew of a safety manager whose immediate boss was changed. The previous manager used the performance/consultative styles. He would provide details of what he was doing and the manager would agree after discussion. That manager moved on and the next manager used the style of benevolent authoritarian. The new manager did not want details or discussion, just brief summaries. The safety manager found that the shift in style could not be overcome and shortly left the company.

Artifacts can give an indication of the nature and style of the culture you are dealing with. As an example, in going through the corporate office of a client, Nathan noted that each cubicle was occupied by not one but two employees that shared one telephone and a file cabinet. What would you surmise about the nature of the leadership? Was this possibly an authoritarian/exploitative type of organization? Leadership did not see the need for designing an efficient, pleasant work environment and possibly had a low opinion of the employees. In this culture, since no capital budget is in place for a decent corporate office, could a budget for safety-related materials and programs be expected from this leadership team?

Another organization Nathan visited had an unwritten rule that managers were to come in very early, well before official office hours, and work late every day as well as spend at least half of each Saturday in their office—even if they did not have anything to do! Why? Because the CEO believed that this showed dedication and that long work hours were essential. This is representative of another example of an authoritarian type of organization.

Lesson Learned # 3

James was involved with a team that was tasked with solving general safety-related issues. The team was diverse, consisting of hourly employees from several departments and given a budget to implement its finding. The team did not have to ask permission of the leadership as the employees were trusted to accomplish their mission and act responsibly.

This is an example of a participative or consultative type of culture.

Table 2.3: Getting the Feel for an Organization

1. Visit and observe the organization.
2. Identify artifacts (see discussion on artifacts) and processes that puzzle you.
3. Ask insiders why things are done that way.
4. Identify espoused values that appeal to you and ask how they are implemented in the organization.
5. Look for inconsistencies and ask about them.
6. Figure out from all you have heard what deeper assumptions actually determine the behavior you observe.

Source: Adapted from Organizational culture and leadership by E.H. Schein (Schein, 2004).

From a hazard and associative risk standpoint, the objective is to assure that concerns get to decision-makers as quickly as possible without barriers. Authoritative types may have potential barriers and screening of information as mid-level leaders may not want to pass bad news or problems along, given they might be held accountable or punished in some way.

The design of the risk communications will vary. For example, in the participative type, sharing of information is easier as you would be considered a trusted professional and expected to communicate. Decisions on actions to be taken do not necessarily require a senior level involvement, just get a solution to the problem fixed first and discuss later.

Using the concepts in Table 2.3 as an example, before a plant visit, James would call the contact person (plant leadership) to set an appointment. He would record the date and time he called and when and if the phone call was returned. The time the person took to return the call was a good indication of the level of interest and level of priority that would be given to the safety-related visit. If the leadership was willing to spend time on the telephone discussing and planning the forthcoming visit, the potential for a successful safety visit was improved. Other indicators were that if leadership took a direct interest during the visit and made time to discuss any findings, issues would typically be resolved in a timely manner.

National and Occupational Cultures

Finally, two other aspects of culture must be considered. While organizational cultures can vary from company to company and department to department, a further complication is that with international enterprise now the norm, understanding international cultures is essential. Even with the variations of traditions, language, and mannerisms impacts perceptions held about leadership; we still tend to expect other people to think and behave as we do, assuming that what we say and how we are perceived is similar to the national and business culture we are in. Each region and country has its own overriding culture. How power is used and/or shared, class and hierarchy, relationships allowed, etc. will vary. What may work in one country may not be accepted in another (The Hofstede Centre, n.d.).

In addition, the various specialty areas of an organization also have their own cultures. As we discussed in Chapter 1, "The Perception of Safety", an image, type of personality, ego, perceived beliefs, style of communication, level of trust, desired status, etc. shades how you view that individual. Read the following and see what images immediately pop into your mind—banker, car salesman, politician, cowboy, miner, "blue-collar" worker, oilman, pilot, and military officer. These stereotypes are reinforced by media, stories, and expected norms and can give insights on the behavior that might be expected when safety programs are introduced and deployed (Schein, 2004).

Further, these are not just stereotypes. They provide information about the occupational culture of each of the mentioned groups. A person in a certain role brings that occupational culture into any discussion. You have certain ways of behaving and make assumptions about how a safety process should work based on your experiences. You must be aware of those assumptions, as they can taint your opinions and perception about what must be accomplished and how to go about getting your job done. You must take into account the overall organizational culture if your safety management system is to operate effectively.

Safety Culture as a Mission-Essential Business Priority

Leadership sets the business priorities for the many activities necessary to accomplish its main organizational mission, which is survival in its particular niche in the marketplace. Leadership must focus on "mission-essential" activities to the highest degree possible. You must make an effort to be considered one of those mission-essential activities when establishing or enhancing a safety culture. Your goals and objectives should be based on being mission-essential, as will be discussed in Chapter 4, "Setting the Direction for the Safety Culture".

Mission-essential activities are the designated actions and criteria that ensure the vision or intent of the organization has the highest possible chance of being accomplished.

To see if you meet the meaning of mission-essential, ask the following questions:

- What are you doing each day that moves you toward the organization's vision, goals, and objectives?
- What safety management system mission-essential priorities are being pushed in lieu of other things such as productivity? Why? (Figure 2.3)

Can You Change a Culture?

What activities are you currently doing that are not mission-essential to the safety management system? Refer to Table 2.4 for the "extremes that might be found between positive and negative cultures".

Figure 2.3
Interrelations between Culture and Vision.

Table 2.4: Extremes between Positive and Negative Cultures

Positive Culture	Negative Culture
Mission clearly includes safety	Lack of shared purpose, self-interest
Sense of safety history and purpose	Limited history or memory about safety
Safety seen as core value	Safety as a priority
Positive beliefs and assumptions about safety	Safety separated from core operations
Positive safety traditions and ceremony reinforce safety as a value	Few if any traditions for safety—not on agenda!
Strong bias toward excellence	Norm of acceptance of mediocre
Shared sense of responsibility for safety process	Little sense of community
Safety seen as a professional discipline	Safety as an add-on duty not a profession
Cultural network fosters positive communication flow about safety	Network of negative communications—fragmented, antagonistic
Leadership is openly clear and activity seeks safety improvement	No clear leadership from senior management
Sense of interpersonal connection, purpose, and belief in future	Sense of hopelessness, discouragement, despair
Safety viewed as beneficial to profitability and/or success	Safety viewed as expense obstacle

Source: Adapted from Shaping school culture—Deal and Peterson (Deal & Peterson, 1999).

What actions can you immediately take to help increase your potential to shape the safety culture of your organization? To help you, the following questions must be addressed before you to attempt to modify or change the safety culture. You should complete a gap analysis that will identify:

- The current organizational culture, including its characteristics and traits.
- What does the organization value?
- How are priorities established within the organization?
- The status of the current safety culture. You need to begin to gather as much information about the organization as possible, including the perception, effectiveness, and inclusion of safety-related activities.
- Does your perception and assessment of the current safety culture match that of the leadership? Is there a gap between what the employees and leadership believe?
- The mandates from leadership. What is your mission? Is it to develop, sustain, or improve the safety culture?
- What structures and processes are needed to best assure that a culture of safety can be maintained in the organization?
- How are quality, production, finance, human resources, etc. managed within the organization?
- How are employees trained, educated, and oriented to the desired norms and values of the organization?
- Are safety concepts considered part of these norms and values?
- How are responsibilities and expected behaviors communicated and enforced within the organization?
- Are critical uncontrolled hazards or associated risks communicated to the leadership team in a format that matches their current communication style?

Your gap analysis should include a thorough understanding of the organization's current state and what the desired future state of the culture is. From this assessment, you can begin to develop your strategy for change.

You need a strategy to present a safety management system to the leadership team. You can do this by using a basic structured safety management system model that can be customized for the organization and not just viewed as a forced compliance-driven approach. Examples of a structured model include those from a government entity such as the Occupational Safety and Health Administration (OSHA) or professional organization, such as the American National Standards Institute (ANSI). Refer to Chapter 5, "Overview of Basic Safety Management Systems", which provides an overview of several safety management systems. If your organization already uses ANSI standards and certifications, follow the appropriate safety management system from one of these groups.

Depending on the leadership mandate that may have been given, before making serious safety process changes, an understanding is needed on the current state of programs and why or why not they have been implemented effectively.

Nine Warning Flag Factors That Defeat Control

According to a study by the Institute for Nuclear Power Operations (INPO) as cited in the Department of Energy (DOE) Human Performance Manual (Volume 1: Concepts and principles, human performance improvement handbook, 2009), nine common weaknesses can serve as "warning flags" that may degrade a safety culture. The INPO concluded "that these latent conditions are conducive to the degradation and accumulation of flawed controls and human-performance-related events".

You can use the nine flags as another way to evaluate your organization. Look for indications or signs of the following:

1. Overconfidence—The history of injury and/or loss-producing events is good as compared to the industry, and no significant loss events have occurred. The organization has begun to believe that since there are no losses, the risk is low. Safety meetings, inspections, policies and procedures, etc. are not followed as routinely as they once were. Indications of safety issues and problems are not reviewed or analyzed.
2. Isolationism—Interactions with other organizations, professional groups, and regulatory and industry groups are rare or nonexistent. Benchmarking is seldom done. Changes in best management practices are not implemented from incident investigation reports, employee feedback, and/or hazard and associated risk issues that have been identified at other organizations. The organization lags the industry in many areas of performance and may be unaware of the real scope of hazards and associated risk. No risk assessment is in place to detect high probability of a loss-producing event.
3. Defensive and Adversarial Relationships—The mind-set toward the regulatory agencies or professional groups is defensive. Recommendations are met with the comment, "Is this required by OSHA (or other regulatory body)?" The main thrust of safety activity is to do the minimum. Employees are not involved in the safety management system and are not listened to when they communicate potential concerns of hazardous conditions. The safety function is considered to be adversarial to productivity and profitability, as poor relationships limit open communications.
4. Informal Operations and Weak Engineering—Standard operating procedures, formal approaches to maintenance, housekeeping, and quality are weak. Discipline is lacking and a lackadaisical atmosphere toward safety is present. Other organizational factors such as changes in leadership, new initiatives, and/or special projects overshadow the intended vision/mission, goals, and objectives. Engineering is weak and design criteria do not consider safety criteria as an organizational priority. Maintenance reporting systems do not put unsafe conditions as a high priority.
5. Production Priorities—Important machine and equipment problems, such as machine guarding defects, electrical system corrections, and other repairs, are postponed to keep production going. Focus on safety is not clearly stressed in employee interactions and site communications.

6. Inadequate Change Management—Organizational changes, staff reductions, retirement programs, and relocations are initiated before their impacts or unintended consequences are fully considered. New hire recruiting, orientation, and training are not used to compensate for the changes. Processes and procedures have not been adjusted following management and operational changes.
7. Plant Operational Events—Loss-producing events are unrecognized or underplayed. Reactions to such events and unsafe conditions are low level or limited in scope and do not reach the attention of the leadership team. Causal factors of loss-producing events are not explored in depth and in a systematic manner.
8. Ineffective Leaders—Leadership is poor, defensive when problems are identified, limited in team skills, or are weak communicators in all areas. Leadership has limited operational knowledge or experience and is not actively learning. The leadership team is not involved in operations and does not exercise accountability or follow up on safety-related issues.
9. Lack of Self-Criticism—The audit or oversight of the organization is limited, with a lack of an unbiased outside views. The oversight system is not formal and any results are not part of the leadership team's internal review. These observations can detect disconnects between operational conditions and the existing self-assessment. Problems remain unidentified, unreported, or not addressed. Adapted from US Department of Energy, Human Performance Handbook (Volume 1: Concepts and principles, human performance improvement handbook, 2009)

Summary

The implementation, sustaining, or improvement of a safety culture requires an almost Sherlock Holmes-like ability to use observation and logic to identify what, where, and how underlying organizational characteristics and traits impact the safety management system. Just as Sherlock Holmes (Sherlock Holmes, n.d.) searches a crime scene, you must be sensitive and aware of subtle clues that may not be quite discernible in the work environment. The existing controls over hazards and risk provide a way to analyze the interactions of job requirements.

If the leadership team desires to improve the safety management system, as we discussed in the chapter, it must ensure that it has clearly defined its overall intent and expectations. Safety is the interaction that results from:

- its beliefs, values, and structure,
- the tools, equipment, and materials in use,
- the people that it hires and interacts with, including the leadership team, and
- the social and physical environment necessary to reach its stated goals—the reason for its existence.

A culture is the result of the long-term experience of the organization in regard to successful problem solving, various personalities, and conditions related to both past and present.

The safety management system should be evaluated for overall effective design and structure that matches the organization's overall operational structure with its inherent hazards and associated risks.

You must understand how "culture" is defined and apply this understanding to the organization. With this knowledge you will have an in-depth understanding of how all aspects of the organization fit together. This would include the safety rules, compliance, and other criteria that are essential for a strategy of improvement of the safety culture over the long term.

In Chapter 3, we highlight the importance of organizational networking in the development of an effective safety culture and how it cannot be overstated. We will provide some concepts on how we are all considered part of a complex web of personal and professional networks. We will demonstrate how your personal network allow you to send and receive information about yourself, the safety culture, the organizational environment, and its related issues, problems, or concerns as well as your successes.

Chapter Review Questions

1. Explain why understanding the culture of your organization is essential to increase the potential for the effective development and implementation of a safety management system.
2. What is organizational culture? Why is it considered essential to the development of an effective safety management system?
3. Discuss the components of organizational culture. Explain why they are important to the safety management system.
4. Compare and contrast the three levels of culture. Explain why it is important to understand these concepts.
5. Explain some area that you can assess regarding your current safety culture. Provide an example from your facility or other facilities in which you may be involved.
6. Explain the patterns of behavior that create habits in a safety culture.
7. To determine the nature of a culture that you work in, consider each aspect of the culture that is involved in driving personalities and characteristics of the organization and how they are interconnected.
8. Refer to Table 2.3 and list the elements that are discussed. Explain how these influence the safety management system and the safety culture.
9. While organizational cultures can vary from company to company, a further complication exists with international enterprise. Explain why problems in communication may develop across culture boundaries.
10. Leadership sets business priorities for the many activities necessary to accomplish its main organizational mission, which is survival in its particular niche in the marketplace. Explain why this is important to "mission-essential" safety management activities.

11. Discuss ways that you can help to change the culture of an organization. Refer to Table 2.4 in "the textbook for guidance".
12. There are nine warning flag factors that can defeat controls of a safety management system. List the nine warning signs and compare and contrast three that you consider most important and state why.

Bibliography

Columbia Space Shuttle Accident and Davis-Besse Reactor Pressure-Vessel Head Corrosion Event. (July 2005). U.S. Department of Energy Action Plan Lessons Learned, Public Domain. Retrieved from http://1.usa.gov/YOWYft.

Craig, D. P. (1978). *Hip pocket guide to planning & evaluation*. Austin, TX: Learning Concepts.

Crutchfield, N. (December 2010). *Nine warning flags, factors that defeat controls* Retrieved from http://bit.ly/r45Vuz.

Deal, T. E., & Peterson, K. D. (1999). *Shaping school culture: The heart of leadership*. ERIC, Jossey-Bass, San Francisco, CA.

Duhigg, C. (2012). *The power of habit: Why we do what we do in life and business*. Canada: Doubleday.

Galloway, S. M. (September 2012). *Zero incident goals motivate risk-taking, not excellence* ProactSafety.com. Retrieved from http://bit.ly/UYdfkn.

git r done. (n.d.). Urban Dictionary. Retrieved from http://bit.ly/UY86c2.

Gladwell, M. (2008). *Outliers: The story of success*. New York: Little, Brown.

Habit. (n.d.). Dictionary.com. Retrieved from http://bit.ly/XUTpm1.

Likert's Management Systems. (n.d.). Wikipedia, the free encyclopedia. Retrieved from http://bit.ly/136rsiY.

Management of Safety, Safety Culture and Self Assessment. (2000). *International conference nuclear energy in Central Europe 2000*. Retrieved from http://bit.ly/15HpFQT.

Schein, E. H. (2004). *Organizational culture and leadership. The Jossey-Bass business & management series*, New York: John Wiley & Sons.

Schein, E. H. (n.d.). Schein's organizational culture model. Jossey-Bass, San Francisco, CA.

Seidman, D., & Clinton, B. (2011). *How: Why how we do anything means everything*. Wiley.

Sherlock Holmes. (n.d.). Wikipedia, the free encyclopedia. Retrieved from http://bit.ly/ZsVvMP.

The Hofstede Centre. (n.d.). Retrieved from http://geert-hofstede.com/.

Volume 1: Concepts and Principles, Human Performance Improvement Handbook, Public Domain. (2009). U.S. Department of Energy. Retrieved from http://1.usa.gov/WdIqoP.

Westrum, R. (2004). A typology of organisational cultures. *Quality and Safety in Health Care*, *13*, ii22–ii27.

CHAPTER 3

Analyzing and Using Your Network

Finally, it is important to consider culture—both occupational and organizational—because it can override collaborative behaviors encouraged by the formal design. In one sense, culture is generated and transmitted through social networks.

—Cross and Parker (Cross, 2004)

I defined a leader as follows: anyone who influences anyone else in a social setting, such as a team or organization.

—Daniel Mezik (Mezick, 2012)

Introduction

The importance of organizational networking in the development of an effective safety culture cannot be overstated. We are all part of a complex web of personal and professional networks. Your personal network establishes how you send and receive information about yourself, the safety culture, the organizational environment, and its related issues, problems, or concerns as well as your successes. You will use this information to establish how you will operate as part of your organization.

In this chapter, we will discuss the importance of defining your personal and professional network in the development of a safety culture. We will introduce concepts for assessing your network through network mapping and the roles that must be understood to effectively move information to the leadership team and throughout the organization.

At the end of this chapter, you should be able to:

- Define the importance of networking,
- Analyze the organizational chart,
- Define your organizational network,
- Develop your organizational network map,
- Define methods that transmit safety information,
- Define different roles associated with the network map.

> You have developed a well-thought-out safety management system, yet efforts to sustain the system are stalled. Safety information, however, is not being used by the leadership team or employees and is diluted or blocked as it moves throughout the organization. Pockets of employees or specialty groups are not getting the message about safety goals, criteria may have been inadvertently bypassed. The safety message may not be getting the consideration needed and is lost within the volume of other information being transmitted.

The Importance of Networking

To avoid the issues described in the above scenario, you must begin to think in terms of your network of personal and professional connections. Your network will determine the speed and quality of the information you provide to the leadership team and employees in your organization. The safety management system is built on how well the safety vision/mission is communicated throughout the organization and influenced by your expertise and efforts. Daniel Mezick defines influence as " … the various processes that affect the thoughts, words, and actions of others"(Mezick, 2012). To continue expanding your influence, you must understand the various internal and external resources, as well as the interconnection between the components of the organization.

The nature, style, and type of information that is being sent must be designed to engage and "stick" with the employees and leadership team that is your target audience. Begin by considering yourself a "node", to be discussed later, in a network consisting of many individuals. Your primary objective is to show that your expertise is of value to your network and the organization as discussed in Chapter 1, "The Perception of Safety."

> *Networking: the exchange of information or services among individuals, groups, or institutions; specifically: the cultivation of productive relationships for employment or business.*
>
> **(Networking, n.d.)**

Duncan Watts tells us that "networks are dynamic objects and not just because things happen in Network Systems, but because the networks themselves are evolving and changing in time, driven by the activities or decisions of those very components" (Watts, 2004). Personal networking has been used by humans for centuries and is not a new concept. It has been used to bypass bureaucracy, gain access to influence or favors, and search for jobs, expertise, and resources for as long as society has existed. Informal networks occur whenever people gather in one place, whether it is at a party, company dinner meeting or picnic, town hall meeting, etc.

> *The reality is that employees feel the need to socialize—and will do so whether it's around the water cooler or online.*
>
> **(Federman, 2013)**

Old sailors would gather around a cask, known as a butt, of water and exchange gossip. This led to the term "scuttlebutt" (Scuttlebutt, n.d.), which is now the modern equivalent of the office watercooler or break room, where employees engage in the same exchange of business and other gossip. The use of social media such as Facebook, Twitter, LinkedIn, etc. has taken a major role in developing and expanding personal networks well beyond the traditional watercooler.

Informal discussions form the basis of the traditional "grapevine" that moves information through an organization, in many cases faster than formal communications systems allow. Organizations spend considerable effort attempting to outpace this informal network to forestall bad information and control issues that may impact morale, production, labor relationships, etc. Adding to the technical side of networking, the use of electronic devices that capture videos and pictures has dramatically increased the transfer of information, with the increased risk that such information goes public or goes global before it is clearly vetted or understood by an organization.

In this environment of connectivity, it is essential that you have an understanding of the nature of social media, networking, how well you are connected within the organization's network, and its impact on the quality and effectiveness of the safety efforts.

Analyzing the Organizational Chart to Assess Your Network

Where are you in your organization's network? Your organization has developed a structure that is the official network. This is normally presented in the form of an organizational chart showing the various departments, leaders, and reporting relationships. It defines the hierarchy of the organization by showing who reports to whom and defines the various job titles, and the specific roles of each member of the organization. The organizational chart is linked in turn to job descriptions. Refer to Figure 3.1 for an "Example Organizational Chart". Refer to Chapter 1, "The Perception of Safety" and Appendix A, "Safety-Related Job Descriptions" for examples of job descriptions for specific safety roles.

Where you are located in this chart tells you where you stand in the organization. This reminds of us a Dilbert cartoon where Dilbert describes where he thinks he is in his organization:

> The Clever Salesman evaluates his prey. "I hope he's an important decision maker." Dilbert: "Take any seat. I call the good chair" Salesman: "Warning! Cubicle! Low-ranking employee!" Dilbert: "Here's our organizational chart: President ... Senior Vice President ... Vice President ... Ok, Lift your foot. Do you see the coffee stain on that carpet?" Salesman: "That's you?" Dilbert: "No, that's my boss. I would be under the carpet" (Adams, 1995).

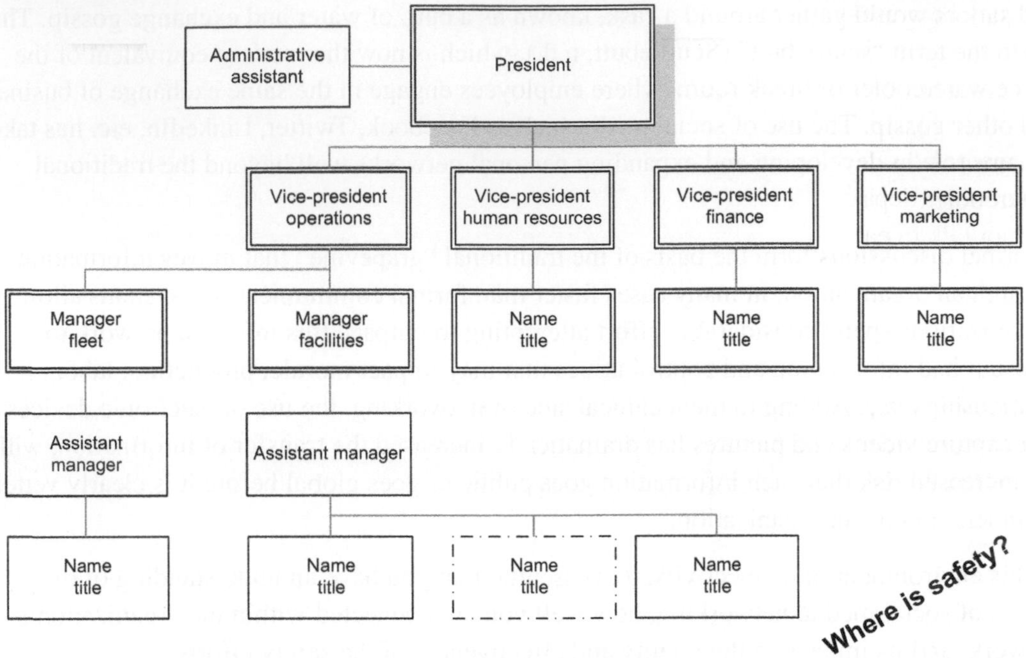

Figure 3.1
Example Organizational Chart.

You have specific assigned responsibilities and authority based on what leadership desires the role of safety to accomplish, and your position may or may not be considered part of the leadership team. The organizational chart provides a map that tells you where you are in the organization, who you report to, and into what area of the organization you report. This establishes your official organization network position.

In reviewing the organizational chart as part of the analysis of reporting relationships, additional questions to ask yourself include:

- Do you reside in an operational or a support function? In many organizations, the role of safety is found in support of operations and is not considered an operational function. From a networking point of view, the potential exists for you being viewed as "not one of us".
- Who do you report to? Does the organizational structure act as an obstacle to safety, impeding your goals and objectives? You may report as a "solid line" to one manager but have multiple "dotted line" reports to different levels of the organization. For example, the reporting structure could be to a number of different levels from the general manager, human resources, engineering, and maintenance—just about any spot in the organization.
- What is the relationship between the various departments and the one where you are located? Do you perceive any communication barriers between departments? Barriers

may be present due to conflicts between personalities, rank of the individuals within the organization, and perceptions about skills sets.

- Are there duplicate safety roles? In our experience, we have found the role of safety split between departments. In one organization, both the corporate risk management department and safety department had assigned safety responsibilities. However, the managers of these departments did not have a high opinion of each other, to the point that they did not talk to each other which was a major obstacle! During a merger or acquisition, duplicate safety roles may find themselves in conflict.

The organizational chart speaks only to the formal structure of the organization and may not function as intended. A senior manager may have allowed his or her responsibilities to be taken over by a strong personality and be in a position that is only a figurehead. A key administrative person may be the actual "go to" person to get a message to a specific leader. The term "power behind the throne" has been historically used to describe who is actually running an organization.

Defining the Organizational Network

Malcolm Gladwell's book *The Tipping Point* (Gladwell, 2000) discusses his three rules for reaching a tipping point when you are attempting to develop what might be called an epidemic for change. He defines three rules that are associated with a tipping point—the "Law of the Few," the "Stickiness Factor," and the "Power of Context".

> A tipping point is the point where an epidemic explodes after growing slowly at first. The goal is to have your safety culture move from a static process to one that goes "viral" through the organization, hence the need for working towards a tipping point.

What Gladwell calls the "Law of the Few" is divided into three sectors—"Connectors," "Mavens," and "Sales". Mavens are the specialist or experts in a given field. As a professional in your discipline, you would be considered a maven in the area of your expertise. This is the value you bring to the organization. A connector is just what it appears to be, a connection or node, and what this chapter discusses. An overriding goal is to be well connected throughout the organization in multiple dimensions in order to access all areas of the organization with information about the control of hazards and risk. As a salesman, you finally have to have the ability to sell your ideas to the organization (Gladwell, 2000).

The Stickiness Factor is how well your message will stick with your intended audience. This factor involves an understanding of what media style and content your message must have to be remembered and utilized by your employees. Your message must be designed and presented so that employees understand and can demonstrate their knowledge of the content (Gladwell, 2000).

Power of Context is concerned with the nature of the environment and culture of the intended audience. The power of context determines how receptive the audience will be to the message. Therefore, your message must be presented at the right time and place, and it must take into account the conditions under which employees are working. Your material and message may be great, but if it is not presented at the proper time and place, your message will not stick. For example, if production is behind and you request time for training, the leadership team will be puzzled about your priorities, especially if the training is on a topic where no visible problem exists or does not aid production (Gladwell, 2000).

Tahl Raz and Keith Ferrazzi make a strong case for the importance of networking in their book *Never Eat Alone*. "If you strip business down to its basics, it's still about people selling things to other people." Raz and Ferrazzi comment, "Ultimately, making it your mark as a connector means making a contribution—to your friends and family, to your company, to your community, and most important, to the world—by making the best use of your contacts and talents." They stress the importance of being constantly aware of and increasing your network of people that can be called upon for assistance, expertise, or advice (Raz & Ferrazzi, 2005).

For example, how many people do you know within your organization? How many employees' names do you know as well as any additional (personal and/or professional) information about them? Do you know about their issues or problems as related to the organization? Do employees feel they can approach you with issues or problems without fear of retribution? We discussed how you are perceived by individuals in your organization in Chapter 1, "The Perception of Safety". Have you ever heard in conversation, "Do not go to a certain individual for any advice or help, he is only talk!" Comments such as these tell you that the individual has restricted or limited his network and put up barriers to communications.

We know of a security/safety manager who makes a point of routinely taking supervisors either as a group or as individuals to lunch. He is able to get them out of their normal work routine to better establish important rapport. He benefits from these lunches because he is viewed as someone who cares and understands the issues and concerns of his staff.

Who do you go to lunch with routinely? Who do you sit by at meetings? Who associates with whom?

When studying your organizational culture, this provides an "artifact" about the culture. These may be viewed as odd questions, but we want you to stop and think about this for a moment. Do you associate with the same employees every day or do you make an effort to meet with diverse employees, non-peers, and others outside your immediate social circle? If it is a multi-shift operation, how often do you visit with the various shifts? See Figure 3.2, How Important Is Your Social Networking?

- Personal
  Who you are?
  Where you are?
  What you are?
  Who do you associate with?

- Marketing
  How you get your message out?
  When do you send your message?
  Who responds to you?, how do they respond?

Figure 3.2
How Important Is Your Social Networking?

Reality Check Indicator

When you go to the break room or approach a group of employees, what response do you get? Do employees talk to you? Or does the room go silent? The type of response can give you an indication of whether you are tied into the informal network of employees or considered an outsider.

Lesson Learned #1

An industrial hygienist we knew was completing a noise study at a processing plant. He noticed that every time he entered the employee cafeteria, all the employees stopped talking. When he asked "why", an employee told him, "Everyone thinks you are recording their conversations." It was found that the safety department also had security responsibility and had recently completed a theft investigation by giving lie detector tests.

This example points out the perception employees had about the leadership team and safety staff. This case is another example of an artifact concerning the safety culture and the impact of fear in the workplace. William E. Deming talks of driving out fear in the workplace. "Deming sees management by fear as counterproductive, in the long term because it prevents workers acting in the organization's best interest" (Cohen, n.d.).

Defining the Basics of Networking Theory

Nicholas Christakis and James Fowler's book, *Connected* (Christakis & Fowler, 2009), provides definitions that are useful in organizing and understanding a network analysis. Christakis and Fowler define a network community as "a group of people who are much more connected to one another than they are to other groups of connected people found in other parts of the network" (Christakis & Fowler, 2009).

> *A social network is an organized set of people that consist of two kinds of elements: human beings and the connections between them.*
>
> **(Christakis & Fowler, 2009)**

To give further insight into the discussion on whom we associate with, Christakis and Fowler have several rules that they believe determine the nature and scope of a network. These include:

- Rule One—"We shape our network". We as individuals decide how many people we want to connect with, what we desire to influence, and how central we are to the overall network (Christakis & Fowler, 2009).
- Rule Two—"Our network shapes us".—Where you are in a network and how central you are within the network. "Being more central make you more susceptible to whatever is flowing within the network" (Christakis & Fowler, 2009).
- Rule Three—"Our friends affect us".—Think about your personal and professional associations. The authors suggest that our connections "offer opportunities to influence and be influenced" (Christakis & Fowler, 2009). You become similar to what your friends are and in turn, you impact them.
- Rule Four—"Our friends', friends', friends affect us". This is what they call their Three Degree Rule that says our influence can go out to approximately three levels or degrees. (Christakis & Fowler, 2009) When we talk about spreading influence throughout a network, this is another important concept to understand. When you use the activity-based safety system (refer to Chapter 11), hold a meeting, or interact with employees throughout a facility, you are not just talking or dealing with one or two people or even a small group of individuals. Refer to Chapter 11, "Developing an Activity-Based Safety System" for more details. You are involved in multiple layers of interactions, not just the one-on-one with the person immediately in front of you. This makes it crucial in how you develop a personal style and a sense of presence (i.e. your personal communication techniques, mannerism, image, etc.). Essentially, you must take the position that you are "on camera" at all times as your influence goes out three degrees. It is both essential and crucial for your safety culture that you are perceived in a manner that is professional and trustworthy. You can destroy your network in one brief unguarded moment. Refer to Figure 3.3, "Three Degrees of Influence".
- Rule Five—"The network has a life of its own".—As we discussed in Chapter 2, "Analyzing the Organizational Culture", organizations have emerging properties that you need to understand. The network evolves and changes as the organization changes.

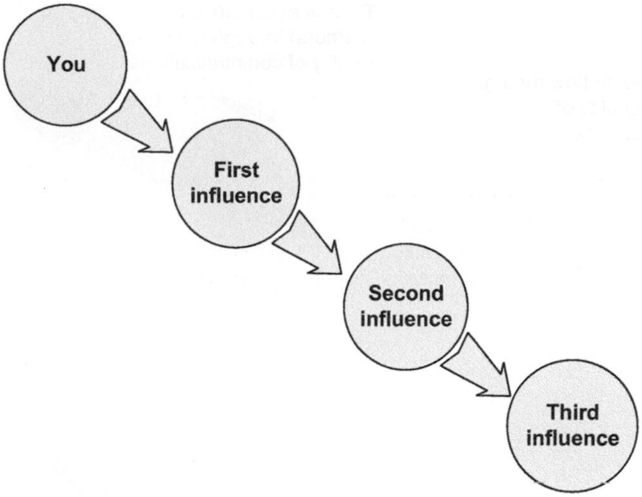

Figure 3.3
Influence Moves Three Degrees. *Source: Christakis, N. A., & Fowler, J. H. (2009). Connected, Back Bay Books.*

> *Emergent properties are new attributes of a whole that arise from the interaction and interconnection of the parts.*
>
> **(Christakis & Fowler, 2009)**

The Safety Information Packet

Your goal is to be able to transmit reliable, proven, and validated safety concepts and techniques through the organization in a manner that enhances the safety culture. An analogy is to view your communications as taking the form of a virus that can go viral and is carried throughout the organizational body. In *The Tipping Point*, (Gladwell, 2000), Gladwell discusses thinking in terms of starting an epidemic, as discussed earlier. Richard Brodie discusses the concept of thinking in terms of a virus in his book *Virus of the Mind* (Brodie, 2009). This concept was developed by Richard Dawkins, who coined the term "meme". You need to begin to think in terms of packets that are considered short bursts of information is easily understood by the intended audience. This process involves new thinking in terms of developing a "meme" that carries your packet of information, which can be accepted and further transmitted to others instead of being rejected, and not going anywhere, causing your meme to disappear.

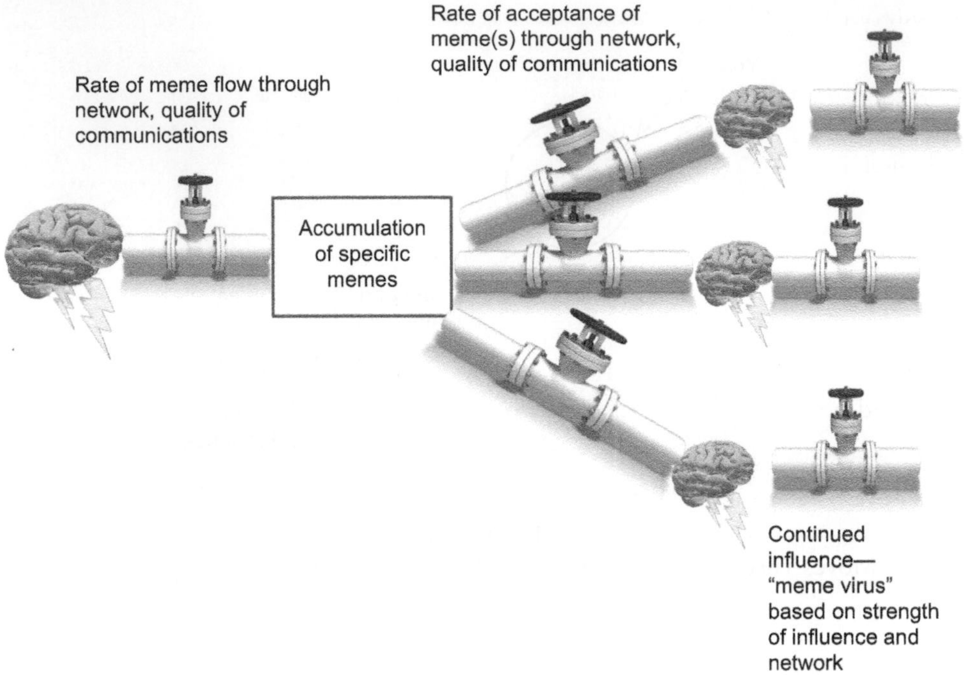

Figure 3.4
Memes Moving from Mind to Mind.

What memes about safety can you think of? Do we have strong, resilient safety memes that persist even though they have not been proven and/or validated? See Figure 3.4, "Memes Moving from Mind to Mind".

> *A meme is a unit of information in a mind whose existence influences events such that more copies of itself get created in other minds.*
>
> **(Brodie, 2009)**

In safety management, we have memes that tell us to do or use certain procedures and methods. We continue to use memes even when they do not work in our current culture. In addition, a meme that worked well in one organization may be rejected in another location.

Richard Brodie outlines several types of memes. A "Distinctive Meme" is one that describes, "categorizes", or "labels" something (Brodie, 2009). We have a safety meme and/or at-risk behavior meme that labels something as safe versus unsafe. The meme called "safety" carries with it information about its meaning which includes perceptions about safety.

Because safety is such a strong meme (everyone knows the definition of safety, right?), we normally do not define the term, assuming everyone knows what we mean. However, safety may or may not have the same connotation with individuals, groups, or organizations. If it is to have influence and replicate itself in other minds, it must be clearly defined. We discussed this in Chapter 1, "The Perception of Safety."

A second type of meme Brodie discusses is the "Strategy Meme". This type of meme tells you what to do (Brodie, 2009). If you behave properly and act in a certain way, then you create a "cause and effect". If I do "X" in this way, then I will remain "safe" as defined by the safety meme. The impact of the strategy meme is that if the overall environment has changed, the strategy meme that is being followed may emphasize behavior that is no longer "safe". An example would be changing equipment, starting a new process, or getting a new vehicle; the old meme may still be applied. For example, after moving into a new house, you go into a room and reach for a light switch only to find it is not there. In the new house, the switch is in a different place. Your light switch meme from the old house is still in active use in your mind.

Brodie's "Association Meme" links various memes together (Brodie, 2009). When the information of one packet of information is activated, it activates other information. An example of how this can create an unsafe situation is the meme, "Where there's smoke, there's fire." When we hear people making assumptions, this is a meme. When investigating the source of "smoke" do you immediately think, "Where's the fire?" However, what if it is not smoke but a harmful chemical vapor? You would be seriously injured from inhalation of the chemical. You observed what you assumed was smoke and this activated the meme, "smoke equals fire." However, all "smoke" is not fire related.

Brodie suggests " … a good meme … easily spreads throughout the population … " (Brodie, 2009). He suggests three ways to get a meme to penetrate and take hold:

- Conditioning—Repetition of the same message (Brodie, 2009). We suggest a form of this type of conditioning in the use of the activity-based safety system in which daily, weekly, and monthly activities become habitual behaviors. Behavioral-based programs, following safety procedures, and completing preshift or work activities are all methods to get conditioned responses necessary to ensuring hazard and associated risk controls are effectively and consistently used.
- Cognitive Dissonance—"Creates mental pressure and resolving it" (Brodie, 2009). Building a case for an improved safety culture creates a dissonance between where the organization is now as an undesirable condition and a vision of where it could be.

> Cognitive dissonance—"In modern psychology, cognitive dissonance is the feeling of discomfort when simultaneously holding two or more conflicting cognitions: ideas, beliefs, values, or emotional reactions" (Cognitive Dissonance, n.d.).

Defining and establishing a vision/mission of what the safety culture could be and setting goals and objectives creates a cognitive dissonance (mental pressure). If the safety meme is strong enough, it replaces the old memes that accepted lower standards and at-risk habits.

- Trojan Horse—This is a hidden meme that is encased within another meme (Brodie, 2009). In Chapter 2, "Analyzing the Organizational Culture", we discussed how the Alcoa CEO desired to improve and change the overall culture of quality, creativity, and customer relations (Duhigg, 2012). He used a safety meme to change other existing memes of the organization.

Changing Reality versus Perception

To change memes, we as professionals have to start thinking about reality versus perceptions. We use a toolbox of concepts to aid us, as discussed throughout this book.

In real life we see smoke, and if we see smoke the meme (perception) tells us that it's fire!

For example, if a production meme (distinction meme) says "keep equipment running", then a machine operator will try to work on equipment that is running and will be subjected to a higher probability of getting caught in a piece of equipment. Introducing a risk perception meme of "stop, think, and ask before reaching into equipment" replaces the "OK to put hands into running equipment to unjam" meme. Refer to Chapter 9, "Risk Perception—Defining How to Identify Personal Responsibility". Not only is a risk perception meme needed in this example, but strong memes can encompass employee involvement, training, job hazard analysis, and the overall safety management system.

Changing a meme requires ensuring that your communications and their content stick with your audience. Chip and Dan Heath discuss how to make a message stick in their book *Made to Stick, Why Some Ideas Survive and Others Die* (Heath & Heath, 2007). They suggest using a blend of the following elements when communicating a message:

- Simple—To the degree possible, keep the message simple. Can the safety vision/mission statement be readily quoted by employees? Most organizations have a long policy statement signed by a senior manager, sometimes more than a page, that most employees cannot remember. Refer to Chapter 4, "Setting the Direction for the Safety Culture." A short and simple, brief statement that conveys the overall policy is needed to activate an association meme. Simplicity applies to safety rules and guidelines as well. Are they designed for ease of use by employees so that they can readily understand and remember them?

- Unexpected—This opens up a gap in knowledge or finds a way to close a gap in knowledge. The objective is to get and keep the attention of the intended target audience.
- Concrete—All the concepts and ideas about the safety management system and what is desired in a safety culture must be made visible and provide a clear understanding of ongoing actions to be taken.
- Credible—The message must have credibility. The threat of "OSHA may visit us" is used too often as a crutch and sets the perception of safety being only minimum regulatory compliance instead of using a hazard and associated risk approach. Building on data and valid research improves the potential for your message to be accepted. The perception of safety and about your professional image and abilities plays into the development of the safety culture.
- Emotional—The Heath brothers suggest that for a message to carry merit it must view people in a compassionate way. It is not just the engineering and design that is important, but ensuring all employees that the organization is sincere in its approach. Messages about being safe for family and community do work when combined with a total approach using the safety management system. Caution must be taken as too much use of emotion in your message can be counterproductive and short lived.
- Story—Use stories to explain or share experiences. Not long-winded stories, but stories that are direct, personal, and paint a picture of what you want to accomplish. (Adapted from *Made to Stick* by Heath and Heath (2007).)

Improving how you integrate and communicate your personal message can ensure that the safety culture meme "sticks" inside the minds of the leadership team and employees.

Now that the concepts of how to devise and develop your safety culture message have been covered, the next step is to properly launch it into the organization. By launch we do not mean an e-mail broadcast, newsletter, or other media, but by beginning to understand how communication travels through your organization.

Silence is a form of communication and you must be alert to its meaning. A report from Vital Smarts gives the results of its research and finds silence in various forms leads to a project failure rate of 85%:

- "*Silence Fails* study demonstrates that when even one of these crucial conversations fails, a silent crisis plays out in a deceptively simple dynamic that results in failure to execute an initiative 85% of the time. The general effects are going over budget, deadlines, and failing to meet quality and functionality specs. Team morale is inevitably damaged in the process. Alternately, when these conversations succeed, the failure rate is reduced by 50–70%."

(Executive Summary, The Five Crucial Conversations for Flawless Execution, 2006)

Continued

Consider the following:

1. *Don't mistake silence for agreement.* When you pitch ideas to people, look closely at their body language. Are they "open" to your idea, making eye contact and sitting slightly forward, or are they "closed", looking away, doodling, or otherwise not connecting with you?
2. *Learn to ask, "What am I missing?"* This simple open-ended phrase will help begin a conversation and shows that you are open to the notion that your idea may not be fully drawn.
3. *Reward differing viewpoints.* If someone offers a differing view, begin with, "I have not thought of it that way ... " Ask follow-up questions such as, "Have you considered ... " and "What has to happen in order for you to get comfortable with this idea?"
4. *Reflect on past interactions.* If your team is normally forthright and now they have clammed up, something about the situation has changed. Reflect on how this current "silent" situation is different from times past. Is there a new dynamic that that is causing the silence?
5. *Ask for feedback.* The best way to get people to open up is to ask for, and then neutrally listen to, feedback. If this is new for you, then you may need to ask several times before people will step up. Be sure to sincerely thank the first person who ventures feedback—even if it is misguided, poorly worded, or completely irrelevant. You want to reward the act of speaking up. Later, you can work on coaching people to give constructive feedback (Miller, 2011).

Silence tells you much about your personal network and how well you are connected into the overall organization's structure. When you make a presentation or complete a training session, what feedback do you get—many questions or dead silence? When you send a newsletter to your personal network, do the articles start positive conversation or do you get only limited feedback? With respect to the leadership team, are your issues on their meeting agenda or are safety items routinely cut from the agenda, deferred, or not covered due to time limitations? Silence is an indicator of the depth of the safety culture and the perception held about safety.

The review of your personal network as well as the organization's network can provide crucial insights on the nature and quality of your connections and whether or not you are fully wired in to the organization.

Social Networking Analysis

Social networking analysis has not, in our opinion, been effectively used to the degree needed to better ensure that important information is communicated about a safety management system. We believe that a more formal approach to networking analysis can lead to more effective safety culture development.

Social networking analysis is a method used to better understand the dynamics of an organization. As all organizations consist of networks of individuals, departments, units, or groups, it can provide insights on how the connections affect the behavior of the communications under study.

It can be used to assess, over a wide range of levels, how the organization sends and receives information.

> *Social network analysis (SNA) ... allows analysts to identify and portray the details of a network structure. Its shows how a ... networked organization behaves and how that connectivity affects its behavior. SNA allows analysts to assess the network's design, how its member may or may not act autonomously, where the leadership resides or how it is distributed among members, and how hierarchical dynamics may mix or not mix with network dynamics.*
>
> **(Petraeus & Amos, 2006)**

To improve and enhance the safety culture requires understanding how to establish and use your personal network to allow your safety meme to be accepted and move rapidly through the organization.

> The formula for determining the number of connections is N(N-1)/2. If you are communicating with five other people, the number of connections is six (including you) × six minus one divided by two, which equals 15 lines of connections. Since these can be two-way communications, 30 two-way directions of communicating exists. See Figure 3.5 on "Lines of Connections from Six People or Nodes". (Metcalfe's Law, n.d.)

Social networking analysis can assist in determining the quality and scope of personal and professional interactions. These interactions can be graphically displayed and used to assess a variety of communication elements and map such items as:

- Who talks to whom
- Direction of the communication (one-way or two-way)
- Quality of the communication
- Whether the communication is positive or negative
- How the message is sent and received internally and externally (face to face, e-mail, memos, newsletters, etc.)
- Structure and style of your message to reach different audiences

While social networking analysis can be very detailed, even a rudimentary analysis can uncover gaps, not just in the message and to whom it is sent but in the quality and nature of how the message is being conveyed.

> *Networks of social information ... are important not just because they help us make better individual decisions but also because they allow things that have caught on in one setting to spill over into another.*
>
> **(Watts, 2004)**

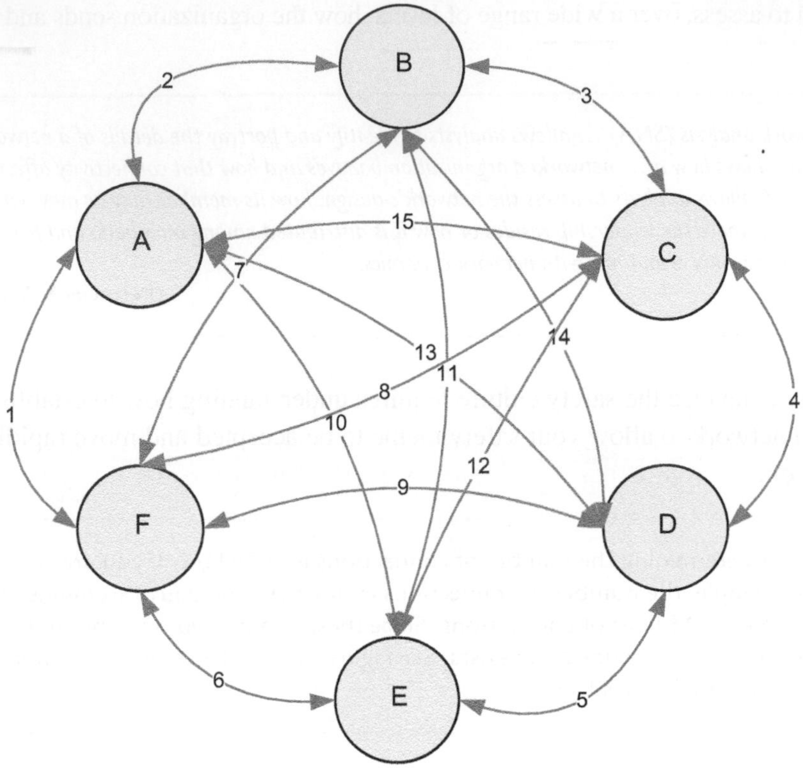

Figure 3.5
Lines of Connections from Six People or Nodes.

Social Network Mapping

A social network map is similar to mind mapping in that it connects nodes such as topics, individuals, ideas, and relationships. A map is developed using three components as building blocks. A social network map shows individuals or groups and the connections between them. Basic definitions for social network mapping include:

> *Individuals in a network are called* actors *or* nodes. *(Actor and node are often used interchangeably.) The contacts between nodes are called* links. *The basic element of a social network graph is the* dyad. *A dyad consists of two nodes and a single link. In the simplest form of a network, the two nodes represent people and the link represents a relationship between them.*
>
> **(Petraeus & Amos, 2006, Figure 3.6)**

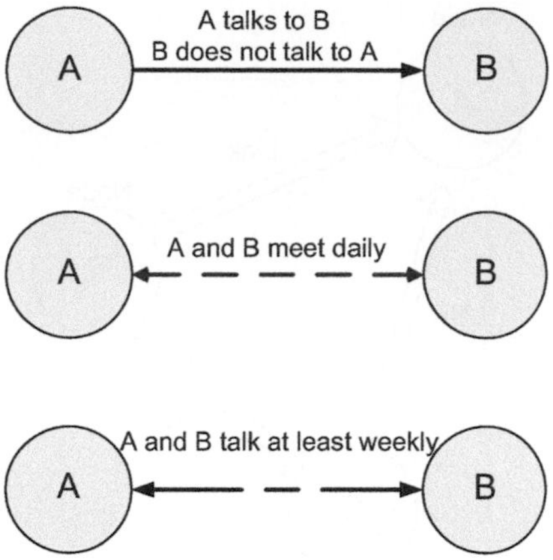

Figure 3.6
Dyads—Relationships.

- A node represents one entity. It could be defined as an individual or you could define a node as a specific group, department, or whatever you are trying to map. You are a "node" within an interconnected group of nodes. The people you associate with are nodes and the people they associate with are nodes. If you are interested in how departments interrelate, you can develop a map that shows how nodes are connected and the communications between them. Information that you send and receive goes and comes from other nodes that you are connected to.
- Connection between nodes is a "link." Your connection to other people is a link. For example, when you talk to another person you "link" with them. When you as part of the leadership team have a face-to-face discussion, hold a daily preshift review, a weekly or monthly meeting, area walk-throughs, and/or review the equipment/machine-specific checklist with your employees, you are directly linked with your employee(s). Refer to Chapter 11, "Developing an Activity-Based Safety System." Through your links, your message is directly communicated to the employees in each of these activities. Where you are not directly connected is where information may not be flowing. In addition, the employees you are linked with are in turn linked to other employees.

As discussed earlier, your influence can go across three degrees of separation (i.e. the employees you talk to will talk to other employees who in turn talk to employees). If you are more than three degrees away from other nodes, your message will have faded away (Christakis & Fowler, 2009).

This spread of influence and your message is why the perception of safety and your professional image is vital.

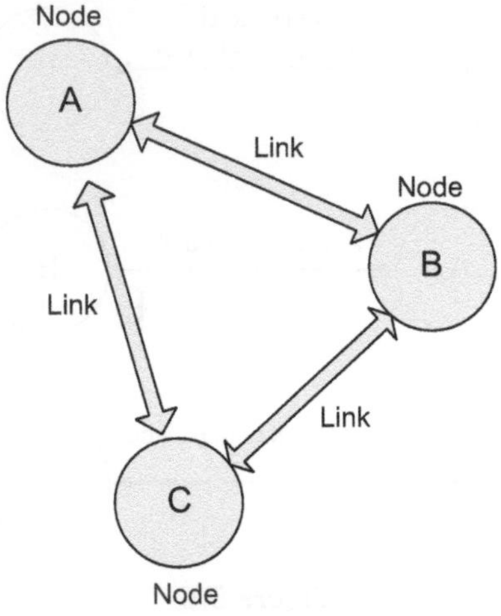

Figure 3.7
Social Network Terms.

- The connection between two nodes is called a "dyad," which consists of two different nodes with a link that represents the relationship between the two nodes.

The lines between the nodes of the dyad as shown in Figure 3.7, "Social Network Terms", have arrowheads that represent the direction of the flow of information between three nodes. In this example, we show two-way communication between the three nodes. In addition, the link can be used to specify other characteristics of the link between the nodes. For example, you can show if the communication is daily or weekly, if it is a positive or negative relationship, or other characteristics that you may want to evaluate.

Building your network map is done by simply connecting your nodes together. Your network map will represent internal and external connections. In addition to your direct personal network, begin the process of identifying the next levels of links. If you talk to the Vice President of Human Resources, try to determine that person's links with members of the leadership team and employees. Continue to look for the links and the characteristics of the links throughout the entire organizational network.

A "hub" is a node that links between multiple nodes. It is the same concept as a bicycle wheel that has a hub connected to the rim by many spokes. Without the nodes successfully connected together, the network, such as a bicycle wheel, does not work. Refer to Figure 3.8, "Social Networks". Without nodes linked together, information does not flow through the network or from and to you.

Your network analysis, like a road map, shows the destination of your information, how well the "road" (a link) is designed, and where intersections are showing alternative routes that can be established that allow you to move around obstacles (leaders or employees resistive to change and obstacles to improving the safety culture). Your network map identifies positive and

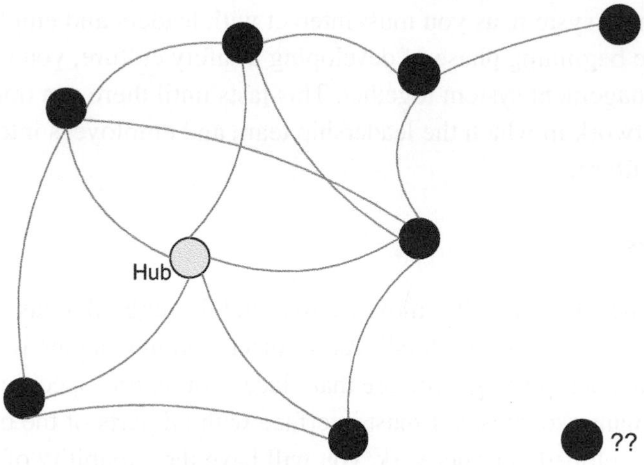

Figure 3.8
Social Networks.

influential leaders and employees who can contribute in getting the safety message transmitted throughout the organization. As discussed earlier, Christakis and Fowler's rule number 4 (Christakis & Fowler, 2009) suggests that if you can get within three degrees of the senior leadership team and have structured your information or message in the style used by leadership, the probability for acceptance of your message increases. Refer to Chapter 2, "Analyzing the Organizational Culture", for a detailed discussion on ways to approach the different types of corporate cultures.

Defining the Roles Identified by the Network Map

As you develop and analyze your personal network and the organizational structure, you will find that your role requires you to cross many other dimensions of the organization. Your role should be in a position to provide in-depth knowledge of an array of subjects and topics. You will also find leaders and employees that can assist you, as they are nodes that occupy a position in the network, making them important linkages in the movement of information. Their role as nodes may act as a hub that can push your message to all levels of the organization. Cross and Parker defines types of roles that your function must fill—"Central Connectors", "Boundary Spanners", "Information Brokers", and "Peripheral People" (Cross & Parker, 2004).

Central Connectors

"Central Connectors" are leaders and employees who act as hubs with multiple links to many other nodes. The safety management system requires you to become the central connector for the safety management system, as you must interact with leaders and employees (Cross & Parker, 2004). In the beginning phase of developing a safety culture, you are the "glue" that holds the safety management system together. This lasts until there is a transition to a trusting and participative network in which the leadership team and employees internalize the value of the desired safety culture.

Boundary Spanners

The Boundary Spanner (Cross, 2004) moves across and through all areas of the organization without restriction (Cross & Parker, 2004). As we discussed in Chapter 1, "The Perception of Safety", you may have a reporting structure that places you in one specific department, such as operations or human resources, yet must interface with all parts of the organization. If you have successfully developed your network, you will have the capability of being a boundary spanner and have higher probability of establishing positive relationships within the organization.

Lesson Learned #2

Nathan was approached by a management consulting firm that wanted to discuss the role of safety managers. The issue they were having was that safety managers were being promoted into senior management positions. Unfortunately, these safety managers were perceived as being only knowledgeable in regulatory compliance, with limited understanding of the complexity of an organization and its management. Senior leadership was concerned that they had made a mistake in promoting the safety manager with whom they could not adequately communicate their needs and concerns.

Information Brokers

The safety function is a specialized discipline, and as such you have knowledge and information that can be of benefit to the entire organization. As both a boundary spanner and a central connector, you can become an essential "Information Broker", (Cross, 2004) an information provider, mentor, resident expert, and advisor for the organization (Cross & Parker, 2004). You are also a "go between" node who connects people in order to share information that is specific to the safety management system and the safety culture. For example, you have gathered information about a serious safety-related topic. You find a solution and transmit your findings to not just the affected employee or department but ensure that the information gets to other areas that have a similar exposure to the identified hazard.

Peripheral People

As you complete the networking map, look for "Peripheral People" (Cross & Parker, 2004)—Individuals that are outside your network or are mavericks that operate independently outside the organizational network norm. These individuals may not be using up to date risk information and may be following the outdated safety criteria.

The basic nature and type of questions that can be asked include:

- Does your network map show communication gaps?
- Are you positively linked with each level of the leadership team?
- Are all employees able to quickly link with you?
- Can a critical hazard or associated risk message rapidly get to the level of leadership appropriate to the potential severity that must be addressed? (as an example–Challenger, Columbia, BP Deep Water Horizon)
- What are the primary types of communication used by the organization?
- What is the strength of your relationship with the leadership team and employees?
- Are all your communications positive and two-way? What actions can you take to improve relationships with the leadership team and employees?
- Are you centrally located as a hub in the network and directly connected to key leadership with no more than three degrees of separation?

Lesson Learned #3

A social network analysis was used with a safety professional who had good personal and technical skills and was beginning the revitalize a safety program. A network map was developed that showed the links he had with other managers and supervisors in his organization as well as their other links. In a matter of minutes, simply using a white board, a number of serious communications gaps were discovered with linkage issues that he immediately began to address. We found gaps where key leadership was not linked to him, as well as a styles of communication that needed to be modified.

Summary

The scope and depth of a safety culture is dependent on how well the safety management system information is communicated. Begin to think in terms of your personal and professional network. Your network will determine the speed that your safety management system information moves through your organization. The quality of your information and its value to the leadership team and employees establishes the perception of safety, and in turn, your influence on the safety culture.

The organizational chart provides a map that tells you where you are and into what area of the organization you report. The chart defines the official structure and your formal position in the organization. Keep in mind that the organizational chart is in constant change and must be checked routinely.

Your personal and organizational network must stress using information that increases your influence with the leadership team and employees and in turn increases the potential for a quality safety culture.

In Chapter 4, we will discuss that when trying to improve, develop, or ensure that the safety management system can be sustained, you have to develop a planning process that will move the organization forward. This planning process ranges from daily, weekly, and monthly activities to annual and multiyear planning. The overall big picture is to ensure that the safety culture is consistently maintained and that your safety management system is effective and efficient in the identification and control of hazards and associated risks.

Chapter Review Questions

1. Explain why you are all part of a complex web of personal and professional network.
2. Why is it important to develop your network? Explain in detail how this will affect your interaction with the organization.
3. Discuss ways to analyze an organizational chart to assess your network.
4. Define an organizational network. Explain the importance of the organizational network.
5. Discuss methods that you can use to check the reality of where you fit in your organization.
6. Discuss the basics of networking. Describe the term "social networking".
7. Understanding how the safety information flows in an organization is important. Discuss how the network is able to transmit reliable, proven, and validated safety concepts and techniques through the organization in a manner that enhances the safety culture.
8. Discuss the difference in changing safety reality and perception. Provide some examples.
9. We discussed the concepts of "meme" in this chapter. Explain what "meme" means to your organization. Discuss methods to change a "meme".
10. Your social networking analysis is important in an organization. Discuss in detail why this is important. Provide several examples.
11. Define social network mapping as described in this chapter. Provide several examples.
12. Define the roles identified by a network map. Provide examples.

Bibliography

Adams, S. (1995, November). *Dilbert cartoon, organization chart*. Dilbert. Retrieved from: http://bit.ly/ViMnYb.

Brodie, R. (2009). *Virus of the mind: The new science of the meme*. Hay House, Carlsbad, CA.

Christakis, N. A., & Fowler, J. H. (2009). *Connected: The surprising power of our social networks and how they shape our lives*. Little, Brown, New York, NY.

Cognitive Dissonance. (n.d.). Wikipedia, the free encyclopedia. Retrieved from: http://bit.ly/105nAsv
Cross, R., & Parker, A. (2004). *The Hidden Power of Social Networks* (1st ed.). Harvard Business School Press, Boston, MA.
Cohen, P. (n.d.). Deming's 14 points. HCi Consulting, Realisation Consulting.
Duhigg, C. (2012). *The power of habit: Why we do what we do in life and business*. Doubleday Canada, New York.
Executive Summary, The Five Crucial Conversations for Flawless Execution. (2006). *Vital smarts*. The Concours Group. Retrieved from: http://bit.ly/14hN2Pj.
Federman, E. (2013, January). *Internal social networks increase workplace productivity*. Socialmedia Today Website. Retrieved from: http://bit.ly/Wxcn29.
Gladwell, M. (2000). *The tipping point: How little things can make a big difference*. Current Affairs/Business. Little, Brown.
Heath, C., & Heath, D. (2007). *Made to stick: Why some ideas survive and others die*. Random House Publishing Group, New York.
Metcalfe's Law. (n.d.). Wikipedia, the free encyclopedia. Retrieved from: http://bit.ly/XUtTPX
Mezick, D. J. (2012). *The culture game: Tools for the Agile manager*. Amazon Kindle. Retrieved from: http://amzn.to/14dizmj.
Miller, J. V. (2011, March). *Psychological safety in the workplace: 5 tips for creating a "Speak up" culture*. Human Capital League. Online Community for Workplace Professionals. Retrieved from: http://bit.ly/X7y0FW.
Networking. (n.d.). Merriam-webster, incorporated. Retrieved from: http://bit.ly/WYIbMF
Petraeus, D., & Amos, J. F. (2006, December). *The US Army/Marine Corps Counterinsurgency Field Manual, FM 3–24, MCWP 3-33.5. Approved for public release*. Headquarters Department Of The Army, Public Domain. Retrieved from: http://bit.ly/105MvMj.
Raz, T., & Ferrazzi, K. (2005). *Never eat alone: And other secrets to success, one relationship at a time*. Doubleday.
Scuttlebutt. (n.d.). Wikipedia, the free encyclopedia. Retrieved from: http://bit.ly/Vyg6hA
Watts, D. J. (2004). *Six degrees: The science of a connected age. Science*. W. W. Norton.

CHAPTER 4

Setting the Direction for the Safety Culture

A goal without a plan is just a wish.
—Antoine de Saint-Exupéry

Do not go where the path may lead, go instead where there is no path and leave a trail.
—Ralph Waldo Emerson

Plans are nothing; planning is everything.
—Dwight D. Eisenhower

Introduction

When trying to improve, develop, or ensure that the safety management system can be sustained, you have to develop a planning process that will move the organization forward. This planning process ranges from daily, weekly, and monthly activities to annual and multiyear planning. The overall big picture is to ensure that the safety culture is consistently maintained and that your safety management system is effective and efficient in the identification and control of hazards and associated risks.

> *Let's go places.*
>
> *Not just the ones you can find on a map.*
>
> *But the ones you can find you in your heart!*
>
> *Let's go beyond everything we know and embrace everything we don't.*
>
> *And once we have reached our destination, let's keep going, because inspiration does not favor those that sit still.*
>
> *It dances with the daring and rewards the courageous with ideas that excite, challenge, and even inspire.*
>
> *Ideas that take you a place that never you imaged …*
>
> **(Toyota, Let's Go Places, TV Commercial, 2013)**

We routinely use planning in our daily lives. When we plan a vacation with our families, we have some idea of where we want to go and a vision of the family having an enjoyable time. We set a goal that addresses how we will achieve this vision and establish objectives for what we want to see when we get to our final destination.

But before we start the journey, we discuss with our family what we want to do and reach an agreement about the trip. Then we compare our various options, based on our time and resources. After we have allocated money for our budget, we now know how much we are willing to spend and the length of time that we plan on traveling. Once we understand the associated cost and how long we will be staying at our destination, we then make our reservations for lodging, car rental, entertainment, the places we would like to see, etc.

> *If you don't know where you are going, chances are you will end up somewhere.*
>
> **Yogi Berra**

Before we begin our trip, we make arrangements for someone to look after our house, stop our mail, and make arrangements to board our pets. The last thing we typically do is program our Global Positioning System (GPS) and print out a map of our destination as a backup. Then, the trip begins.

We spend a considerable amount of time planning our vacation to ensure that we have covered everything as it is a complex event that requires much detail and coordination.

The objective of this chapter is to demonstrate concepts that can help to define a structured format that keeps the safety management system on track. To accomplish this, you must write a safety policy that outlines the vision/mission, goals, and objectives that are needed to pull together all of the critical program elements that support the safety culture. This policy ensures that all individuals involved and the required activities, materials, and methods of communication are properly considered and are aligned with other organizational drivers.

After completing this chapter you will be able to:

- Discuss methods for developing a vision/mission statement.
- Compare and contrast personal and organizational scope drift.
- Discuss how to develop and communicate a safety policy statement.
- Define how to develop goals and objectives using S.M.A.R.T. principles.
- Define the difference between a goal and an objective.
- Discuss why it is important to develop a plan.

Charting Your Course—The Planning Process

Taking the vacation planning analogy one step further and applying it to your organization, you will find that planning and developing a safety management system is no different. You will have a similar coordination, planning, and budgeting process.

When developing and planning your safety management system, a number of things must be accomplished before you can start your journey:

- You must define your overall vision/mission of what you want to accomplish.
- You must decide on what you want to accomplish and then put your reasoning in writing. This will be your safety policy statement that defines and outlines your vision/mission and overall purpose, or your intent.
- You must determine your destination. What do you want as your end result? This will be your overall desired goals for the safety culture.
- You must map out your intended path toward your goal and what actions it will take to get to your destination. These are your objectives.

The primary goal for any safety management system to be successful is to clearly articulate the overall vision/mission, safety policy statement, and stated goals and objectives to everyone in the organization. When this is done properly, the probability of acceptance and gaining greater support and commitment is improved.

Figure 4.1
Charting Your Course. (*Source: Presenter Media, n.d..*)

Nothing can stop the man with the right mental attitude from achieving his goal; nothing on earth can help the man with the wrong mental attitude.

Thomas Jefferson

Lesson Learned # 1

Are there words or phrases used by your organization that effectively turn people off to projects? James has experienced certain terms routinely being used that had become unknowingly negative with employees. He noted that when employees heard the statement, "We are rolling out a new XXX next month and … ", their eyes would roll back and they would begin glancing around with strange looks on their faces. Their reception was, "Here comes another flavor of the month, do not get too excited, it will go away in six months if we just wait. Do just enough to keep them happy."

The term "roll out" had become a toxic statement. James readily saw the need to find better ways to introduce initiatives that would have a positive perception and communicate the value of any change or endeavor. He found that by replacing "roll out" with "introduce", an initiative had a higher probability of acceptance.

Vision/Mission, a Major Trait of Leadership

Much has been written on the importance of developing an overall vision/mission. In starting the development of your safety management system or correcting its course, you must have a vision/mission of what you want to achieve (i.e. what is the desired intent or end results). Steven Covey has as one of his seven habits that we should "begin with the end in mind" (Covey, 1990). Establishing a clear vision/mission of what is to be accomplished brings a sense of purpose to the

You must first have a clear understanding of what you want to accomplish or what has been mandated. You begin developing a meaningful safety management system by answering the following questions:

- What is the current state of the safety culture?
- What are the desired end results that I want to achieve?
- What does management expect to be accomplished? In many cases, management may not have an understanding of hazards and the associated risk and control requirements. An assessment of the leadership team's understanding has to be taken into account.
- What regulatory compliance activities are ongoing and need to be supported?

- What activities are needed to address risk management efforts for areas not covered by regulatory compliance?
- What time constraints of key individuals must be considered?
- What type of budget for resources is needed for implementing the right plan?
- Given budget, time constraints, skill levels, etc. what are the critical essentials necessary for meeting the desired goals and objectives?
- How will the plan be communicated? (Adapted from Safety and Health Management Basics, Module Eight: Evaluating The Safety Management Systems, n.d.)

safety process that can now be communicated throughout the entire organization. As you work toward creating your vision/mission, you need to ask yourself two questions:

- Do I just want to be in compliance with regulatory requirements or
- Do I want to build a true safety culture that will sustain itself and have long-term impact on my organization?

Refer to Figure 4.2, "Which Way Do I Want to Go?".

A compliance-driven vision/mission results in meeting the mandated guidelines and regulations but may be limited to only meeting minimum requirements. This approach may lead to not identifying all hazards and associated risk if only targeting elements based on the regulatory compliance standards.

The vision/mission is a point of reference that is a check and balance to see if your efforts are on the right course to your ultimate destination. This is just like using GPS. If you change your direction, the GPS will attempt to reroute you to get you back on your intended path. The key is, if you do not have a clear vision of where you are going, then how do you know when you have reached your designation (i.e. when your goal and objectives has been achieved)?

Figure 4.2
Which Way Do I Want to Go?

Organizational Scope Drift

As you develop your safety management system through planning your goals and objectives, one of the issues that you may face is "scope drift".

> Scope drift is a rifle sighting term describing the condition where your telescope is not aligned with the rifle barrel. When aiming at a target through an unaligned rifle scope, you will miss your target every time.

Using this analogy, organizational scope drift creates a gradual drift away from the safety culture as your safety management system goals and objectives are not aligned with the vision and mission that have been established. We have experienced this gradual change as different responsibilities were assigned, business conditions shifted, new leadership had different priorities, and mergers with different organizations created turmoil in the culture due to different structures, personalities, operating procedures, etc. Once you are off target, it will be hard to realign the desired goals and objectives.

Personal Scope Drift

We have experienced how easy it is to fall into a trap of working on our own unaligned projects as well as those mandated by leadership. The following are examples:

- You do not have a written plan with defined goals and objectives for improving the safety culture. As organizations are in constant change, you can lose sight of what you want to accomplish unless you write down your goals and objectives. You should create your own personal plan about what you want the safety culture to become as well as plans for the organization. If your daily routine does not allow for time to assess your network and the nature of the organizational culture, you need to meet with your leadership and make the case for change.
- You enjoy working on cool projects that have nothing to do with your vision/mission. Steven Covey described these types of projects as "unimportant and not urgent". This type of project gives you and the organization the perception that you are very busy even if those projects have no relevance to the safety management system. Always ask the question, does this project add value to the safety culture or the safety management system? (Covey, 1990).
- You may have to work at the direction of agendas that are not your own. In this case, ask questions about what you have been assigned to do. Use a questioning attitude. It will be necessary to understand the internal and possible external political and

leadership team reasoning to get the full intention and desired objective of the leadership's agenda (Volume 1: Concepts and principles, human performance improvement handbook, 2009).

Lesson Learned # 2

In our experience, first, do what you are asked to do and meet the desires of leadership by completing the basic requirements of your job description. In many cases, your perception and leadership's perception may be different. Do what is asked in order to meet the mandates of leadership. If a memo with bullets is desired, do not write a thesis just because you think that is a good idea. We are not saying be mediocre. Be professional and always maintain your ethics—Do the right thing!

You will routinely be pulled in many directions as your job requires responding to leadership and employee safety concerns that you are handed or that routinely develop. How many times have you set a schedule for the day and then your manager stops by your office and states "Oh, by the way, I just came from a meeting and you need to do 'X' immediately"? This is a case where you have to change your agenda and do what is asked. Consider this as the typical interaction that can alter your daily course.

You must develop a strong personal discipline to maintain the view that your response and how you handle these requests can directly impact the perception about you and the role of safety in the organization as discussed in Chapter 1, "The Perception of Safety", and in the maintaining your network and its influence on the safety culture as discussed in Chapter 3, "Analyzing and Using Your Network".

The Safety Policy Statement

The safety policy statement is different from the formal overall safety policy. The safety policy provides in-depth details about the who, what, and where of safety administration, defines responsibilities, and describes the approach to be taken with regard to safety. The safety policy can be considered an artifact that points to the espoused values of the organization. How well the policy is implemented and followed will give an indication as to the true values held by the organization. The safety policy statement is a brief, concise sentence that conveys the essence of the safety policy and drives immediate action and behavior.

For any safety culture to become entrenched into the organization, the leadership team must demonstrate its commitment to all employees and the safety management system. A well written safety policy statement establishes leadership's intent and makes clear to everyone in the organization what to do when a conflict arises between a safety issue and immediate operational priorities. Refer to Appendix F for a sample safety policy.

To be effective, the safety policy statement should contain the following elements:

- A declaration of the leadership team's commitment to safety efforts.
- Goals and objectives of the safety management system (Building an Effective Health and Safety Management System, 1989).

Any action used to introduce the safety policy statement must be done after consideration of current and past policies, actions, and events that have been implemented. If other policies are not enforced, then the safety policy statement may also follow the same fate. In addition, be aware of resistance to following policy in the form of cynical comments by employees that can indicate lack of acceptance or rejection. The key to a successful safety policy statement is to periodically involve all employees in its review.

> *Once a policy is developed, management should set a goal for safety and health, and then build objectives that will allow employees to reach the goal. The goal should be a realistic one, so as not to discourage employees from striving for the goal. Each employee should be able to see his or her work activities moving toward the goal, thus allowing him/her to meet the objectives.*
>
> **(Safety Pays, n.d.)**

Once the safety policy statement is signed and endorsed by the leadership team, it must be communicated to all employees by a series of organization-wide meetings such as the pre-shift, weekly, and monthly meetings as discussed in Chapter 11, "Developing an Activity-Based Safety System". As part of keeping the policy statement visible, it should be posted on hard copy and electronic bulletin boards throughout the organization and incorporated into informational materials such as newsletters, memos, and the safety manual. The safety policy statement should be presented and discussed during orientations with new, transferred, or relocated employees and contractors.

Communicating Your Safety Policy Statement

No matter the size of the organization, the safety policy statement must be easily readable and understood through the spoken word. The safety policy statement must be:

- Presented in a brief, clear manner in the language of the employees.
- Consistent in its transmission of the intended message. Everyone should receive the same content and message regardless of who delivers it and how it is delivered.

- A point of reference to be used when safety-related issues or activities are in conflict with operational priorities.
- Supported by the leadership team through involvement in the safety management system by routine guidance and the enforcement of established hazard and associated risk control, safety rules, and work practices.

The following examples provide simple, brief policy statements that convey an easily remembered leadership message:

- "Our most important responsibility is the safety of our employees".
- "Everyone in our organization will have a safe place to work".
- "Everything we do will be done in a safe manner".
- "If a task is not safe, do not do it".

Safety policy statements that are multiple pages long are too complex to be remembered. Trying to convey the intent of the leadership's message so that everyone will be on the same page is very difficult when you have a lengthy safety policy statement. You should use the "Kiss" principle, "Keep It Simple, Stupid", for a statement to be remembered (KISS principle, definition, n.d.). This acronym was first coined by Kelly Johnson, lead engineer at the Lockheed Skunk Works (creators of the Lockheed U-2 and SR-71 Blackbird spy planes, among many others) (Johnson, n.d.). We have used "Keep It Simple and Streamlined" as a variation on the same theme.

Communicate by Action

In communicating the safety policy statement, the manner in how it is presented is crucial. What you do, or fail to do, speaks louder than words! A nod or gesture at a critical point in the discussion of the statement can communicate the perception that what is being said is not really to be taken seriously. The tone and method used to introduce the statement signal whether the message will be received well by employees.

Aligning the Organization

For the safety culture to evolve, demonstrating commitment and concern is not enough. The safety culture improves or degrades depending on the commitment of the essential resources for control and improvement of the workplace. It is also dependent on how well leadership is organized and capable of addressing issues beyond just regulatory compliance. When we view the safety policy statement from the standpoint of how influence expands and travels throughout an organization, we see how imperative it is to manage and constantly review how positively and clearly the safety policy statement is received.

The bottom line is that the written statement is really not the policy. It communicates the intent of the leadership team's safety policy. The real policy is the attitude of the leadership team and employees in dealing with safety issues. The policy demonstrates commitment by the actions of the leadership team necessary to become ingrained in the organizational culture.

Defining Goals and Objectives

> In most organizations, a planning process is in place for quality improvement and annual setting of personal and professional goals and objectives. If your organization has planning tools to develop goals and objectives, use them! Do not reinvent the wheel, become an expert in that process. This ensures that the format you use is in line and consistent with the rest of the organization.

Now that the safety policy statement has been developed, you are ready to define your goals and objectives. A goal defines where you want to be, your final destination. It provides insight on the direction that you need for your safety management system (Safety Pays, n.d.). Refer to Figure 4.3, "Setting Goals".

Goals and objectives are derived from your vision/mission and form the basis for a plan that becomes a "living" document that should be flexible enough to allow change as organizational conditions change.

> *It's never too late to be who you could have been.*
>
> **George Eliot**

Using the desired planning process, you are in the same alignment as the organization.

Figure 4.3
Setting Goals.

Defining Goals That Improve the Safety Management System

> *Goals are general guidelines that explain what you want to achieve. They are usually long term in nature and represent your overall vision.*
>
> **(Define Goals and Objectives, Step 2, 2003)**

Your task is to provide the leadership team with guidance that provides timely feedback on decisions and actions that are not in tune with the safety culture.

> "Would you tell me, please, which way I ought to go from here?"
>
> "That depends a good deal on where you want to get to," said the Cat.
>
> "I don't much care where … " said Alice.
>
> "Then it doesn't matter which way you go," said the Cat.
>
> " … so long as I get SOMEWHERE," Alice added as an explanation.
>
> "Oh, you're sure to do that," said the Cat, "if you only walk long enough."
>
> *Alice's Adventures in Wonderland*, (Carroll, n.d.)

According to Oregon OSHA's document, "Safety and Health Management Basics", "Goals are easy to write. They are nothing more than wishes" (Safety and Health Management Basics, Module One: Management Commitment, n.d.).

This may be simply stated but until an action plan is written down and implemented, goals are really nothing but a wish, similar to a New Year's resolution. The tendency is to just start doing something without a formal or even rudimentary outline or plan on what is to be accomplished. This leads to the "scope drift" discussed earlier and meandering from unfocused activity to activity. Without goals and objectives, "bias to action" can develop that results in "busy work", where you can get caught up in an activity trap or ineffective multitasking.

Paul J. Meyer's book *Attitude Is Everything: If You Want to Succeed Above and Beyond* describes a format for developing and gauging effective goals that he calls "S.M.A.R.T." He suggested that goals be *Specific, Measurable, Attainable, Realistic,* and *Timely* (Meyer, 2003).

The following is an overview of the S.M.A.R.T. principles:

Specific

Use information that clearly establishes the five Why's and one How, as discussed below. Being specific requires the use of designated individuals, locations, reasoning, and time frames that are tangible.

Using selected data such as reduction of the total case incident rate (TCIR), or workers' compensation experience modification rate (EMR), etc. does not add "value" to an organization, as described in Lesson Learned #3.

Lesson Learned # 3

When James was working on his Six Sigma Black Belt Certification, he conducted research on goal setting in order to map out his required certification project. The project was originally designed using 5 years of TCIR as a measuring device as calculated from the OSHA 300 log. As he conducted his research and tried to validate the data collected, he found that this TCIR data did not capture all the information needed.

He challenged the project team to find out why the data could not correlate. Although a traditional goal for overall TCIR reduction had been set, it was invalid as it did not provide the essential information to clearly identify where to focus the project. Use of TCIR does not take into account all the different types of injuries that are not recorded on the OSHA log. Only after developing a more in-depth analysis was the team able to target a specific injury reduction goal.

This above lesson learned provides insight on an issue that directly affects goal setting. In the absence of systematic data collection, analysis of hazard, risk assessment, and related safety management system data, you may not have a clear picture of the real situation. Poor data collection leads to an erroneous gap analysis that in turn results in setting goals that do not address the right underlying safety-related issues.

Lesson Learned # 4

As James continued to research goal setting, he referenced W. Edwards Deming's classic book *Out of Crisis*, in which Deming described his "14 Points". Deming noted that "Goals focus on the outcome rather than the process". Deming stressed that numerical goals can be in conflict with other goals and do not give any information about the underlying process of activities that generates the numbers (Deming, 2000).

Refer to Appendix G for an overview of numerical and descriptive goals.

TCIR and EMR data and measurements are subjected to influence from conditions and elements beyond the safety management system's control and can negate what on the surface may appear to be a successful accomplishment. The reduction of the TCIR and EMR may or may not be an indication that the safety culture improved.

■ Lesson Learned # 5

We have both experienced what looked like major improvements in both TCIR and workers' compensation injuries only to find that organizational changes had reduced exposures with shifts from permanent employees to contractors or temporary employees, technology changes, etc. The overall safety culture remained unchanged as well as the basic hazards and associated risk. ■

To better define and establish a goal, answer the five "Why's" and one "How" question:

Who:	Who is involved?
What:	What do I want to accomplish?
Where:	Identify a location.
When:	Establish a time frame.
Why:	Specific reasons, purpose, or benefits of accomplishing the goal.
How:	Define the actions to be taken.

After developing answers to these questions, you move on to the next goal setting stage and determine if the goal is measurable.

Measurable

In this stage, you establish a clearly defined set of criteria for measuring progress toward the goal that you have established. To determine if a goal is measurable, you should ask the "how" question: "How will I know when the goal has been?" Once a goal is defined as measurable, you then decide if it can be achieved or is attainable.

> *Measurable—Establish concrete criteria for measuring progress toward the attainment of each goal you set. To determine if your goal is measurable, ask questions such as … How much? How many? How will I know when it is accomplished?*
>
> **(Meyer, 2003)**

The bottom line is that what is measured must be important to both the organization and to those individuals who will directly and indirectly impact what is being measured.

Attainable

One of the things that we often forget to ask is if the intended goal is attainable. When goals have been established and are considered to add value to the process, you begin to visualize in your mind ways you can make goals come true.

> *Attainable—When you identify goals that are most important to you, you begin to figure out ways you can make them come true. You develop the attitudes, abilities, skills, and financial capacity to reach them.*
>
> **(Meyer, 2003)**

If you set a goal to reduce the number of injuries by 10% and/or reduce your TCIR by 25% this year, what does this really mean? Is this measurable and attainable? To reach a goal, you have to have some control over the outcome.

Many organizations base their safety performance data on incident rates as the sole measurement of the safety process. As noted before, TCIR can be impacted for many reasons. Therefore, if you are using TCIR as a goal measurement tool then you should consider the following:

- Incident rates, frequency (total recordable rates, which includes recordable injury cases), and severity (the number of restricted lost days due to workplace injuries) are downstream measurements. This is similar to trying to inspect quality of a finished product after the fact rather than trying to control the quality during the process. As stated before, for example, if you set a goal to reduce injuries by 10% and/or reduce your TCIR by 25% this year, what does this really mean? The TCIR can change based on changes in the numbers of employees, contracting actions to others, increased use of temporaries, and decisions made for operational purposes. A goal may appear successfully attained and yet no change has occurred in the safety culture.

To put this into context, there is a Dilbert cartoon that portrayed Dilbert's boss reporting:

> *Our Goal this year is Zero disabling injuries. Last year, our goal was 26 disabling injuries. In retrospect, this was a mistake. We had to injure nine employees to meet the goal ...*
>
> **(Adams, 1998)**

This cartoon example sends the message that you expect employees to get hurt and some level of failure in the safety management system. Is this the message that you want to convey to the organization? Using the TCIR alone without using upstream activities presents a false sense that hazards and associated risk are in control when they are not.

Lesson Learned # 6

Nathan had a safety manager state in a conversation that the corporate safety audit content and scoring criteria was increased each year. No matter how hard his plant tried to meet the target criteria, passing the audit became unattainable. He was never able to improve and finally gave up trying. Raising the bar too often leads to disillusionment and lowers the motivation to remain involved.

For example, if goals are set too high, they may be impossible to reach. If goals are set too low, they may be attainable, but achieving them results in no improvement in the safety management system or the safety culture. The organization will lose interest if the goal is too easy and become frustrated if the goal is too high.

If a goal is not attainable and pressures exist for its completion (bonuses, merit raises, awards, incentives, etc.), then human nature will find a way to reach the goal. Finding ways to get credit for attaining the goal may result in unintended consequences of "pencil whipping" of reports, hiding of claims, exaggeration of data, and other negative activities that degrade the safety culture.

After the goal is defined as achievable, the question becomes, "Is it realistic to think that the organization can achieve this goal?"

Realistic

To be realistic, a goal must represent something of value that leadership and employees can fully support and are willing to find the time and put in the effort for its success. Each goal must be relevant to both the organization and the safety management system. Each goal must be evaluated to ensure that it brings value to the organization's safety culture if it is achieved.

> *Realistic—To be realistic, a goal must represent an objective toward which you are both willing and able to work. A goal can be both high and realistic; you are the only one who can decide just how high your goal should be. Your goal is probably realistic if you truly believe that it can be accomplished.*
>
> **(Meyer, 2003)**

Consideration must be given to available time, budgets, materials costs, skills required, and communication needs. A goal becomes unattainable if the requirements for its success are not addressed.

Lesson Learned # 7

Nathan once encountered an issue where the recommendation was made to establish a goal for providing literacy classes to employees. The goal's intent was that given the low literacy rates at the facility, if people could learn to read, they could read the safety handouts and manuals. This was a worthy goal; however, it was unrealistic given the nature of the organization, employee turnover, the production process, etc. It could not have been attained due to many organizational constraints.

Once your goals are determined to be realistic, now determine if they can be achieved in a timely manner.

Timely

Time is one of the most important and probably least appreciated parts of goal setting. A goal should clearly identify the specific time frame needed to complete it. Without a goal being tied to a time frame, no sense of urgency is present and there is no ability to set it as a priority. Setting a time schedule should not be done without close coordination with the leadership team, operations, training, human resources, and other departments that may have previously established their own goals.

> *Time Bound—A goal must have a target date. If you desire to make a million dollars, but don't set the timeline for it, it won't be motivating. A deadline too far in the future is too easily put off. A goal that's set too close is not only unrealistic, it's discouraging.*
>
> **(Meyer, 2003)**

Lesson Learned # 8

A safety/security manager we know keeps a "collision calendar". The calendar tracks all the various organizational and personal events, training, holidays, meetings, mandated activities, etc. and prevents overlap with other departments that may have already established and scheduled events. Once "collisions" are identified, he has the option to use alternate dates or try to negotiate for the date or time with the person with the schedule conflict. This calendar allows him to review with the leadership team the potential for over scheduling employees.

An additional issue that must be factored into setting a time for completing a goal is to consider the intent of the goal. Does it involve a hazard control issue or associated risk? Flexibility is needed to assure crisis-related goals are quickly reviewed and the time requirement set to ensure that corrective action gets a high priority (Meyer, 2003).

Taking individuals away from their defined jobs and duties requires that consideration be given to what resources are needed to cover their absence. Understanding time constraints and conflicts prevents goals from becoming unattainable. Good time planning increases the probability that goals will be accepted and supported by all those involved in the process.

For example, to improve the safety culture, one goal may be to implement a structured safety management system as discussed in Chapter 5, "Overview of Basic Safety Management

Systems". The goal begins by determining what activities will be necessary to activate safety management system elements. Using the S.M.A.R.T. format, each element is reviewed for its specific criteria and a measurement method is determined. American National Standards Institute (ANSI) Z10-2012 notes that " … because the risks, organizational structure, culture, and other characteristics of each organization is unique, and each organization has to define its own specific measures of performance" (ANSI Z10, Overview and History, Module 2, Lesson 1, n.d.; Occupational Health and Safety Systems, 2012).

The following is an example of setting a goal:

Goal: Each supervisor in department X will complete job hazard analyses on two designated jobs by January 31. (The objectives on how to achieve this goal will be discussed later.)

Refer to Table 4.1 for "additional suggestions on specific goals" that you can use based on the example that we provide you.

Too often, an assumption is made that the leadership team knows and understands the intent of the goals and objectives. A discussion with the leadership team is essential to communicate the reasoning for each goal, as well as how each goal can be attained and its relevant value to the overall organization.

Defining Objectives

For each established goal, an objective must be developed that is tangible and provides the essential steps or actions necessary to accomplish the stated goal.

According to Michigan.gov, "Objectives define strategies or implementation steps to attain the identified goals" (Define Goals and Objectives, Step 2, 2003).

In addition, "How" must be asked as well: How will it happen? How will it be known that the goal is achieved?

Table 4.1: Additional Suggestions for Specific Goals

- Inspections (facility, equipment, materials, etc.) will be conducted each week in high risk or hazardous areas.
- Maintenance work orders will be analyzed for valid safety priority for correction.
- Conduct one-on-one safety communications with employees.
- Establish communication links with all employees.
- Hold safety meetings (2 minute shift review. weekly meeting, and monthly meeting).
- Machine-specific checklists to be completed by the machine or equipment operator, similar to the daily sign off required by forklift/powered industrial truck operators.
- Develop a machine-specific checklist to validate all equipment safe operating procedures.
- Review the orientation training programs for new, temporary, and transferred employees to ensure hazards and associated risk and safety information are included.

> *Goals can only be achieved by setting objectives. Objectives are the specific paths you will follow to achieve a goal. They are statements of results or performance. They are short, positive steps along the way to your organization's goals.*
>
> **(Safety Pays, n.d.)**

Objectives take a little more thought than writing goals. "Well written objectives should have the following elements present:

- Starts with an action verb (decrease, increase, improve, etc.).
- Specifies a single key result to be accomplished.
- Is quantifiable. Uses numbers to measure a desired change (i.e. 50% increase).
- Specifies a target date for accomplishment" (Safety and Health Management Basics, Module One: Management Commitment, n.d.).

Objectives should be integrated as a permanent feature of the safety management system. Objectives can be quantified and provide information on what works and does not work.

Example Goal: Area manager will work with each supervisor in department X to complete job hazard analyses on two designated jobs by January 31.

Objective 1—Each supervisor will select two designated jobs based on the departmental loss-producing events analysis provided by the safety department, initial hazard/risk assessment, and employee interviews.
Objective 2—Area manager will discuss with supervisors how the analyses will be developed, their time requirements, and the resources they will need.

Writing Your Objectives

Each goal and each goal's objectives should be clear and concise: what is to be achieved, by whom, to what degree, and by when. Goal and objectives must be communicated throughout the organization so that all employees understand how each goal was developed. Employees should understand their level of involvement, the reasoning, and purpose of the goal and how they fit into the overall achievement of the safety management system.

To be effective, goals and objectives must be in writing and in a structure that allows for their tracking and monitoring. Depending on the complexity of the goal, tracking can be completed using any number of tools: project management software, mind maps, spreadsheets, etc. The key is to write down short goals and each objective to reduce the potential for confusion or misunderstanding. Refer to Table 4.2 for "some general guidelines on how to write your objectives". In addition, cross-check your goals and objectives against the safety policy statement and whether they bring value to the safety culture. Goals and objectives should target areas with the highest hazard severity potential and associated risk to achieve the maximum benefit from investment in time and resources. Refer to Chapter 7, "Leadership and the Effective Safety Culture".

Table 4.2: Guidelines for Writing Objectives

Objective	Sample Action Words
Starts with an action verb	"Conduct"
Specifies a single key result to be accomplished	"One preshift daily talk"
Specifies a target date for its accomplishment	"To be completed by (Insert Date/Time)"
Relates directly to the role of all individuals in the organization	"To deliver your daily message about hazard or associated risk in area"
Is readily understandable by those who will be contributing to its attainment	"Discussed message in a common language to address comprehension"
Is realistic and attainable but represents a significant challenge	"Safety professional mentor supervisors to develop own topics with employee suggestions"
Is consistent with available resources and organizational policies and practices	"Use employee feedback discussions, preshift review documents, checklist, monthly meeting notes, surveys, and appropriate HR forms". Refer to Chapter 11, "Developing an Activity-Based Safety System".

Source: Adapted for use from Illinois OSHA, Managing Worker Safety and Health (n.d.).

To review the status of the goals and objectives, milestones or benchmarks are set to ensure that each goal and objective is in process and whether it has been attained and properly implemented. A combination of formal reviews and discussions with the leadership team and employees is used to confirm that objectives have been completed and the goals reached.

Communicating Your Goals and Objectives

To be effective, once you have defined your goals and objectives, they must be communicated to all employees to ensure that everyone understands the level of expectation desired. As with the safety policy statement, goal communication will reaffirm the leadership team's commitment to safety (Roughton & Mercurio, 2002; Safety Pays, n.d.).

If your goals and objectives have been achieved, but discussions and safety management reviews indicate that conditions have not changed (i.e. loss-producing events have not declined or little improvement in hazardous recognition and control), then you should review your goals and objectives for their attainability. In this case, your goals and objections may have been set too low, wrong goals were set, introduction of the goals was not effective, buy-in from leadership team and employees was not strong enough, etc. Furthermore, organizational conditions may have changed after the goals were established.

If you find that the goals and objectives are not being met, conditions, resources, production requirements, etc. may have changed and are impacting the potential for achievement. (Roughton & Mercurio, 2002; Safety Pays, n.d.). An immediate, direct "face-to-face, sit down" communication meeting with the leadership team must be completed to correct/refine and/or reestablish commitment of the goals and objectives, reassessing time requirements, resources, employee involvement, etc.

When writing your objectives you must state in specific measurable terms what is to be achieved and to what degree it should be accomplished. Remember, the key is to try and keep your objectives easily understood, tangible, and measurable (Managing Worker Safety and Health, n.d.).

One issue is keeping goals and objectives visible and consistent. Putting your objectives in writing will clarify the specific meaning and intent of what you want to accomplish. With the newer technologies, videos can be quickly and economically customized into brief presentations for daily review and communicating your message. As objectives are completed and goals achieved, the videos can be updated to show their status. If questions develop, the objectives are in a format that you and others can review in various media formats.

Reviewing Your Objectives

To be effective, you must periodically review the status of objectives to ensure that you are getting the desired results. For example, as already stated, if supervision meets the goals and objectives, but the department continues to have loss-producing events, then a reassessment of objectives is needed (Roughton & Mercurio, 2002). Sometimes the required changes will be obvious. But in other cases, you may need to carefully explore the reasons for the objective's not being met. The key is to identify possible solutions with employee involvement. The wrong employee may have been assigned a particular responsibility without an explanation of the requirements or expectations. In this case a simple change in the assignments may fix the issue. The level of authority and responsibility may need to be changed or increased so all employees know who is assigned to accomplish what.

The evaluation can either be verbal or written and can include the following elements:

- Unacceptable performance should be identified and changed as quickly as possible. The evaluation is an opportunity to provide encouragement and ongoing training and allow for more employee involvement.
- As the goals and objectives should be clear and understood, the employee being evaluated should not be surprised regarding expectations. If it is found that misunderstanding has developed or there has been personal or organizational "scope drift", it will be necessary to modify the goals and objectives to ensure that they are still specific, measurable, attainable, realistic, timely. Employees may need further explanation of what is expected with additional training required.
- The evaluation is an opportunity for the leadership team and employees to explore new ways of improving the safety management system.
- The outcome of the evaluation session should be to encourage employee responsibility. During these evaluations, provide positive reinforcement and feedback for each goal and objective that has successfully been completed (Managing Worker Safety and Health, n.d.; Roughton & Mercurio, 2002).

Once the evaluation has been completed, all of the agreed upon changes must be incorporated into the revised objectives. Many evaluation systems break down when leadership fails to incorporate and implement new or modified agreed upon tasks.

Task and activity should be monitored based on events to support the evaluation. For example, you may need to monitor supervisors' incident investigation processes after each loss-producing event until the supervisors have developed the necessary investigation skills.

Keep in mind that the complexity and formality of your evaluations should be consistent with the safety management system (Managing Worker Safety and Health, n.d.; Roughton & Mercurio, 2002).

Goals and objectives can be measured in variety of ways that can be either numerical or descriptive. Refer to Appendix G for a description of numerical and descriptive goals.

Resistance to Goals and Objectives

While establishing goals and objectives may appear to be a simple process, they require change in the organization that will affect the norms, beliefs, or habits currently in place. When making or suggesting any type of change, unless the need for change has been clearly defined and its importance communicated, resistance to change must be expected. The safety management system may come into conflict with other departments that do not understand the scope of risk-related importance of the goals and objectives.

Individuals may feel that their time and resources are being diverted to projects or activities that are not of value to them or their area of responsibility. They may get defensive if objectives are required in a time frame they have not been fully agreed upon.

Lesson Learned # 9

We have both experienced the situation where a corporate initiative is "rolled out" to locations without it having been clearly communicated to leadership and employees as to why it is important. During one meeting that Nathan experienced, one of the participants told the meeting facilitator, "I have many goals being set for me by various corporate departments. Every one of them labeled critical and to take priority over all the others. Could we have a moratorium on initiatives because I can either try to do your stuff or do my real job?" The lesson is that as you set goals and objectives, it is crucial to ensure that consideration is given to all the goal and objective criteria we have outlined because you will probably find yourself in similar situation at some time.

Understanding interactions with your social network can identify the nature and scope of your influence and aid in finding the best approach to establish goals and objectives. Involving all levels of the organization to participate in setting objectives creates a more dynamic safety culture by establishing an atmosphere of buy-in, acceptance, and commitment. In turn, this can create a stronger level of trust in which objectives have an increased potential of attainment and what is working and what is not working can be determined. Our experience has found that very candid discussions can be held if individuals know that they are actively involved in the process.

The Plan

After the safety management system and safety culture have been assessed and a gap analysis is completed, the stage is set for developing an overall plan outlining specific goals and objectives for the organization.

As a starting point, the initial plan is typically developed with the leadership team. Once the initial plan is completed, you begin to solicit input from the widest possible network to gain the necessary buy-in. This check within the network ensures that concerns, obstacles, and communications from diverse opinions are considered and incorporated into the plan. The more individuals involved in reviewing your plan, the better the chance of getting it accepted when it is finalized and introduced to the organization.

A plan pulls together the actions needed by various diverse safety programs and activities. As an example, ANSI Z10 outlines 21 core elements that define a safety management system (Occupational Health and Safety Systems, 2012). The basic format suggested by OSHA's Voluntary Protection Program includes six basic core principles that OSHA considers the starting point for a safety culture (Voluntary Protection Program (VPP), n.d.). For discussion on these systems, refer to Chapter 5, "Overview of Basic Safety Management Systems" and Chapter 6, "Selecting Your Process."

A plan serves as the tangible visual road map, the physical representation of your intent to improve the safety culture from its current state to a new desired state. A good plan defines the steps to be taken and determines the order in which each goal is to be approached and in what order each objective is to be undertaken.

The Critical Part of Planning

Sustaining a system is sometimes an overlooked aspect of planning. A safety plan must take into consideration all of the safety management system elements that are in place and identify essential areas that may require change and improvement.

The consequences of assigning a goal without establishing the means of sustaining it, waste resources, energy may harm the morale of those individuals who achieved the goal. Maintaining a process requires that the safety culture has the necessary norms and beliefs in place that provide the long-term buy-in needed to sustain the safety management system.

Communicate Your Plan

Lesson Learned # 10

Nathan was once asked by a regional manager to evaluate a safety director. "All he does is wander around the plant. I see him sitting on the line just talking to workers or in supervisors' offices, I never see him in his office working. I do not think he is doing any safety stuff. Tell me whether or not he is doing the job." Nathan met with the safety manager and, while in his office, saw a notepad that had a number of pages torn away. At the top of the remaining page was written "101" with comments about a safety issue. Curious, Nathan asked what this was about. The safety manager replied, "That is my to-do list." He was asked, "Why 101?" "Well, I finished the first 100 items I had listed for improving the safety program. I have been out in the plant visiting with workers and they have given me a lot of insights and the supervisors are starting to come by and tell me their concerns. I had to get their trust first".

The lesson learned is to not assume that the leadership team knows what you have accomplished. You need to let them know by regularly communicating what has been accomplished. Ideally, you will be able to provide your accomplishment and feedback in leadership team meetings. Take these meetings seriously and be prepared to quickly and clearly articulate the status of goals and objectives. Use the written objectives in discussions with your supervisors and employees. During conversations, spot check the status of goals that are not directly under your authority but are part of your plan.

Summary

The chances of influencing the decisions of the leadership team in shaping the safety culture are improved with an understanding of goals and objectives.

Goals should be designed to bring value to the overall strategic plan of the organization. The leadership team is dependent on the guidance you provide in establishing properly designed goals and objectives for the safety management system.

Goals define the desired end results or targets necessary to close gaps in the safety management system. Objectives are the steps and activities designed to reach goals. It is critical that as much data and information as possible is gathered about the current state of the organization safety culture in order to compare it against a desired future state.

Your review of goals and objectives should address the following:

- Are the goals and objectives specifically chosen to improve the safety management system and the safety culture?

- Have the S.M.A.R.T. concepts been used to assess goals and objectives?
- Has responsibility for each goal and its objectives been assigned?
- Has the time requirement for the assigned actions been reviewed based on an assessment of time and budget constraints?
- Have resources been assessed—people, money, media, and equipment?
- How will the goals and objectives be communicated, tracked, and evaluated?
- How will "course corrections" be made to keep the plan effectively implemented?
- Have the requirements and needs for sustaining achieved goals been assessed?

A safety culture is the end result of the safety management system and the many other functions of the organization fitting together. It is the result of all organizational goals and objectives incorporating hazard control and associated risk into the decision-making process. A successful organization places the safety of the organization and employees not ahead of priorities of production, service, sales, and quality control but includes concepts of hazard and associated risk control into the overall strategic planning process.

In Chapter 5 we will discuss an "Overview of Basic Safety Management Systems" and in Chapter 6 we will continue our discussion on "Selecting Your Process". In these chapters, we will discuss approaches for developing a foundation for a successful safety process and define what a safety process should look like structurally. Do not assume people know what is being done!

Chapter Review Questions

1. Explain the planning process and how it can enhance the improvement of the safety culture in your position at your company.
2. Discuss why vision/mission are the major traits of leadership as defined in this chapter.
3. Compare and contrast personal and organizational "scope drift". What does "scope drift" mean to you? How can you apply this to your position?
4. What is a safety policy statement? If your organization has a written safety policy statement, how does it compare to the discussion in the chapter? Write a safety policy statement that you can use in your organization.
5. What is the difference between a safety policy statement and a safety policy? Provide an example.
6. What is the best method to communicate your safety policy statement?
7. What are the factors that determine whether the safety culture improves or degrades?
8. What is the method that you would use to define your goals and objectives?
9. What are some of the goals that could be developed to improve the safety management system?
10. Describe the elements of S.M.A.R.T. and how you can apply it to your planning process.

11. How would you define objectives?
12. Write five objectives for a safety management system. Use Table 4.2 as a guide.
13. Review the objectives that you have written. Is each objective clear and concise?
14. Explain why there is resistance to goals and objectives.
15. What are the critical parts of planning? Discuss your thoughts.

Bibliography

Adams, S. (December 1998). Dilbert Cartoon, Zero A Number. Dilbert. Retrieved from http://bit.ly/X3OcrF.

Carroll, L. (n.d.). Alice's Adventures in Wonderland, Chapter 6. New York: Dover Publications.

ANSI Z10, Overview and History, Module 2, Lesson 1. (n.d.). NC State University. Retrieved from http://bit.ly/T0e7UD.

Building an Effective Health and Safety Management System. (1989). Partnerships in Health and Safety program, Partnerships in Injury Reduction (Partnerships), Reprinted/Modified and/or adapted with Permission. Retrieved from http://bit.ly/WsHteC.

Covey, S. R. (1990). *The 7 habits of highly effective people: Powerful lessons in personal change*. A Fireside Book, Simon and Schuster, Inc., Retrieved from http://amzn.to/WsCA5g.

Define Goals and Objectives, Step 2. (2003). *Michigan State Government, public domain*. Michigan State Government, Public Domain. Retrieved from http://1.usa.gov/WlIGEa.

Deming, W. E. (2000). *Out of the crisis*. MIT Press. Retrieved from http://amzn.to/14o8bbI.

Johnson, K. (n.d.). KISS principle. Wikipedia, the free encyclopedia. Retrieved from http://bit.ly/Yc5Snw.

KISS principle, definition. (n.d.). Business Dictionary.com. Retrieved from http://bit.ly/X8DRfs.

Managing Worker Safety and Health. (n.d.). Illinois OSHA Onsite Safety & Health Consultation Program, Public Domain, Adapted for Use. Retrieved from http://bit.ly/WTsneh.

Meyer, P. J. (2003). *What would you do if you knew you couldn't fail? Creating SMART goals. Attitude is everything: If you want to succeed above and beyond*. Meyer Resource Group, Incorporated.

Occupational health and safety systems. (2012). The American Industrial Hygiene Association. ANSI/AIHA Z10.

Presenter Media. (n.d.). Retrieved from http://bit.ly/XLpgGC.

Roughton, J. E., & Mercurio, J. J. (2002). *Developing an effective safety culture: A leadership approach, Adapted for Use*. Butterworth-Heinemann. Retrieved from http://amzn.to/X8Gaz8.

Safety and Health Management Basics, Module Eight: Evaluating The Safety Management Systems. (n.d.). Oregon Occupational Safety and Health Division (Oregon OSHA), Online Course 100, Public Domain, Permission to Reprint, Modify, and/or Adapt as necessary. Retrieved from http://bit.ly/UVLJj0.

Safety and Health Management Basics, Module One: Management Commitment. (n.d.). Oregon Occupational Safety and Health Division (Oregon OSHA), Public Domain, Permission to Reprint, Modify, and/or Adapt as necessary. Retrieved from http://bit.ly/Yztx5p.

Safety Pays. (n.d.).*Oklahoma Department of Labor*. Oklahoma Safety and Health Management, Public Domain. Retrieved from http://bit.ly/14o9Duz.

Toyota, Let's Go Places, TV Commercial. (December 2013). Toyota. Retrieved from: http://bit.ly/XXFQm9.

Volume 1: Concepts and Principles Human Performance, Improvement Handbook, Public Domain. (2009). U.S. Department of Energy. Retrieved from http://1.usa.gov/WdIqoP.

Voluntary Protection Program (VPP). (n.d.). Occupational Safety and Health Administration (OSHA), Public Domain, Adapted for Use. Retrieved from http://1.usa.gov/T0j6EW.

PART 2

Safety Management Systems Defined

CHAPTER 5

Overview of Basic Safety Management Systems

Management is doing things right; leadership is doing the right things.
—Peter Drucker

Introduction

When you are in the beginning stage of developing and implementing a safety management system for your organization, communication is the key to success, as discussed in Chapter 3, "Analyzing and Using Your Network". To get all employees involved, at all levels of the organization, the quality and depth of your network is crucial to ensuring that your messages get transmitted. "Both the leadership team and employers will gain from a shared, collaborative effort and the system will be better as a result of everyone's involvement." (Building an Effective Health and Safety Management System, 1989).

A safety management system "is a term used to refer to a comprehensive business management system designed to manage safety elements in a workplace" (Safety Management Systems, n.d.). A basic safety management system's main purpose is to accomplish the following elements:

- To ensure everyone in the organization can recognize and understand real or potential hazards and associated risk.
- To prevent or control operational hazards and associated risk.
- To train employees at all levels of the organization so they can demonstrate the importance of correcting potential hazards they may be routinely exposed to as well as how to protect themselves and others.

The objective of this chapter is to discuss the basic elements recommended for a safety management system and why selecting and implementing a structured format benefits the organization's safety culture. After completing this chapter, you will be able to:

- Discuss the common link between safety management systems.
- Contrast leadership and employee involvement in a safety management system.
- Discuss why defining the roles and responsibilities is critical to the safety management system.
- Discuss the importance of hazard and risk assessment, prevention, and control.
- Discuss how you must interact with other departments regarding training.

- Discuss why you should keep safety performance simple.
- Discuss and provide your opinion on why a safety management system is important to the development of a safety culture.

The Common Link between Safety Management Systems

A safety management system must clearly define how the organization intends to execute the stated goals and objectives and to measure the effectiveness of actions to reach or maintain the safety-related vision/mission. Refer to Chapter 4, "Setting the Direction for the Safety Culture" for a detailed discussion. A literature review found that current safety management systems revolve around several basic criteria. As an example from one source, a safety management system must include at a minimum the following:

- "Express the leadership team's commitment to safety and clearly state the policies, objectives and requirements of the safety management system.
- Define the structure of the safety management system as well as the responsibilities and authority of key individuals of the leadership team.
- Define each element of the safety management system.
- Communicate the expectations and objectives of the safety management system to all employees" (Safety Management System Toolkit, 2007).

Refer to Figure 5.1, "Applying the Normal Business Approach to a Safety Management System".

After reviewing many types of safety management systems and comparing their elements side by side, it is evident that six basic core principles are the vital link to developing and sustaining a safety culture.

Occupational Safety and Health Administration's (OSHA) Voluntary Protection Program places management and employee involvement together as complimentary establishing the core of an effective safety management system. We will separately discuss management and employee involvement as individual elements, as a clear and distinct difference is needed in the approach to the leadership team and in gaining employee involvement. Refer to Figure 5.2, "Journey to a Successful Safety Management System with a Sustainable Process". If one part of the system is missing, it represents a serious gap that must be addressed. Refer to Figure 5.3, "Identifying Gaps in a Safety Management System" (Roughton & Crutchfield, 2008).

> *Leadership is not magnetic personality—that can just as well be a glib tongue. It is not 'making friends and influencing people'—that is flattery. Leadership is lifting a person's vision to high sights, the raising of a person's performance to a higher standard, the building of a personality beyond its normal limitations.*
>
> **Peter Drucker**

Business	Safety Management System
Vision/Mission about what the organization desires and espouses	Vision/Mission that aligns with the business mission
Corporate goals both long and short term	Safety goals that aligns with corporate goals
Corporate objectives to reach goals	Safety objectives considered as part of corporate objectives
Policies and procedures for overall management and administration of the organization	Safety policy statement, other safety policies, and procedures in similar format and incorporated in business policies and procedures
Business processes • Identify noncompliance with policies and procedures (i.e., HR compliance, accounting, operational, quality, etc.)	Safety processes • Identify and control hazards and associated risk as well as compliance to corporate and regulatory policies and procedures
• Implement solutions to organizational issues	• Advise and assist implementing solutions on hazard and associated risk reduction and control
• Measure performance of the organization	• Measure performance of the safety management system
• Lessons learned developed and communicated	• Lessons learned developed and communicated
• Repeat the above process—plan to check and act	• Repeat the above process—plan to check and act

Figure 5.1

Applying the Normal Business Approach to a Safety Management System. *Source: Adapted from Safety Management System Toolkit. Developed by the Joint Helicopter Safety Implementation Team of the International Helicopter Safety Team (Safety Management System Toolkit, 2007).*

The following are the basic core safety process elements that can be used to design and implement a structure that provides the foundation for a safety culture:

- Management leadership
- Employee involvement

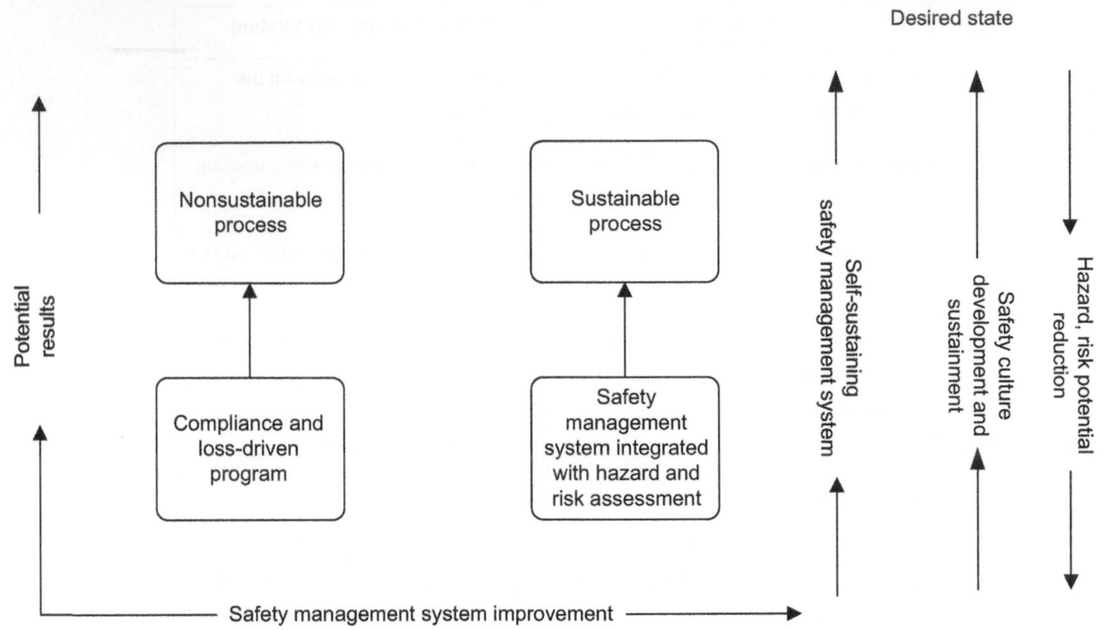

Figure 5.2
Journey to a Successful Safety Management System with a Sustainable Process. *Source: Adapted from Job hazard analysis: A guide for voluntary compliance and beyond (Roughton & Crutchfield, 2008).*

- Risk and hazard identification and assessment
- Hazard prevention and control
- Education and training
- Performance and measurement (Injury and Illness Prevention Program (I2 P2), n.d.; Managing Worker Safety and Health, n.d., Figure 5.4).

If a safety management system structure is not used, decisions on valid hazards and associated risk assessments controls may not be given the correct priority that is needed to ensure that the organization and all employees are protected.

Management Leadership

> *The job of management is not supervision, but leadership.*
>
> **Dr W. E. Deming**

Management commitment filters down from the top leadership through all levels of the organization. For our discussion, we will be using the term leadership team in order to incorporate all levels of management and supervision. The leadership team commitment is considered the driving force found in all successful safety management systems. The leadership team provides the vision/mission and direction as well as adequate resources to ensure

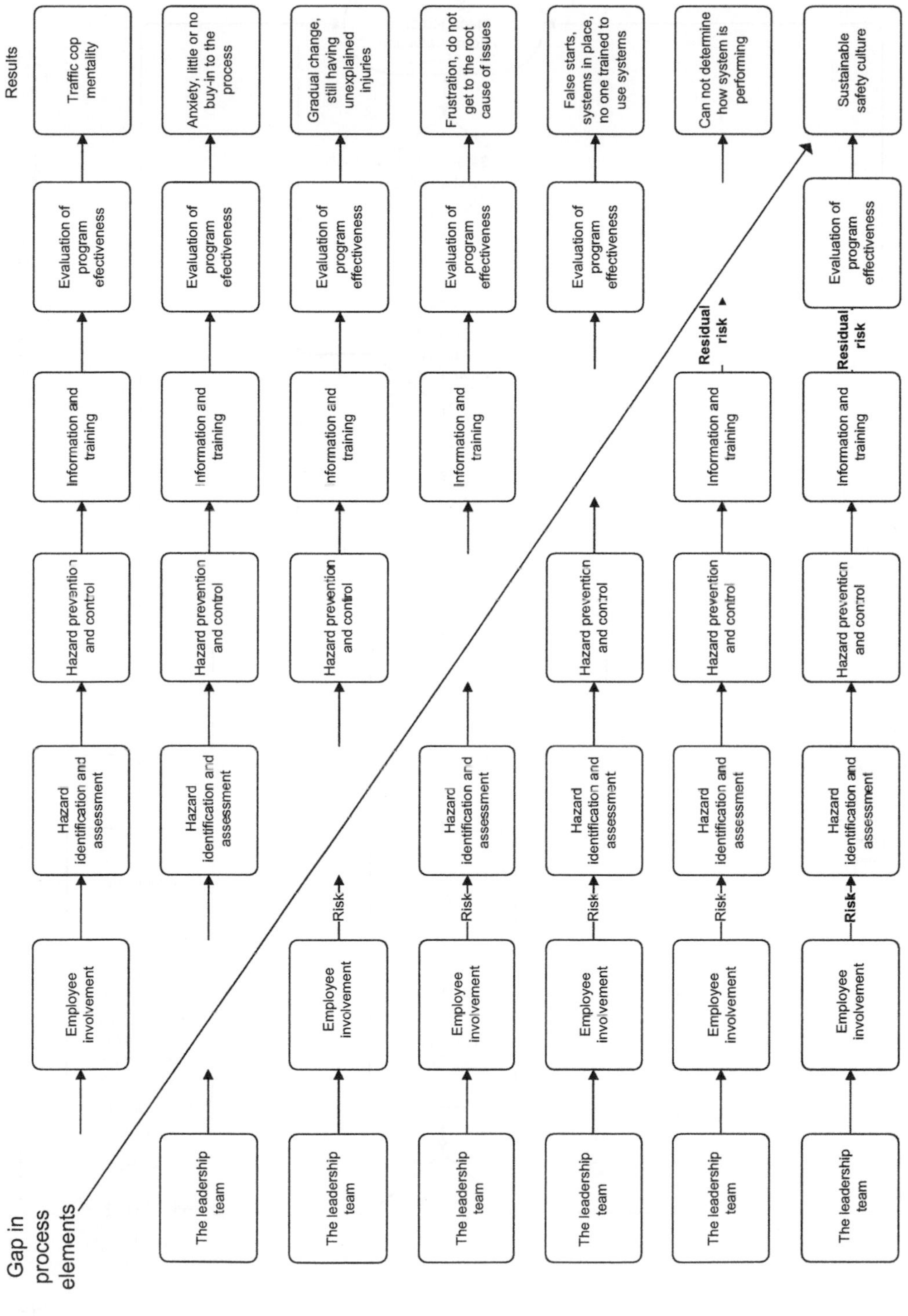

Figure 5.3
Identifying Gaps in a Safety Management System. *Source: Adapted from Job hazard analysis: A guide for voluntary compliance and beyond (Roughton & Cructhfield, 2008).*

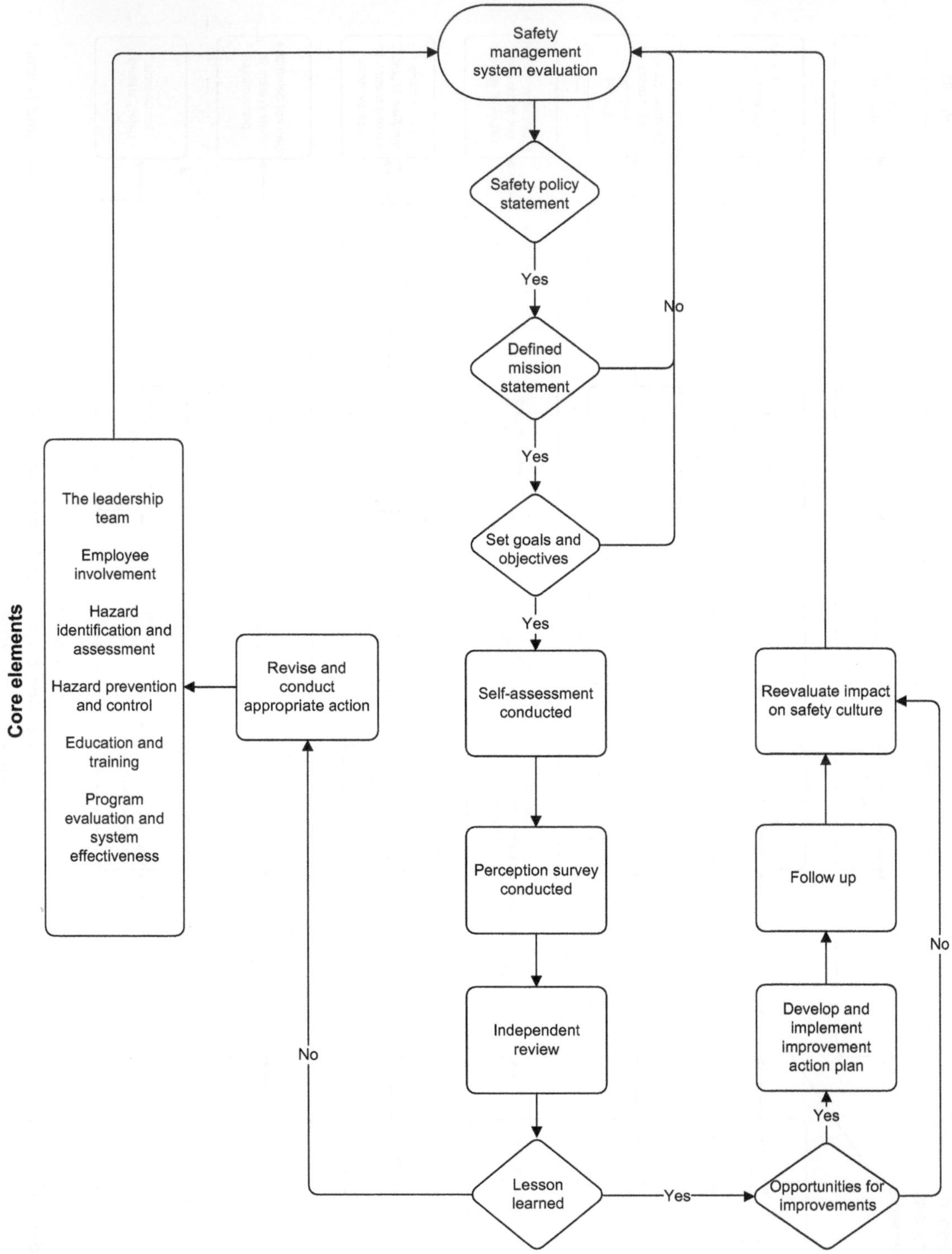

Figure 5.4

Overview of the Safety Management Review Cycle. *Source: Adapted from Job hazard analysis: A guide for voluntary compliance and beyond (Roughton & Cructhfield, 2008).*

that controls are in place. If the leadership team demonstrates commitment by providing the motivating dynamics and the needed resources, an effective safety management system can be developed and sustained. This demonstration by the leadership team should include the following elements that are consistent with an effective safety management system:

- Establishing the responsibilities of all managers, supervisors, and employees for safety-related goals and objectives, and holding them accountable for carrying out their assigned responsibilities.
- Providing managers, supervisors, and employees with authority, access to relevant information, education and training, and resources needed to carry out their safety responsibilities.
- Identifying a qualified employee to receive and respond to reports about safety conditions and, where appropriate, to initiate corrective action (Roughton & Mercurio, 2002; Safety and Health Management Basics, Module One: Management Commitment, n.d.).

If all employees can see that the leadership team puts emphasis on safety, they will then more likely mimic these efforts in their own activities. An indicator that a safety culture is effective is when the leadership team follows all established safety rules, guidelines, and work practices without fail. As an example, when a leader walks through a workplace and does not follow the expected rules and guidelines, leadership sends a strong message that it is not serious about a safe work environment and that it is somehow outside the required rules. Too often we have seen leadership not actively follow safety procedures.

Lesson Learned # 1

A colleague tells the story of having a manager light up a cigarette under "No Smoking" signs. When asked why, the manager said the signs had been placed there by his predecessor who was a nonsmoker and was oblivious to the existing fire hazards present.

There are many other ways available for the leadership team to show its support for the safety management system. For example:

- Participate in facility safety inspections.
- Personally support corrective actions and alternative solutions for conditions and activities that are hazardous until the hazards can be minimized, corrected, or controlled.
- Personally provide and review the tracking of safety performance, etc. (Managing worker safety and health, n.d.).

We cannot direct the wind, but we can adjust the sails.

Bertha Calloway

> *The first responsibility of a leader is to define reality. The last is to say thank you.*
>
> **Max DePree**

Traditionally, a designated person or group may have the initial responsibility for the safety process. The overall "safety" of an organization cannot be the sole responsibility of one person or group if the definition of safety is an emerging property that develops from the interactions of all parts or elements of the organization. Your objective as part of the leadership team is to influence and shape the organization in its control programs for hazards and associated risk.

> *Every sale has five basic obstacles: no need, no money, no hurry, no desire, no trust.*
>
> **Zig Ziglar**

Being too stringent by relying only on rule-based compliance can create a climate that regards the safety function as more of a police force for regulatory compliance than as a mentor who understands the needs of the organization and demonstrates how hazards and associated risks develop and create loss-producing events that must be controlled. The following are basic self-evaluation questions that you can ask that may impact the safety culture:

- Is there a written safety policy statement for the organization?
- Is the safety policy statement signed by the current leadership team?
- Is the safety policy statement readily available to everyone in the organization?
- Has the written safety policy statement been communicated to all levels of the organization?
- What access do you have to the senior leadership team?
- Is everyone in the organization aware of their specific safety-related responsibilities?
- Are all employees evaluated on their individual safety performance?
- Does the leadership team regularly communicate with all employees in the organization about their commitment to safety?
- Does the leadership team participate in on-site walk-throughs to reinforce their commitment to safety practices and behaviors?
- Is there a process in place that addresses on-site contractor safety?
- Does the leadership team provide the resources needed (budget, time, equipment, materials, etc.) to implement and improve the safety management system? (Building an Effective Health and Safety Management System, 1989).

A strong management commitment to the safety process, as with any successful management system, forms the core of all efforts. Refer to Figure 5.4, "Overview of the Safety Management Review Cycle" (Roughton & Crutchfield, 2008).

Employee Involvement

Getting employees involved in the development of the safety management system is particularly important to create ownership and buy-in. Employee involvement in the development of the safety management system will help to ensure that it fits within the existing organizational culture.

Many opportunities can be used to get employees involved in the safety process. For example, to promote employee involvement, you can ask them to be involved in the development of hazard assessment, inspections, preventative maintenance, training, emergency response, and loss-producing reporting systems.

> *Effectively engaging workers to actively participate in health and safety requires thoughtful planning and implementation processes/policies that will build an atmosphere of trust.*
>
> **(Occupational Health and Safety Systems, 2012)**

Employee involvement follows closely with the leadership team's commitment. Your task is to provide the means for employees to develop and/or express their commitment to safety for themselves and their fellow employees. Most employees want to do the right thing with regard to their assigned duties and tasks. They know that they must safely complete their specific task; however, they must in turn balance what they feel is safe vs. perceived risk that can be accepted to get the job task completed. This may not stem from any perceived pressure from the leadership team but rather the desire or gratification to complete a task and move to the next activity.

One area that must be considered in ensuring employee participation is that involvement should be as diverse as possible. The following are areas that can become avenues for a two-way communication and a conduit for enhanced involvement:

- Ensure that two-way communications are in place and that allow rapid response is provided on any reports and recommendations. The system should ensure immediate correction of issues and allow employees to make recommendations as appropriate.
- Easy and practical access to safety-related information with quick and reliable communications methods that do not filter or block concerns about identified hazards.
- Techniques that allow each individual to assess specific perceived risk and hazard identification with a practical method to prioritize issues. Refer to Chapter 9, "Risk Perception—Defining How to Identify Personal Responsibility".

If employee communication is blocked or filtered, the safety culture is diminished. What appears trivial at one level of the organization may be a serious issue at another level until fully analyzed and assessed.

Defining Roles and Responsibilities

Clearly defined and well-communicated safety roles and responsibilities for all levels of the organization are critical in the beginning phases of developing a safety management system. These outline all expectations about performance and accountability among the leadership team, all employees, contractors, and site visitors. Once the roles and responsibilities are assigned, everyone in the organization must be made aware of their individual roles and responsibilities.

> *Unless commitment is made, there are only promises and hopes . . . but no plans.*
>
> **Peter Drucker**

Each level of leadership should be held accountable for ensuring that channels of communication remain open in all directions about safety concerns and issues. Refer to Chapter 7, "Leadership and the Effective Safety Culture", for a discussion on defining roles, responsibility, delegation, authority, and accountability for everyone in the organization (Building an Effective Health and Safety Management System, 1989).

Your coordination and influence with the person writing and developing job descriptions is crucial. As we discussed in Chapter 3, "Analyzing and Using Your Network", the network analysis can assist you in determining the actions you must take to get safety criteria built in to the process. In addition, we will discuss activity-based safety systems in more detail in Chapter 11, "Developing an Activity-Based Safety System".

Hazard and Risk Assessment Identification and Analysis

Hazard and risk identification and analysis is a core function of a safety management system. With the leadership team on board and a level of employee involvement, the next phase is to identify hazards and associated risk within all operations.

As discussed in role and responsibilities, it is important to proactively assess all jobs, their steps, and tasks to identify the specific hazards that may be built into the job and the severity of associated risk. If this assessment is performed as a team effort with involvement at all levels of the organization, it will begin to promote hazard awareness to both the leadership team and employees (Building an Effective Health and Safety Management System, 1989).

Without a comprehensive approach to hazards and associated risk, the safety management system and the safety culture cannot be consistently maintained. Your guidance and expertise drives the process. The concerns over inherent operational hazards and the conditions where new hazards may be potentially created must be translated into terms that are understandable at all levels of the organization. The following items should be reviewed for the level and scope of their implementation throughout the organization:

- Routine development and use of job hazard analyses as well as education of employees about each job's hazards and associated risk. The process should include involvement or comprehensive communication with the leadership team so that this understanding is present throughout the organization. For a more detailed review of job hazard analysis, refer to Chapter 12, "Developing the Job Hazard Analysis".
- Providing a reliable system for employees to notify and receive timely feedback from the leadership team about conditions that appear hazardous.
- Investigating all real/potential loss-producing events such as injuries, property damage, etc. so that causes and future prevention methods can be determined and implemented.
- Analyzing hazards, associated risk, and injury trends to identify patterns that can be further analyzed for corrective actions (Managing worker safety and health, n.d.; Roughton & Crutchfield, 2008).

Refer to Chapter 9, "Risk Perception—Defining How to Identify Personal Responsibility", and Chapter 10, "Risk Management Principles", for a detailed discussion on a method to help identify risk perception and controlling risk.

Hazard Prevention and Control

Effective planning and design of the workplace and job tasks are ongoing concerns as technology, facilities, materials, equipment, and operations are constantly under change. As part of the overall assessment, the design of any work order system should be reviewed to ensure that it can prioritize requests for hazardous corrective actions.

Prevention by Design

The concept of prevention by design or safety by design recommends approaching hazard and hazard identification and control at the beginning stages of design to essentially engineer safety concepts into processes, equipment, machinery, and facilities. Various global initiatives have been launched and the US NIOSH mission is as follows:

"The mission of the Prevention through Design National initiative is to prevent or reduce occupational injuries, illnesses, and fatalities through the inclusion of prevention considerations in all designs that impact workers. The mission can be achieved by:

- Eliminating hazards and controlling risks to workers to an acceptable level 'at the source' or as early as possible in the life cycle of items or workplaces.
- Including design, redesign and retrofit of new and existing work premises, structures, tools, facilities, equipment, machinery, products, substances, work processes and the organization of work.
- Enhancing the work environment through the inclusion of prevention methods in all designs that impact workers and others on the premises" (Prevention through Design, 2012).

Once the hazard and associated risk assessments are completed, the next step in the development of a safety management system is the implementation of control measures to eliminate or reduce the risk of harm to employees.

Information and Training

The gathering of information and training materials is the linchpin that holds the safety management system together. Establishing a structured method to gather and maintain information and data specific to your operational environment increases your ability to influence and design safety performance requirements and/or job practices. By becoming a "curator" of information, you assemble the tools and materials necessary for the education of the leadership team and employees. Refer to Chapter 15, "Becoming a Curator for the Safety Management System" for an overview of concepts and thoughts to help you with data collection.

Training Programs

Employee training and education programs should be designed and implemented to ensure that all employees demonstrate and understand that they are fully aware of the hazards and associated risk inherent in the operations. Refer to Chapter 13, "Education and Training, —Assessing Safety Training Needs".

As part of any assessment, how the organization has defined its training structure and methods must be addressed. Training may reside in a separate department and not be part of your responsibility and authority. Coordination and a strong rapport with that department's leadership allows for the sharing of resources and the proper determination of how training is to be delivered. However, your role may also include that of safety training. This can leave the perception that safety training is not a real part of the overall operation and separate from the "real world". Understanding your network to provide a uniform approach to employee and leadership training.

Evaluation of Program Effectiveness

The safety management system must have a feedback loop that communicates to the leadership team the status of the various programs and systems required to control hazards and associated risk. The safety management system should be revised as gaps or deficiencies are identified by comprehensive evaluations. An annual review of system elements with periodic spot checks to verify that programs have remained in place and are fully utilized is necessary. The organization should be able to keep a schedule of reviews and ensure evaluations are not delayed or pushed back, as scope drift can be gradually occurring.

> System reviews are focused on management system elements and are not specific to operations. Examples include:
>
> - Determination of resource levels and expertise
> - Effectiveness of communication and employee participation
> - Review of change management time lines and effectiveness
>
> System reviews may be input for decisions made during management reviews (Occupational Health and Safety Systems, 2012).

Safety performance measurements should be kept simple and be able to generate information or data that clearly provides valid evidence on the effectiveness of the safety management system.

Summary

In this chapter we discussed the basics of a safety management system's process elements. The basic safety management system consists of six core elements:

- Management leadership
- Employee involvement
- Risk and hazard identification and assessment
- Hazard prevention and control
- Education and training
- Performance and measurement

By using these process elements, you should be able to establish a safety management system structure that:

- Recommends visible leadership team involvement in implementing and sustaining the safety management system so that all employees understand that the management's commitment is serious.
- Encourages employee involvement in the structure and operation of the safety management system and in decisions that affect the employee's safety.
- Defines the importance of and how the leadership team and employees should be accountable for assigned safety responsibilities.
- Identifies levels of authority and the resources required for responsible parties so that assigned duties can be met.

A safety management system must clearly define how the organization intends to execute the stated goals and objectives and to measure the effectiveness of actions to reach or maintain the safety-related vision/mission.

Management commitment and leadership filters down from the top leadership through all levels of the organization. The leadership team provides the vision/mission and direction as well as adequate resources to ensure that controls are in place.

Getting employees involved in the development of the safety management system is particularly important to create ownership and buy-in. Employee involvement in the development of the safety management system will help to ensure that it fits within the existing organizational culture.

Clearly defined and well-communicated safety roles and responsibilities for all levels of the organization are critical in the beginning phases of developing a safety management system. These outline all expectations about performance and accountability among the leadership team, all employees, contractors, and site visitors.

Hazard and risk identification and analysis is a core function of a safety management system.

Effective planning and design of the workplace and job tasks is an ongoing concern as technology, facilities, materials, equipment, and operations are constantly under change.

The gathering of information and training materials is the linchpin that holds the safety management system together. Establishing a structured method to gather and maintain information and data specific to your operational environment increases your ability to influence and design safety performance requirements and/or job practices.

Employee training and education programs should be designed and implemented to ensure that all employees demonstrate and understand that they are fully aware of the hazards and associated risk inherent in the operations.

The safety management system must have a feedback loop that communicates to the leadership team the status of the various programs and systems required to control hazards and associated risk.

There are many types of safety management system structures. In Chapter 6, we will discuss various types of safety management systems have been used by organizations for many years. These systems can range from voluntary standards established by industry or trade groups, consulting organizations, and international standards organizations to required governmental legislation.

Chapter Review Questions

1. What is a basic safety management system's main purpose? List the main purpose and discuss.
2. Discuss the normal business approach to a safety management system. Use Figure 5.1 in the textbook as a guide.
3. What are the six basic core safety process elements that can be used to design and implement a safety management system structure to support a safety culture?

4. Discuss each of the six basic core safety process elements of a safety management system in detail.
5. Management commitment and leadership filters down from the top leadership through all levels of the organization. Discuss why this is important to understand.
6. According to OSHA, management leadership and employee involvement are the most important elements of a safety management system. Review the descriptions of each role and provide your opinion as to which roles would be the most important. Support your opinion with some examples from your personal experience and/or research.

Bibliography

Building an Effective Health and Safety Management System. (1989). Partnerships in Health and Safety program, Partnerships in Injury Reduction (Partnerships), Reprinted/Modified and/or adapted with Permission. Retrieved from http://bit.ly/WsHteC.

Injury and Illness Prevention Program (I2 P2). (n.d.). Occupational Safety and Health Administration (OSHA), Adapted for Use. Retrieved from http://1.usa.gov/lIszWK.

Managing Worker Safety and Health. (n.d.). Illinois OSHA Onsite Safety & Health Consultation Program, Public Domain, Adapted for Use. Retrieved from http://bit.ly/WTsneh

Occupational Health and Safety Systems. (2012). The American Industrial Hygiene Association, ANSI/AIHA Z10.

Prevention through Design. (June 2012). National Institute for Occupational Safety and Health (NIOSH), Pubic Domain. Retrieved from http://1.usa.gov/11f7iTQ.

Roughton, J. E., & Crutchfield, N. (2008). *Job hazard analysis: A guide for voluntary compliance and beyond. Chemical, petrochemical & process*. Elsevier/Butterworth-Heinemann. Retrieved from http://amzn.to/VrSAq5.

Roughton, J. E., & Mercurio, J. J. (2002). *Developing an effective safety culture: A leadership approach, adapted for use*. Butterworth-Heinemann. Retrieved from http://amzn.to/X8Gaz8.

Safety and Health Management Basics, Module One: Management Commitment. (n.d.). Oregon Occupational Safety and Health Division (Oregon OSHA), Public Domain, Permission to Reprint, Modify, and/or Adapt as necessary. Retrieved from http://bit.ly/Yztx5p.

Safety Management System Toolkit. (2007). Developed by the Joint Helicopter Safety Implementation Team of the International Helicopter Safety Team, *The international helicopter safety symposium, Montréal, Québec, Canada*. Retrieved from http://bit.ly/YaSCOg.

Safety Management Systems. (n.d.). Wikimedia Foundation, Inc., Retrieved from http://bit.ly/15lZVIH.

CHAPTER 6

Selecting Your Process

The secret of success is learning how to use pain and pleasure instead of having pain and pleasure use you. If you do that, you are in control of your life, if you don't, life controls you.

—Anthony Robbins

Introduction

Many types of safety management systems have been used by organizations for many years. These systems can range from voluntary standards established by industry or trade groups, consulting organizations, and international standards organizations to required governmental legislation. In the end, these systems have the same intent—to provide guidance to organizations on how to manage and structure a successful safety process.

Based on many years of experience, we are convinced that selecting or developing a safety management system is essential, as it acts as a stabilizing structure to keep the focus on operational hazards and associated risk as conditions change in the organization. In turn, it provides a framework for greater employee involvement that can lead to continued employee commitment and increased productivity.

A safety management system's main purpose is to accomplish the following:

- To educate and train employees at all levels so there is an understanding of the real and potential hazards.
- To provide methods to identify and understand real and potential hazards and associated risk.
- To avoid, prevent, and/or control hazards and associated risk.

The objective of this chapter is to discuss the basic elements recommended for a safety management system and why selecting and implementing a structured format benefits the organizational safety culture. After reviewing this chapter, you should be able to:

- Identify the commonalities of safety management systems.
- Discuss and compare plan–do–check–act (PDCA) and define, measure, analyze, improve, and control (DMAIC).
- Identify how the safety management system can enhance and benefit your safety culture.
- Compare and contrast the difference between voluntary standards established by industry, international standards organizations, and government-related safety programs.

Safety Culture.

What Do All Safety Management Systems Have in Common?

After reviewing a number of safety management systems from around the world and comparing their elements side by side, it is evident that a basic theme forms the overall core foundation of these systems. Refer to "Appendix H", "Comparison of Governmental-Related Safety Management Systems Process Elements".

> *According to Awwad J. Dababneh, "In a note produced by the Royal Society of Chemistry, London (2005), it was concluded that there are no fundamental differences between any of these management systems, and there are large degrees of overlap in system elements and most are common requirements.*
>
> *The differences between BS 8800 and ISO 9001 and ISO 14001 relate largely to scope and the specific needs of OH&S requirements compared to quality and environment specifications. The distinction between BS 8800 and OHSAS 18001 and ILO-OSH 2001 is mainly in the order which the elements are addressed. The intent and basic requirements are common to all three documents"*
>
> **(Dababneh, n.d.)**

The PDCA Cycle

At the core of these elements is the simple concept of a PDCA methodology, called the Shewhart cycle (Shewhart, 1931), that can equally be applied to all safety management systems. PDCA has been at the root of quality control and safety programs. The Shewhart cycle has been adapted within the framework of many voluntary compliance standards such as ANSI Z10-2012 "Occupational Safety Process" (Occupational Health and Safety Systems, 2012). Refer to Figure 6.1, "Shewhart, PDCA Cycle for Learning and Improvement".

Safety management systems are a parallel process to quality management processes. Frank Bird, one of the pioneers in loss control, advocated using the book *Quality is Free* (Crosby,

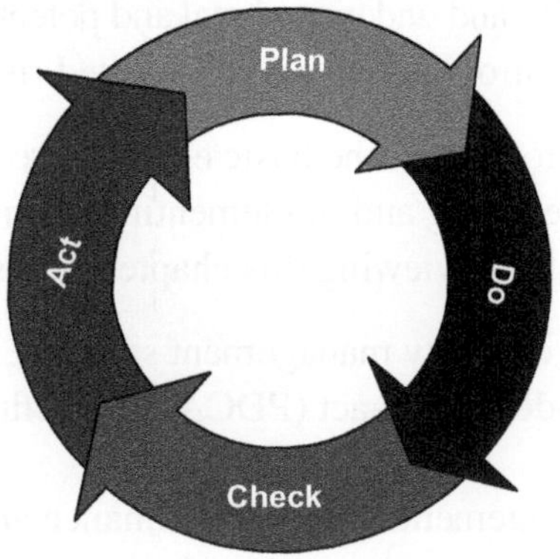

Figure 6.1
Shewhart, PDCA Cycle for Learning and Improvement.

1979) as a guide to developing a safety process by simply replacing the word "quality" with "safety". Your networking analysis within the organization should identify the leadership roles with responsibility for quality management due to this close relationship between the tools of quality management and safety. It is important that you parallel existing programs and processes as much as possible to ensure that the safety role be viewed as an integral part of the organization.

> *Total Quality Management promotes continuous improvement in all aspects of an organisation's activities. It emphasises identifying the key processes, setting performance standards, measuring achievement against these standards and then taking corrective action and identifying opportunities for improvement—all in a continuous cycle.*
>
> **(Successful Health and Safety Management, hsg65, 1997)**

> *PDCA (plan–do–check–act or plan–do–check–adjust) is an iterative four-step management method used in business for the control and continuous improvement of processes and products. It is also known as the Deming circle/cycle/wheel, Shewhart cycle, control circle/cycle, or plan–do–study–act.*
>
> **(Deming, n.d.; Shewhart, 1931)**

> *The elements of a Safety management system (SMS). A number of important elements are specified that have to do with the setting of policy and creation of plans and organizational capacity to realize that policy (PLAN), the analysis of hazards and effects leading to planning and implementation of those plans in order to manage the risks (DO), and the control on the effective performance of those steps (CHECK). A number of feedback loops are specified to see where the information gained should be sent (FEEDBACK)*
>
> **(Pearse, Gallagher, & Bluff, 2001)**

The following steps use the PDCA process and show how it is used to devise a structure for use in developing a safety management system. By having a specific process, a structure is provided from which a map can be developed for the organization to follow. In addition, the PDCA provides a dynamic format that reflects how the process is not static and is in continual movement and renewal.

Step 1: Plan: Define the goals and objectives of the safety management system (SMS). As discussed in Chapter 4, "Setting the Direction for the Safety Culture", the goals and objectives will define specific conditions and behaviors desired from the implementation of the elements of the SMS. This will establish the criteria desired for the safety culture.

The intended results expected are defined in this phase and are discussed with and approved by the leadership team.

Step 2: Do: Begin the implementation of the SMS goals and objectives.
Begin the implementation starting with the education, training, and communication of the SMS and its importance to all employees. This phase represents a transition from the past efforts to a new process and its procedures. The implementation must first ensure that all employees are on board with the desired changes. Using the activity-based safety system, Chapter 11, "Developing an Activity-Based Safety System", quickly spreads the information about the SMS and how all employees will be involved.

Step 3: Check or Study: As the SMS and its elements are implemented, periodic review and surveys are made to ensure that resistance to change is reduced and any concerns or issues are addressed. Study may find where buy-in needs to be improved, concerns with time requirements and budgets need to be addressed, and hazard assessments improved. In this phase, the parallel to quality management is apparent as the use of statistical process analysis, surveys, questionnaires, and interviews provide a clearer picture of what the current process is achieving.

Step 4: Act: The findings of the Check/Study phase defines changes and modifications needed. Corrections are made to the safety management system by integrating what has been found to work into the safety management system and things that do not improve the safety culture abandoned (How to Effectively Assess and Improve your Safety and Health Program through Safety and Health Evaluation, n.d.; Roughton & Crutchfield, 2008; Shewhart, 1931).

Define, Measure, Analyze, Improve, and Control

If your organization uses the Six Sigma approach to quality management, the use of a variation of the PDCA known as DMAIC can be used. DMAIC is a more advanced approach than the traditional PDCA.

As issues with the current safety culture and the safety management system are identified, the PDCA/DMAIC concepts provide a framework for administration of the safety management system elements. Refer to Figure 6.2 for an "Overview of the DMAIC and the PDCA Processes Relationship".

The following sections will provide a brief summary of the 15 DMAIC steps in conducting a Six Sigma type review of the safety management system.

Defining the Project

Refer to Figure 6.3, "Basic Six Sigma Process Model Aligned to SMS" diagram mapped to a successful safety management system.

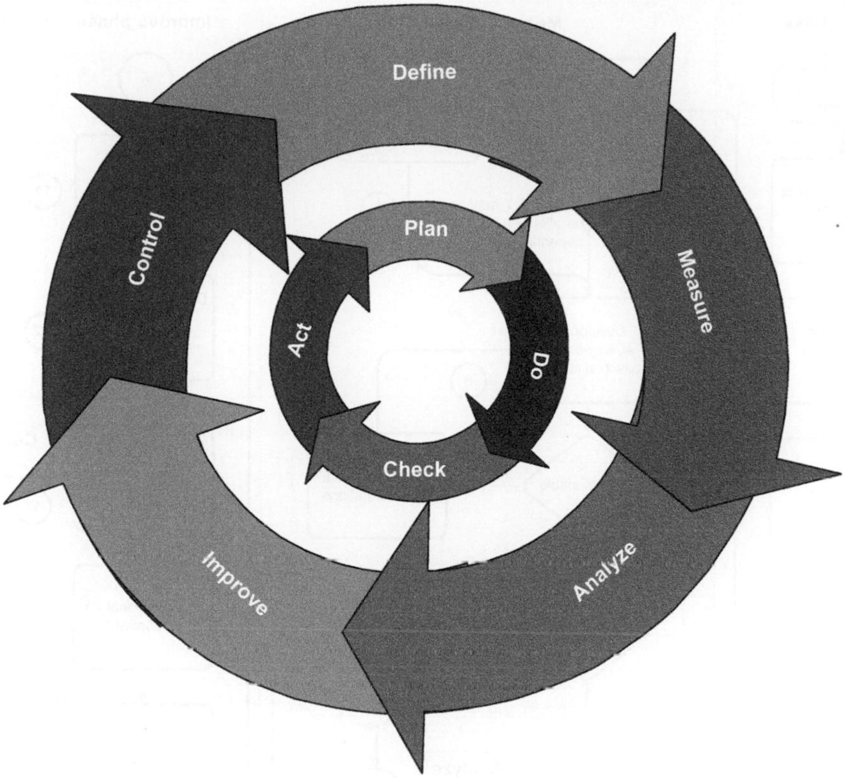

Figure 6.2
Overview of the DMAIC and the PDCA Processes Relationship.

Step 1: Define the goal of the safety management system.

Step 2: Define current state of the safety culture: What is to be measured and in what format?

Step 3: Define the appropriate metrics: How will the safety culture and safety management system be measured?

Step 4: Develop the objectives of the safety management system: Put an "evergreen" (a living method) process in place that can stand the test of time in an ever-changing organizational environment.

Step 5: Select and organize the safety management system team: Working with the leadership team, define the roles, responsibilities, and authorities needed for the safety management system. Establish employee involvement.

Measuring the Project

Step 6: Develop a macro overview using organizational charts, a network map, and flow diagram of the current safety management system process.

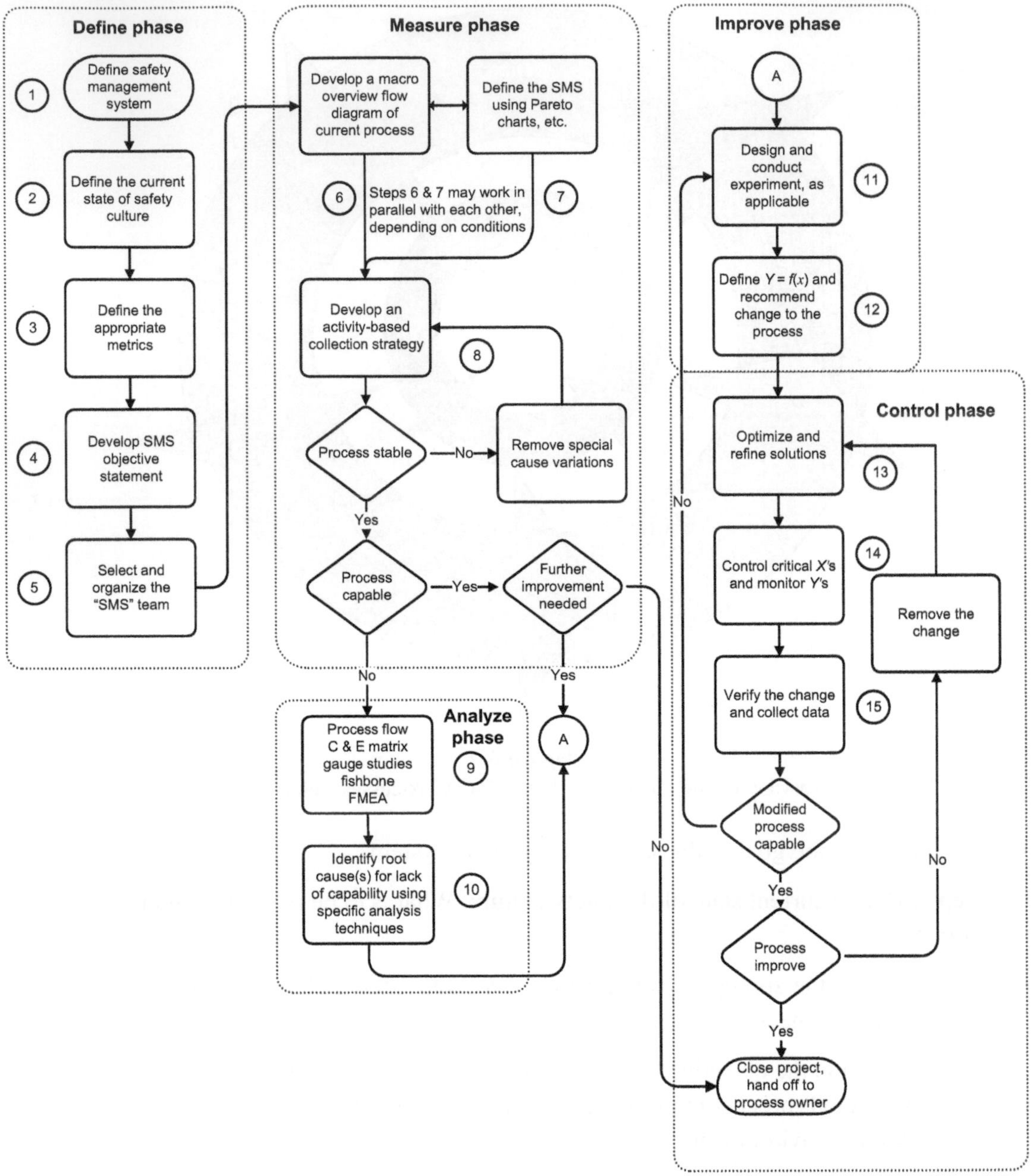

Figure 6.3

Basic Six Sigma Process Model Aligned to SMS.

Source: Adapted from Black belt training manual, define phase, black belt. Breakthrough Management Group Inc.
Adapted from Handbook for basic process improvement, U.S. Navy, May 1996, Public Domain, http://1.usa.gov/Y7U2ZX.
Adapted from Roughton, J., & Crutchfield, N. (2008). Job hazard analysis, a guide for voluntary compliance and beyond. Butterworth-Heinemann.

Step 7: Define the safety management system project with "Pareto" charts, graphs of loss history, "dashboards" showing activity status, etc.

Step 8: Develop an activity-based data collection strategy used to collect specific baseline data about the safety management system. The data collection should be focused on preloss activities that are used to reduce hazards and associated risk.

Analyzing the Project

Step 9: Refine the process using process flow, XY matrix, gauge studies, Fishbone, FMEA, Mind maps, etc. The analysis of the safety management system provides insights into what is or is not working, gaps in communication, and concerns and issues with the implementation.

Step 10: Identify root cause(s) of underlying problems in the implementation: issues with leadership team and employee involvement, materials, data gathering, lack of specific data analysis, and communications. Were the right goals, activities, correct authorities, adequate budgets, etc. in place to support the safety management system? In short, what obstacles are preventing the full application of any of the elements of the safety management system?

Improving the Project

Step 11: Design and conduct an "experiment", as applicable. The safety management system is tested by looking at the overall system. After the analysis, the first question to ask: Is the safety management system model selected correct for your organization? Begin by looking at the measurements used: activities, 2-minute shift reviews, risk guidance tool, meetings, inspections and checklist, etc. Review what is not working, make adjustments to the process, and test for the ability of the leadership team and employees to be able to use the formats chosen.

Step 12: Define the $Y = f(x)$ and recommend changes to the process. Y is the outcome of doing X. This is one of several Six Sigma tools that can be used to look at the relationships between elements of the safety management system process. If a risk guidance tool is used, does it impact the avoidance, reduction, and control of hazards? If no relationships can be defined, then the system must be further reviewed. Refer to Chapter 14, "Assessing Your Safety Management System".

Controlling the Project

Step 13: Optimize and redefine corrective measures and solutions to gaps in the safety management system implementation. This will be a continuous process as changes in the organization's business climate or conditions are constantly ongoing. The safety management system offers a structure but can never be considered permanent. The strength of the safety culture is dependent on the SMS being flexible and resilient enough to adjust to changing conditions.

Step 14: Control critical *X*'s (inputs) and monitor *Y*'s (outputs for effectively solving a problem). By mapping the safety management system process, the relationships are defined, which allows a review of the flow of communication and activity. The implementation of a safety management system is not simple. It requires an understanding of as many of the interrelationships of an organization as possible. The potential for successful implementation of a safety management system is increased if the sequence of safety management system elements is defined. Refer to Figure 6.4 for an overview of "the possible drive toward a safety management system".

Step 15: Verify the change and collect ongoing data about system effectiveness. (Handbook for basic process improvement, 1996).

Benefits of Using a Standardized Safety Management System

Once the basic safety management system elements as discussed in Chapter 5, "Overview of Basic Safety Management Systems", are implemented and deemed successful, your organization has reached a maturity that will allow you to move to the next level. The intent is to ensure that the organization has the initial fundamentals in place before taking on an advanced safety management system.

Safety management systems do more than simply string together various and miscellaneous programs. They provide a set of structured process elements that form the core characteristics of a proven safety management system. These are templates that allow you to implement a more comprehensive approach. By selecting and implementing a process that is nationally and/or internationally recognized, your safety process gains immediate credibility in your organization.

Pros and Cons of a Standardized Safety Management System

A variation in the perception exists about how the correct safety management system is developed or implemented. Each organization has different structures that need a variety of approaches. In Six Sigma, the "Voice of the Customer" or, as a variation, the "Voice of the Process", must be listened to before any safety management system is introduced.

The following comments discuss the potential pros and cons that you must consider and expand on based on your organization. Your objective is to determine what will work best in your organization.

The pros of implementing a specific safety management system can include, but are not limited to, the following. A standardized safety management system should provide:

- A structured system that can be integrated with all other management processes and not just become a stand-alone program.

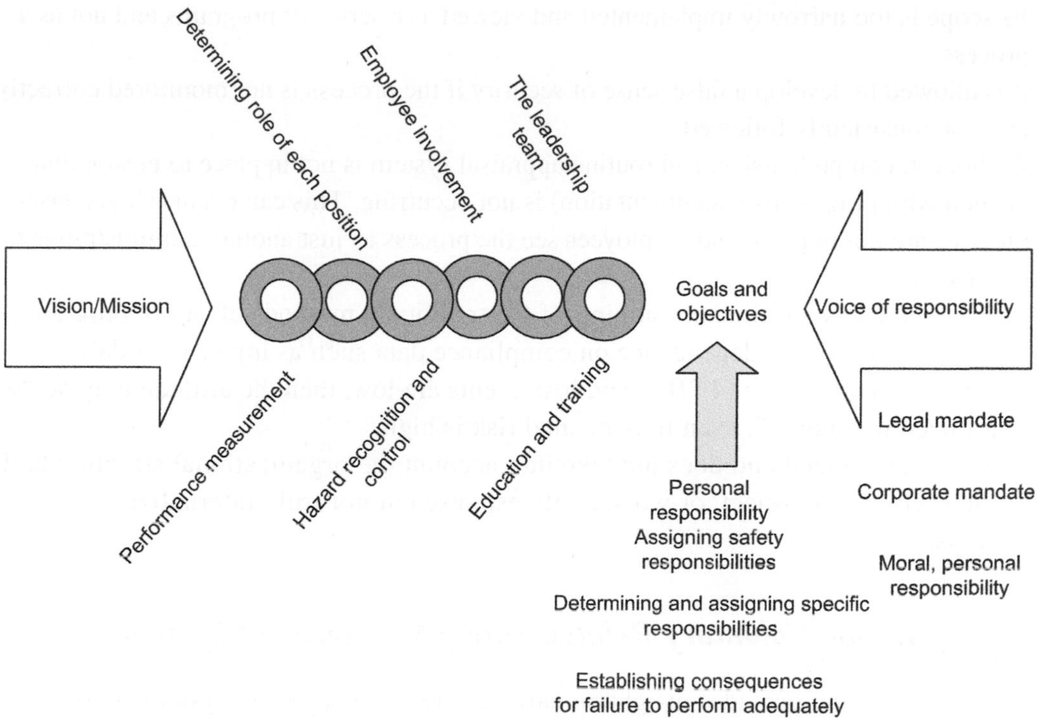

Figure 6.4
The Drive toward a Safety Management System.

- A way to focus your leadership team and your employees and is a "compass" to guide the process. Everyone is on the same page.
- Time-saving by using a structured map or template to ensure continuous improvement.
- Information from the experience of other organizations using a similar process that can include benchmarks to show achievements and progress.
- A way for everyone in the organization to know the status of the system that can encourage discussion, corrective action, and problem solving.
- Assistance in discussing with investors and underwriters that you are managing hazards and associated risk and that a formalized process is in place.
- Internal and external auditors with the same criteria for reviewing the process.
- Regulatory agencies to assurance that the organization is managing hazards and associated risk using a formalized process.

The cons of a safety management system include several areas that must be constantly monitored. The safety management system may have problems if:

- It is not implemented properly and appears to be a "canned" approach that cannot be integrated with other management processes.

- Its scope is too narrowly implemented and viewed as a series of programs and not as a process.
- It is allowed to develop a false sense of security if the process is not monitored correctly and not consistently followed.
- An honest, comprehensive, and routine appraisal system is not in place to ensure that "pencil whipping" (false documentation) is not occurring. This can occur when consequences are not in place and employees see the process as just another administrative task or paperwork.
- It is perceived as too time consuming. If the traditional methods of safety program success are used with dependence on compliance data such as injury recordability (total case incident rate (TCIR)) and loss events are low, then the attitude may be "we don't need to do this!", even if associated risk is high.
- It is not customized and does not take into account the organizational structure and culture which may reject, or worse, only espouse but not truly internalize the process.

Government- and Voluntary-Related Safety Management Systems

The following section provides insight on various safety management systems in order to help you make a more informed decision on what type of management system you and your organization will want to adopt. Two types of safety management systems are available—government-related as both mandated and voluntary and voluntary systems developed by nationally and internationally recognized standards agencies and professional societies.

The list below provides examples of several standardized government-related guidelines safety management systems:

- OSHA's proposed Injury and Illness Prevention Plan (I2P2) (Injury and Illness Prevention Program (I2 P2), n.d.).
- The Voluntary Prevention Program (VPP) (Voluntary Protection Program (VPP), n.d.).
- Department of Energy Voluntary Protection Program (Voluntary Protection Program (DOE-VPP), n.d.).

Various states and provinces may have their own suggested or mandated safety management systems. Examples of a state or province system are:

- The mandated California OSHA Injury and Illness Prevention Program Title 8-§3203 (Guide to Developing Your Workplace Injury and Illness Prevention Program with checklists for self-inspection, n.d.).

- Building an effective health and safety management system (Building an Effective Health and Safety Management System, 1989), Partnerships in Injury Reduction, Alberta.
- The mandated Montana Safety Culture Act (Safety Culture Act, Public Domain, n.d.).

An older model of an OSHA safety management system that was proposed and offers a good process audit format is the Program Evaluation Profile (PEP), proposed in an earlier effort to develop a safety management program (Program Evaluation Profile (PEP), Adapted for Use, 1996).

Refer to Appendix C, Comparison of Government-Related Safety Management Systems Process Elements for an overview of the commonality between existing and proposed government-related safety management systems.

OSHA Proposed I2P2

An example of a proposed regulatory mandated safety process is the US OSHA I2P2. The proposal has been under review for a number of years and, if implemented, the rule will "require employers to develop and implement a program that minimizes worker exposure to safety and health hazards" (Injury and Illness Prevention Program (I2P2), n.d.).

Voluntary Protection Program

VPP has been in effect and has been effective based on OSHA studies in establishing in-depth compliance-based programs.

> *The VPP recognizes employers and workers in the private industry and federal agencies who have implemented effective safety and health management systems and maintain injury and illness rates below national Bureau of Labor Statistics averages for their respective industries.*
>
> *In VPP, management, labor, and OSHA work cooperatively and proactively to prevent fatalities, injuries, and illnesses through a system focused on: hazard prevention and control; worksite analysis; training; and management commitment and worker involvement. To participate, employers must submit an application to OSHA and undergo a rigorous onsite evaluation by a team of safety and health professionals.*
>
> *Union support is required for applicants represented by a bargaining unit. VPP participants are re-evaluated every three to five years to remain in the programs. VPP participants are exempt from OSHA programmed inspections while they maintain their VPP status.*
>
> **(Voluntary Protection Program (VPP), n.d.)**

Department of Energy Voluntary Protection Program Elements

> *The Department of Energy Voluntary Protection Program (DOE-VPP) promotes safety and health excellence through cooperative efforts among labor, management and government at the Department of Energy (DOE) contractor sites. DOE has also formed partnerships with other Federal agencies and the private sector for both advancing and sharing its Voluntary Protection Program (VPP) experiences and preparing for program challenges in the next century. The safety and health of contractor and federal employees are a high priority for the Department.*
>
> **(Voluntary Protection Program (DOE-VPP), n.d.)**

Building an Effective Health and Safety Management System

This is a health and safety management system provided by Partnerships in Injury Reduction, Alberta, Canada, that involves the introduction of processes designed to help reduce Injuries.

> *Successful implementation of the system requires management commitment to be vested in the system, effective allocation of resources, and a high level of employee participation. The scope and complexity of a Health and Safety management system will vary according to the size and type of workplace.*
>
> **(Building an Effective Health and Safety Management System, 1989)**

Montana Safety Culture Act

> *The Safety Culture Act enacted by the 1993 Montana State Legislature encourages workers and employers to come together to create and implement a workplace safety philosophy. It is the intent of the act to raise workplace safety to a preeminent position in the minds of all Montana's workers and employers.*
>
> *Listed … are the six requirements all employers must meet, and the additional three requirements employers with more than five employees must meet, to comply with the Montana Safety Culture Act (MSCA) … from the Department of Labor and Industry offered as guidelines for implementation of the MSCA.*
>
> **(Safety Culture Act, Public Domain, n.d.)**

Program Evaluation Profile

The PEP (Program Evaluation Profile (PEP), Adapted for Use, 1996) was intended to have been used as a safety and health program evaluation for the employer, employees, and OSHA. The PEP was to provide a method to score an organization's safety program. A numerical score was to provide an indicator of where program weaknesses were apparent (Program Evaluation Profile (PEP), Adapted for Use, 1996).

Examples of Advanced Safety Management Systems

The following are examples of voluntary advanced safety management systems. These systems have the same basic formats and use similar concepts in their structure. By researching and studying the underlined fundamentals of these recognized voluntary standards, you can review your current system in your organization and determine the best approach for moving toward a more structured, advanced, and industry-recognized safety management system. Refer to Chapter 4, "Setting the Direction for the Safety Culture".

- Occupational Health and Safety management systems, ANSI/ASSE Z10, 2012 (Occupational Health and Safety Systems, 2012).
- Canadian Occupational Health and Safety Management Standard, Z1000-06 (CSA Z1000-6) (Occupational Safety and Health Management, 2006).
- Occupational Health and Safety Management standard (OHSAS 18001) (Occupational Health and Safety Management standard (OHSAS 18001), 2007).
- Occupational Safety and Health (OSH) Management Systems, ILO-OSH 2001 Guideline (ILO-OSH, Overview and History, Module 2, Lesson 1, 2001).

The purpose of the standard is to provide organizations with an effective tool for continual improvement of their safety performance. Although Z10 is a guidance standard, implementation is not required by OSHA, the use of these guidelines is expected to improve occupational health and safety among American organizations.

(Occupational Health and Safety Systems, 2012)

Guidance standard—A standard that is advisory in nature and is not a regulatory requirement

(Glossary Terms, 2010)

Occupational Health and Safety Management, ANSI Z10-2012

Refer to Table 6.1 for "Overview of Occupational Health and Safety Management, Z10-2012 Five Major Elements".

Table 6.1: Occupational Health and Safety Management, Z10-2012, Five Major Elements

- Management leadership and employee participation
  - Occupational health and safety policy
  - Responsibility and authority
  - Employee participation
- Planning
  - Identify and prioritize OHSMS issues
  - Develop objectives for risk control based on prioritized OHSMS issues
  - Formulate implementation plans to accomplish prioritized objectives
- Implementation and operation
  - Hierarchy of controls
  - Design review and management of change
  - Procurement and contractors
  - Emergency preparedness
  - Education, training, competence
- Evaluation and corrective action
  - Monitoring, measurement, assessment, incident investigations, audits
  - Corrective action
- Management review
  - Annual review of OHSM

Source: Occupational Health and Safety Systems (2012).

Canadian Occupational Health and Safety Management Standard, Z1000-06

> *CSA Z1000-06 Occupational Health and Safety Management, is intended to reduce or prevent injuries, illnesses and fatalities in the workplace by providing companies with a model for developing and implementing an occupational health and safety management system.*
>
> **(Occupational Safety and Health Management, 2006)**

Refer to Table 6.2 for "Overview of Occupational Health and Safety Management, Z1000-06, Six Major Elements".

Table 6.2: Overview of Occupational Health and Safety Management, Z1000-06, Six Major Elements

OHS safety management system
• Responsibility, accountability, and authority
• Management representatives
• Worker participation
• OHS policy
Planning
• Review
• Legal and other requirements
• Hazard and risk identification and assessment
• OHS objectives and targets
Implementation
• Preventative and protective measures
• Emergency prevention, preparedness, and response
• Competence and training
• Communication and awareness
• Procurement
• Contracting
• Management of change
Documentation
• Control of documents
• Control of records
Evaluation and corrective action
• Monitoring and measurement
• Incident investigation and analysis
• Internal audits
• Preventive and corrective action
Management review
• Continual improvement
• Review input
• Review output

Source: Occupational Safety and Health Management (2006).

Occupational Health and Safety Management Standard (OHSAS 18001)

The British standard has established formal consensus criteria for occupational health and safety management systems.

Refer to Table 6.3 for "Occupational Health and Safety Management Standard (OHSAS 18001)".

OHSAS 18001 has been developed to be compatible with the ISO 9001 (Quality) and ISO 14001 (Environmental) management systems standards, in order to facilitate the integration of quality, environmental and occupational health and safety management systems by organisations, should they wish to do so.

The (OHSAS) specification gives requirements for an occupational health and safety (OH&S) management system, to enable an organisation to control its OH&S risks and improve its performance. It does not state specific OH&S performance criteria, nor does it give detailed specifications for the design of a management system.

(Occupational Health and Safety Management standard (OHSAS 18001), 2007)

Table 6.3: Occupational Health and Safety Management Standard (OHSAS 18001)

General requirements
- Establish an OHSMS for your organization

Planning requirements
- Analyze OH&S hazards and select controls
- Respect legal and nonlegal OH&S requirements
- Establish OH&S objectives and programs

Implementation requirements
- Establish responsibility and accountability
- Ensure competence and provide training
- Establish communication and participation
- Establish OH&S communication procedures
- Establish OH&S participation and consultation
- Control your organization's OH&S documents
- Implement operational OH&S control measures
- Establish an OH&S emergency management process

Checking requirements
- Monitor and measure your OH&S performance
- Evaluate legal and nonlegal compliance
- Evaluate compliance with legal requirements
- Evaluate compliance with nonlegal requirements
- Investigate incidents and take remedial action
- Investigate your OH&S incidents
- Take corrective and preventive action
- Establish and control OH&S records
- Conduct internal audits of your OHSMS

Review requirements
- Review the performance of your OHSMS

Source: Occupational Health and Safety Management standard (OHSAS 18001) (2007).

Occupational Safety and Health Management Systems, ILO-OSH 2001 Guidelines

> *The International Labor Organization (ILO) is a specialized agency of the United Nations (UN) with a primary goal to promote opportunities for women and men to obtain decent and productive work in conditions of freedom, equity, security, and human dignity. The ILO developed its voluntary guidelines on occupational safety and health (OSH) management systems in 2001: ILO-OSH 2001 Guidelines.*
>
> **(ILO-OSH, Overview and History, Module 2, Lesson 1, 2001)**

Refer to Table 6.4 for "Occupational Safety and Health Management Systems, ILO-OSH 2001 Guidelines".

Table 6.4: Occupational Safety and Health Management Systems, ILO-OSH 2001 Guidelines

Policy
- Occupational safety and health policy
- Worker participation

Organizing
- Responsibility and accountability
- Competence and training
- Occupational safety and health management system documentation
- Communication

Planning and implementation
- Initial review
- System planning, development, and implementation
- Occupational safety and health objectives
- Hazard prevention
  - Prevention and control measures
  - Management of change
  - Emergency prevention, preparedness, and response
  - Procurement
  - Contracting

Evaluation
- Performance monitoring and measurement
- Investigation of work-related injuries, ill health, diseases, and incidents, and their impact on safety and health performance
- Audit
- Management review

Action for improvement
- Preventive and corrective action
- Continual improvement

Source: ILO-OSH, Overview and History, Module 2, Lesson 1 (2001).

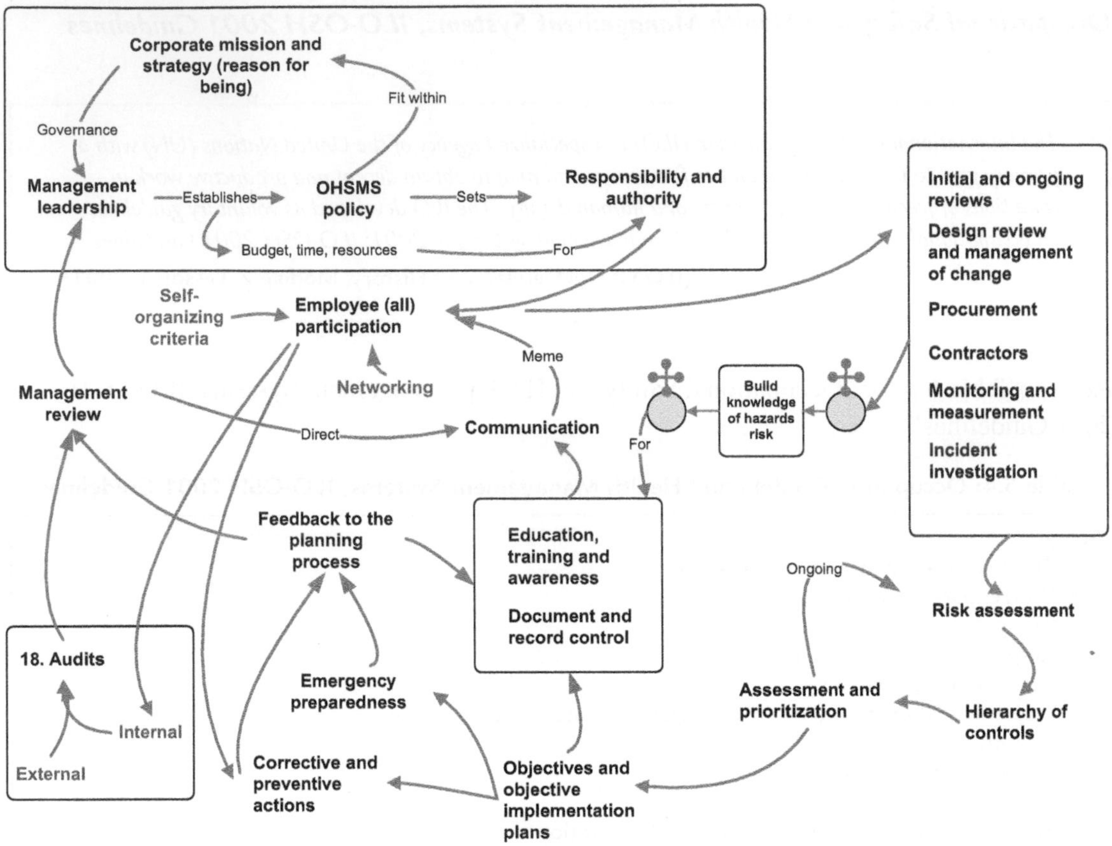

Figure 6.5
An Overview of the ANSI Z10 OHSMS Process.

Refer to Figure 6.5 for an "Overview of the ANSI Z10-2012 OHSMS Process". This diagram provides a view of the interrelationships between the various elements of the Z10 and offers insights on its implementation (Occupational Health and Safety Systems, 2012).

Summary

Many types of safety management systems have been used by organizations for many years. These systems can range from voluntary standards established by industry or trade groups, consulting organizations, and international standards organizations to required governmental legislation. In the end, these systems have the same intent—to provide guidance to organizations on how to manage and structure a successful safety process.

After reviewing a number of safety management systems from around the world and comparing their elements side by side, it is evident that a basic theme forms the overall core foundation of these systems.

At the core of these elements is the simple concept of a PDCA methodology, called the Shewhart cycle (Shewhart, 1931), that can equally be applied to all safety management systems. PDCA has been at the root of quality control and safety programs.

If your organization uses the Six Sigma approach to quality management, the use of a variation of the PDCA known as DMAIC can be used. DMAIC is a more advanced approach than the traditional PDCA.

Once the basic safety management system elements as discussed in Chapter 5, "Overview of Basic Safety Management Systems", are implemented and deemed successful, your organization has reached a maturity that will allow you to move to the next level. The intent is to ensure that the organization has the initial fundamentals in place before taking on an advanced safety management system.

A variation in the perception exists about how the correct safety management system is developed or implemented. Each organization has different structures that need a variety of approaches. In Six Sigma, the "Voice of the Customer" or, as a variation, the "Voice of the Process", must be listened to before any safety management system is introduced.

Two types of safety management systems are available—government-related as both mandated and voluntary, and voluntary systems developed by nationally and internationally recognized standards agencies and professional societies.

In Chapter 7, we will discuss "leadership and how effectively a safety culture" will be enhanced. As we discussed in Chapter 4, "Setting the Direction for the Safety Culture", you have to ask yourself daily: "What is important today and what has to be done now?" When you choose your project of the day, it should be in line with your vision/mission of improving the safety culture. The so-called "real world" will intrude if you have not established personal leadership to counter the scope drift that we have discussed in Chapter 3, "Analyzing and Using Your Network". Your internal vision/mission as a leader provides the guidance on how to adapt and change in all types of situations!

Chapter Review Questions

1. What is the purpose of a safety management system?
2. What process elements do all safety management systems have in common? Refer to Appendix H for a comparison of safety management system process elements.
3. Explain the concepts of the PDCA cycle.
4. Who is the quality guru who developed PDCA?
5. How does the PDCA apply to safety management? Provide several examples.
6. Why is the safety management system similar to the quality management process?
7. Compare and contrast the PDCA and the DMAIC system.
8. What are the benefits of using a standardized safety management system?

9. Identify the pros and cons of a standardized safety management system. Provide some examples from your experience.
10. What is the difference between a government- and voluntary-related safety management system?
11. Identify the difference between OSHA's proposed I2P2, VPP, and the DOE-VPP.
12. Compare and contrast the pros and cons of developing and implementing the ANSI/AIHA Z10 safety management system versus the OSHA VPP program.
13. Discuss the perceived or real benefits of each safety management system that is listed in this chapter and the relative difficulty of implementing such a safety management. Support your discussion with your personal experience and/or resources from other sources.
14. Various states and provinces may have their own suggested or mandated safety management systems. Provide examples of these systems.
15. Name the advanced voluntary safety management systems.
16. What is the major difference in the voluntary safety management systems?

Bibliography

Building an Effective Health and Safety Management System. (1989). Partnerships in Health and Safety program, Partnerships in Injury Reduction (Partnerships), Reprinted/Modified and/or adapted with Permission. Retrieved from http://bit.ly/WsHteC.

Crosby, P. B. (1979). *Quality is free: The art of making quality certain.* (Vol. quality_crosby), New York: McGraw-Hill.

Dababneh, A. J. (n.d.). Effective Occupational Safety and Health Management System: Integration of OHSAS 18001, ILO-OSH 2001, and OR-OSHA. University of Jordan Faculty of Engineering and Technology Department of Industrial Engineering. Retrieved from http://bit.ly/X1YGJY.

Deming, W. E. (n.d.). Plan–Do–Check–Act. Wikipedia, the free encyclopedia. Retrieved from http://bit.ly/14FZb1H.

Glossary Terms. (2010). North Carolina State University. Retrieved from http://bit.ly/XevTjx.

Guide to Developing Your Workplace Injury and Illness Prevention Program with checklists for self-inspection. (n.d.). State of California, Department of and Industrial Relations, Adapted for Use. Retrieved from http://bit.ly/Xewi5m.

Handbook for basic process improvement. (June 1996). U.S. Navy Public Domain. Retrieved from http://1.usa.gov/Y7U2ZX.

How to Effectively Assess and Improve your Safety and Health Program through Safety and Health Evaluation. (n.d.). Oregon Occupational Safety and Health Division (Oregon OSHA), OR-OSHA 116, Public Domain, Permission to Reprint, Modify, and/or Adapt as necessary. Retrieved from http://bit.ly/106AhmR.

ILO-OSH, Overview and History, Module 2, Lesson 1. (2001). North Carolina Safety University. Retrieved from http://bit.ly/XOTbMM.

Injury and Illness Prevention Program (I2P2). (n.d.). Occupational Safety and Health Administration (OSHA), Adapted for Use. Retrieved from http://1.usa.gov/lIszWK.

Occupational Health and Safety Management standard (OHSAS 18001). (2007). Praxiom Research Group Limited. Translated Into Plain English.

Occupational Health and Safety Systems. (2012). The American Industrial Hygiene Association. ANSI/AIHA Z10.

Occupational Safety and Health Management. (2006). Canadian Standards Association, CSA Standard. Z 1000-06.

Pearse, W., Gallagher, C., & Bluff, L. (Eds.). (2001). *Occupational health & safety management systems, proceedings of the first national conference*. Crown Content. WorkCover. Retrieved from http://bit.ly/WLdr5y.

Program Evaluation Profile (PEP) Adapted for Use. (1996). Occupational Safety and Health Administration (OSHA). Retrieved from http://1.usa.gov/VuKM1C.

Roughton, J., & Crutchfield, N. (2008). *Job hazard analysis: A guide for voluntary compliance and beyond. Chemical, petrochemical & process*. Elsevier/Butterworth-Heinemann. Retrieved from http://amzn.to/VrSAq5.

Safety Culture Act, Public Domain. (n.d.). State of Montana Department of Labor & Industry, Public Domain. Retrieved from http://1.usa.gov/ohJMKR.

Shewhart, W. A. (1980). Economic control of quality of manufactured product. Republished by BookCrafters, Inc., Chelsea, Michagan.

Successful Health and Safety Management, hsg65. (1997). Health and Safety Executive, Crown Publishing, Permission to Reprint. Retrieved from http://bit.ly/YKDzRk.

Voluntary Protection Program (DOE-VPP). (n.d.). Department of Energy, Public Domain. Retrieved from http://1.usa.gov/14ob6Rt.

Voluntary Protection Program (VPP). (n.d.). Occupational Safety and Health Administration (OSHA), Public Domain, Adapted for Use. Retrieved from http://1.usa.gov/T0j6EW.

CHAPTER 7

Leadership and the Effective Safety Culture

The successful man will profit from his mistakes and try again in a different way.
—Dale Carnegie

Accountability links responsibility to consequences.
—Harry Truman

Coming together is a beginning. Keeping together is progress. Working together is success.
—Henry Ford

The difference between an unsuccessful person and others is not a lack of strength or knowledge, but rather a lack of will.
—Vince Lombardi

Introduction

When the word "leader" is heard, it immediately conveys an image of someone who is able to step to the front of a group, clearly communicate what he or she desires the group to do. Leaders inspire the group to do more than what is normally expected, and get the group to achieve what others did not think was possible. Real leaders are rare and leadership is something you know when you see it.

Most of what we see is actually management. Mangers get things done. They administrate, ensure rules and procedures are followed, and in general see that established goals and objectives are met. A true leader may or may not be a good manager and a manager may not be a good leader. The difference is that an organization cannot excel without leadership. It can survive through management but not excel.

In the absence of a balance of leadership and management, our professional and personal responsibilities can become so complex that confusion about what is most important allows duties to be neglected that are important to the success and improvement of the organization. Leadership provides the clarity to cut through the confusion and focus the energy of the organization on what is truly important.

As we discussed in Chapter 4, "Setting the Direction for the Safety Culture", you have to ask yourself daily, "What is important today and what has to be done now?" When you choose your project of the day, it should be in line with your vision/mission of improving the safety culture. The so-called "real world" will intrude if you have not established personal leadership to counter the scope drift that we have discussed in Chapter 3, "Analyzing and Using Your Network". Your internal vision/mission as a leader provides guidance on how to adapt and change in all types of situations.

> Business is NOT static; it is dynamic and always changing.

After completing this chapter, you should be able to discuss concepts of leadership and how to better advise your leadership team for the success of the safety management system. You should be able to:

- Compare and contrast the definitions of leadership.
- Discuss the impact of leadership on the safety management system.
- Compare and contrast responsibility, authority, and accountability.
- Discuss why delegation has not been considered part of a safety management system.
- Define your assigned accountability and responsibility for the safety management system and the sustaining of the safety culture.
- Discuss the importance of written job descriptions.
- Discuss why it is necessary to establish consequences for performance.

Leadership Defined

Ask the following questions to assess the effectiveness of the current leaders in your organization:

- Do leaders in your organization work together to create an environment where employees are involved in the safety management system?
- Are assigned accountabilities matched with responsibilities within the leadership team?
- Are goals assigned without consulting with the employees?
- Are goals established without clear objectives on how to meet those goals?
- Does the leadership team provide you with the authority to accomplish the goals of the safety management system?

If you can answer "yes" to these basic questions, then you may have elements of leadership in place. If the answer is "no", then the organization may not be managed under real leadership.

> *Management is doing things right; leadership is doing the right things.*
>
> **Peter Drucker**

The leadership team by its visible presence and behavior sets the tone that drives the development of the safety management system. Your personal leadership goal should be to increase your authority and influence, which in turn allows you to better advise the leadership team on solutions that deeply weave the safety management system into the culture of the organization.

Your responsibilities will be split between program management when developing and implementing the safety management system. You must become the leader who can communicates the vision/mission in a way that inspires the leadership team and employees as to what is expected with regard to the sustaining the safety culture.

> The two major roles you must play: management and leadership:
>
> You should manage the safety management system and develop your leadership skills to support the effort for a safety culture. You must shift between a program state of mind and a mindset that thinks in terms of systems and process.

For example, in many cases, supervisors must double as a supervisor and a production employee as the need arises. This is necessary in some cases but can unfortunately create a vague undefined role for the supervisor who is held accountable for production. The supervisor must be both a leader and producer to ensure that assigned production is completed.

> *Coming together is a beginning. Keeping together is progress. Working together is success.*
>
> **Henry Ford**

In reviewing definitions of leadership, we find that it has many definitions. On one website, "The Happy Manager", noted that, "There is no single definition of leadership. It's one of those truisms that some of the most important things in life seem difficult to define. Yet ... you know when you see or feel it" (Leadership Quotes, n.d.).

The following comments from various sources provide insights that bear out the quote "you know when you see or feel it" (Leadership Quotes, n.d.), giving us an idea of how hard it is to define leadership. Although leadership is hard to define, when you see a leader, you just know it by their actions, their attitude, and their positive responses to any situations.

The military constantly studies the concepts of leadership from which we can draw insight for leading a safety management system. At times, it is essential that leaders go beyond just managing a given situation as sometimes the mission must take on specific risk that must be controlled. The following comments provide a mosaic from which we can better define what and how leadership can be realized:

- The World War II General, Omar Bradley, once stated that "Leadership … means firmness, not harshness; understanding, not weakness; generosity, not selfishness; pride, not egotism."

Refer to Table 7.1 for Bradley's thoughts on "Leadership Expectations" (Army Leadership, ADP 6-22, 2012).

- According to Omar Bradley "along the way, you will make honest mistakes. You will face difficult decisions and dilemmas. This is all part of the process of learning the art of leadership" (Army Leadership, ADP 6-22, 2012).
- "Leadership is characterized by a complex mix of organizational, situational, and mission demands on a leader who applies personal qualities, abilities, and experiences to exert influence on the organization, its people, the situation, and the unfolding mission" (Army Leadership, ADP 6-22, 2012).
- "Leadership and increased proficiency in leadership can be developed. Fundamentally, leadership develops when the individual desires to improve and invests effort, when his or her superior supports development, and when the organizational climate values learning. Learning to be a leader requires knowledge of leadership, experience using this knowledge and feedback" (Army Leadership, ADP 6-22, 2012).

Table 7.1: Leadership Expectations

• Have a vision and lead change.
• Be your formation's moral and ethical compass.
• Learn, think, and adapt.
• Balance risk and opportunity to retain the initiative.
• Build agile, effective, high-performing teams.
• Empower subordinates and underwrite risk.
• Develop bold, adaptive, and broadened leaders.
• Communicate up, down, and laterally; tell the whole story.

Source: Army Leadership, ADP 6-22, 2012.

> *Leadership is the process of influencing people by providing purpose, direction, and motivation to accomplish the mission and improve the organization.*
>
> **(Army Leadership, ADP 6-22, 2012)**

- "Being a leader is not about giving orders, it's about earning respect, leading by example, creating a positive climate, maximizing resources, inspiring others, and building teams to promote excellence" (Army Leadership, ADP 6-22, 2012).
- "To be a good leader, you should learn what types of authority you have and where it comes from. Whenever in doubt, ask. Once you're confident that you know the extent of your authority, use sound judgment in applying it" (Duties, Responsibilities, and Authority explained, n.d.).
- "Occasionally, negative leadership occurs in an organization. Negative leadership generally leaves people and organizations in a worse condition than when the leader–follower relationship started. One form of negative leadership is toxic leadership." (Army Leadership, ADP 6-22, 2012).
- "Toxic leadership is a combination of self-attitudes, motivations, and behaviors that have adverse effects on subordinates, the organization, and mission performance. This leader lacks concern for others and the climate of the organization, which leads to short- and long-term negative effects. The toxic leader operates with an inflated sense of self-worth and from acute self-interest. Toxic leaders consistently use dysfunctional behaviors to deceive, intimidate, coerce, or unfairly punish others to get what they want for themselves" (Army Leadership, ADP 6-22, 2012).

The following is advice from leadership from another industry that has high hazards and associated risk, according to the *Human Performance Improvement Handbook, Concepts and Principles* developed by the U.S. Department of Energy, is defined as:

- "a leader is … any individual who takes personal responsibility for his or her performance and the facility's performance and attempts to positively influence the processes and values of the organization" (Volume 1: Concepts and principles, human performance improvement handbook, 2009).
- "Workers, although not in managerial positions of responsibility, can be and are very influential leaders. The designation as a leader is earned from subordinates, peers, and superiors" (Volume 1: Concepts and principles, human performance improvement handbook, 2009).

> *Performance Indicators—Parameters measured to reflect the critical success factors of an organization. A lagging Indicator is a measure of results or outcomes. A leading indicator is a measure of system conditions or behaviors which provide a forecast of future performance (also known as 'metrics').*
>
> **(Volume 2: Human performance tools for individuals, work teams, and management, human performance improvement handbook, 2009)**

Leadership Impact on the Safety Management System

The safety management system and the safety culture are influenced and shaped by how the organization is managed and how it is led. "Managers and supervisors are in positions of responsibility and as such are organizational leaders" (Volume 1: Concepts and principles, human performance improvement handbook, 2009).

For example, your employees will look to their leadership team for meaning, purpose, direction, and insight. If they feel and clearly see the organization's leaders are providing meaning to what is being done, show that all jobs have purpose, and share the direction desired for the organization, then employees will more than likely take on greater responsibilities and go beyond what is normally expected. However, if employees do not see any leadership qualities, they may be less inclined to get involved.

> How is leadership defined? One comment that is often heard about great leadership is "I would follow that person to the end of the earth and back!"

Leadership and Organization Structure

A strong central structure can have an organizational culture that readily reinforces the safety process at all locations. The organization may have a general culture that espouses a participative style. This style is ideal to enhance the safety culture as it can increase employee involvement in the safety management system.

As discussed in Chapter 1, "The Perception of Safety", your organizational culture must provide the appropriate level of support to the safety management system in order to have a successful safety culture that will stand the test of time.

The values, goals, objectives, business strategies, and beliefs of leadership provide indicators of the nature of their leadership style. These indicators send a message to the organization on what is considered important.

Based on this, the leadership team sends a strong message when it:

- Is visible and active in periodic safety-related meetings such as the preshift reviews, weekly/months, face-to-face, etc.
- Maintains open communications with all employees and reduces or eliminates barriers to reporting safety issues, hazards, and associated risk.
- Provides quick correction of loss-producing activities and/or conditions.
- Keeps safety management system information on meeting agendas.
- Ensures that all safety criteria is fully incorporated into orientation and other related training.

- Tracks the status of the organization's safety performance, both pre- and post-loss data, with emphasis on the former.
- Takes all concerns regarding safety issues, hazards, and associated risk seriously.

In a decentralized organizational structure, one location may have a stronger safety management system and a safety culture while another may be just the opposite because it has a toxic style of leadership that does not allow an in-depth safety culture. Evidence of this style is when the main emphasis is on compliance and minimal compliance at best.

Leadership Expectations

"Achieving safety excellence begins with leadership commitment." This statement tends to be overused. However, while this statement is overused, it is vital for the leadership team to show their commitment to the safety process as they must demonstrate visible positive involvement in daily safety activities. Refer to Chapter 11, "Developing an Activity-Based Safety System" for more details on defined safety activities.

As part of the leadership in your organization, understanding and internalizing the concepts of leadership will provide you with the foundation necessary to expand your influence. The criteria outlined by Omar Bradley as listed in Table 7.1, "Leadership Expectations", are one set of criteria that can guide your growth as a leader.

It is importance to understand your current organization's culture and the types of management style used by your leadership team. Refer to Chapter 3, "Analyzing and Using Your Network" for a detailed discussion on how to use your network.

Lesson Learned # 1

James started out in industry as a new employee with a major corporation. After a few years, he was promoted to a supervision role. During the training orientation into the supervision role, he was told to follow the philosophy that if an employee cannot do their assigned job and keep up with the production line, "just fire them and hire somebody new". So for several years, this is how James played the game. But the problem he had was that he never had a stable workforce because of how he was taught to manage employees. One day James realized that "just firing employees" was not a good strategy.

Even though he was "successful" as a supervisor, he still had the problem of dealing with respect from his employees. He was good at what he did, but he had a reputation of "If you do not do your job, you will be fired". With this reputation, when he would move into another production line, the employee reaction was "I do not want to work for you" and he constantly fought personnel turnover.

To combat this issue, after reflection about his supervisory style, James decided to try to change his personal brand. He started to build a stable workforce by conducting face-to-face discussions with each employee, trying to understand their needs. By conducting discussions in meetings, one-on-ones, and with groups, he made it very clear that his expectations was "All he wanted out of anyone was for them to come to work and just do a quality job" His personal network evolved from being one-way negative to positive two-way communication.

Employees quickly found out that this was truly his new leadership style and they understood in a short period of time that he meant what he said and stood by his decisions. Once all employees understood that he was fair and sincere he had even more success as his production lines became stable and he had employees that wanted to work with him. He moved from just being a supervisor to leader by incorporating core principles. ■

The case study reflects the following insights:

- "Leaders of integrity adhere to the values that are part of their personal identity and set a standard for their followers to emulate. Identity is one's self-concept, how one defines him or herself. Leaders who are effective with followers identify with the role and expectations of a leader; they willingly take responsibilities typical of a leader and perform the actions of a leader. Leaders who are unsure of themselves may not have a strong idea of their identity" (Army Leadership, ADP 6-22, 2012).

Establishing Organizational Priorities

Most organizational leaders and managers have a general and basic understanding of the value of safety. However, as the organization sets priorities for all of its other mandates, it will tend to see safety as an additional priority, on the list of agenda items, especially if loss-producing events are low. While understanding its importance, the leadership may not have an informed understanding of the core concepts of hazard, associated risk, and control.

Management by Walking Around

Leaders must be visible and in constant contact with all levels of the organization. To be a leader, you must remain visible to not just employees but to all levels of the organization.

"Management By Walking Around" (MBWA) was introduced by Peters & Waterman in their book *In Search of Excellence* (Peters & Waterman, 2004). This technique, if properly used, suggests that members of the leadership team routinely and consistently walk around their operation, observing, and conducting one-on-one discussions with their employees.

The leadership team gets the feel and tone of its operation if this is kept routine. MBWA is part of network development as leaders, including you, gain insight about the concerns of employees and the issues they must deal with each day. As discussed in Chapter 3, "Analyzing and Using Your Network", influence can go down to three degrees of contact. Leaders are not just talking to one or two employees but to many others as those directly contacted relay what they experienced to others.

If the walk-around results in negative impacts, the leaders did not follow the safety rules, berated someone, or in general left bad feelings, the influence of leadership is diminished.

Refer to Chapter 11, "Developing an Activity-Based Safety System" for other activities that can be used.

You Are Directly Responsible for Establishing Purpose

In the real world, the problem is that most levels of management are routinely challenged with completion of assigned tasks with limited direction. Management's objective is to complete its tasks as assigned using the resources it have been given. To reduce the potential for rote completion of these tasks, it is up to you to develop the manager leadership skills so that employees understand the intent and know they can question the assigned task if that intent is not being met. Without built in flexibility, you will project a negative attitude to your employee in the form "I have got to do this", regardless of the unintended consequences. A true leader will take these types of challenges and puts a positive spin on any type of situation and looks for opportunities for improvement as "We want to do this because ... ".

> *Human error is caused not only by normal human fallibility, but also by incompatible management and leadership practices and organizational weaknesses in work processes and values.*
>
> **(Volume 1: Concepts and principles, human performance improvement handbook, 2009)**

In most organizations, the traditional pyramid of responsibilities is in place where each level of leadership is held accountable for the goals and objectives of its subordinates. The senior leadership team delegates the activities needed to meet its goals and objectives to individuals whose expertise or areas of authority provide the best path to completion. Each level of leadership retains responsibility for completion of the assigned goals.

For routine, normal operations on a daily basis, the pyramid discussed above generally works well. When problems develop that might need to be quickly resolved because of their potential severity or harm to the organization, the normal lines of communuication and authority may need to be bypassed or circumvented depending on the critical nature of the hazard and scope of the associated risk. Delays in decisions may allow a severe situation to become a catastrophic event.

Open Door Policy

In addition to MBWA, you should make yourself available to your employees through actions such as having an open door policy. If the safety culture is strong, no one will fear bringing an issue or problem to your attention. Being accessible means walking a fine line between encouraging your employees to communicate with you yet not interfering with their normal relationships with supervisors or managers.

The open door is part of the ongoing process of expanding your network of contacts within the organization. The open door is not to bypass the normal chain of command or authority. It offers a "relief valve" that allows anyone to approach you for insights, advice, and information. Care must be taken because as a key leader, these methods ensure that you do not interfere with the authority that has been assigned to the other members of the leadership team.

For hazard and associated risk issues and problems, the regular chain of command may not respond fast enough to respond to the critical nature of the issue. Historical examples of these issues are the Challenger and Columbia space shuttle disasters. An open door policy and procedures that allow unresolved high hazard and risk information to move rapidly to senior leadership should be in place. Essentially, a way to short circuit and bypass normal communication channels is needed. Refer to Chapter 9, "Risk Perception—Defining How to Identify Personal Responsibility".

> *If I had an hour to solve a problem and my life depended on the solution, I would spend the first 55 minutes determining the proper question to ask, for once I know the proper question; I could solve the problem in less than five minutes.*
>
> **Albert Einstein**

Defining Roles, Responsibility, Delegation, Authority, and Accountability

Effective implementation of a safety management system is a leadership team responsibility. As an in-depth system must reach all levels of the organization, the leadership team should actively involve all employees of the organization in the process. To reach the real potential of a system, it is important that roles, responsibility, delegation, and authority for safety-related assignments are clearly defined and coupled with a strong accountability structure. This is accomplished by designing safety activities that are aligned with your organizational goals and objectives.

As a business grows in complexity and the number of employees increases, being responsible for all of the details of an effective safety management system may become less feasible. The assignment of responsibilities throughout the leadership team is more efficient if a safety management system structure has been adopted (Managing worker safety and health, n.d.).

> *A manager alone cannot perform all the tasks assigned to him. In order to meet the targets, the manager should delegate authority. Delegation of Authority means division of authority and powers downwards to the subordinate. Delegation is about entrusting someone else to do parts of your job. Delegation of authority can be defined as subdivision and suballocation of powers to the subordinates in order to achieve effective results.*
>
> **(Delegation of Authority, n.d.)**

Review Your Organization to Determine Safety-Related Tasks for Each Role

To determine the specific tasks leadership and employees will be assigned and how you want these roles to contribute, first conduct a review of your entire organizational structure. The review will help identify the key roles that you deem vital to your safety culture. When conducting this review, ensure that you take into account your internal and external network as discussed in Chapter 3, "Analyzing and Using Your Network".

To define and integrate accountabilities and responsibilities into the organization, begin by developing a description list of job titles that will play a key role in safety for your organization. To accomplish this, review your organizational structure to understand how each position can best support the safety management system. These positions will provide visible evidence that the leadership intends to have a sustained positive safety culture and must be selected with care. These individual are a good resource in the developing further job descriptions for positions to be closely associated and normally involved in the safety process (Managing Worker Safety and Health, Appendix 5-2, Sample Assignment of Safety And Health Responsibilities, n.d.; Roughton & Mercurio, 2002).

The key is to ensure that you list all positions in your organization. All leadership and employees should have assigned activities and responsibilities within the safety management system. Some will have more than others, but everyone should have an assignment. It may be walk-throughs, data gathering, presentations, safety committee, loss-producing event review, risk assessment, etc.—i.e. everyone. Refer to Chapter 11, "Developing an Activity-Based Safety System".

From our experience with job hazard analyses, we have found it's best not to assume that each job, task, or assignment has been clearly defined (Roughton & Crutchfield, 2008). Job descriptions may also be different from what the job title states. The following questions can be asked of leadership and employees:

- What are your specific duties?
- Who do you report to and what is your position and authority?
- Where is that person within the organization?
- What authority do you have and how is it described?

As discussed in Chapter 1, "The Perception of Safety", the problem is that safety job descriptions traditionally include many duties that must be performed along with that of ensuring that various governmental regulatory compliance criteria are met. This can be a tall task for anyone to follow and be successful. The issue we have is that this type of job description assumes that one person or group can take responsibility for all of the safety culture.

The Value of Developing and Implementing Written Job Descriptions

A weak area we have seen is not having the assignment of safety management system responsibilities directly placed into written job descriptions. We have seen statements about maintaining compliance with Occupational Safety and Health Administration (OSHA) or other regulation, but they have been limited in defining involvement in other elements of the safety management system. While compliance is an essential responsibility, additional specific duties must be formalized that go beyond regulatory and injury control statements.

A job description provides the structure of roles in the organization to include responsibilities and authorities within the safety management system and/or other associated program elements. Individual job descriptions describe the most important characteristics and responsibilities of a job title (A Reference for Developing a Basic Occupational Safety and Health Program for Small Businesses, 2008; Roughton & Mercurio, 2002).

Well-written job descriptions enhance communication and coordination among positions and aid in:

- Removing any doubt about the lines of responsibilities and authority for each position.
- Determining that all responsibilities have been accounted for in the organization.
- Identifying how new tasks and responsibilities should be assigned.
- Developing job performance objectives and establishing performance measurements.
- Improving and enhancing communication throughout the organization.
- Developing whether all responsibilities have been accounted for within the organization and whether new tasks and responsibilities should be assigned.
- Developing job performance objectives and establishing performance measurements (Managing worker safety and health, n.d.; Managing Worker Safety and Health, Appendix 5-2, Sample Assignment of Safety And Health Responsibilities, n.d.).

When filling a position that has existed over a period of time, the opportunity may exist to make changes to the existing job description. If you are moving into a new position where the safety culture has been determined to be weak and in need of improvement, you may be able to initiate changes in the existing job description to include specific responsibilities and areas that will enhance the ability to make improvements.

Writing the Basic Job Description

If written job descriptions that have safety-related activities built into them do not exist, begin by writing a short basic "general statement" defining the specific responsibilities and limits of authority and the resources required to reach the specific goals and objectives for which the position will be held responsible. Refer to Appendix I for a "Sample Safety Responsibilities Worksheet" that is a starting point to document safety expectations.

You need to collect as much information as possible that will help to provide guidance on how to accomplish specific safety-related activities. At this point you do not have to define each job's specific safety tasks in detail. All you need is a brief summary of the expectations that can be agreed upon with leadership. The job description worksheet starts to become a living document that must be regularly reviewed and modified as necessary to keep up with activities that may change in time as to what needs to be accomplished. As performance reviews are conducted, the job description can be updated, modified, and responsibilities expanded as warranted.

Defining Clear Goals and Assigning Responsibilities

To have a successful safety culture, all elements of the organizations must work together and every employee in the organization must be held accountable for their role in sustaining the process. A connection must exist between what and how a role is to be held accountable and the end results desired. The responsibilities must be based on preloss activities as discussed in Chapter 11, "Developing an Activity-Based Safety System".

> *When an individual or organization is assigned authority and/or delegated responsibility, they must provide a plan; execute the plan, and measure and report real results relative to that plan.*
>
> **(Artley, 2001)**

For example, if supervisors have little control over maintenance, hiring practices, training, budget, time, quality, etc., then they have little control over the elements that create the potential for a loss-producing event. In this scenario, if supervisors are held accountable for injuries that may occur, they are being held accountable for human and latent errors that are not under his or her control (Volume 1: Concepts and principles, human performance improvement handbook, 2009). The best approach is to have clearly defined, tangible responsibilities for preloss activities and lines of authority.

Lesson Learned # 2

A colleague of Nathan had a situation where he was discussing the cost of workers' compensation with a general manager. The manager remarked that those costs were considered inconsequential as he spent far more money on trinkets, t-shirts, etc. to give to customers. The actual hazards and potential risk were not visible to him.

The lesson is that what appears to be a major problem at one level may not be perceived as such at another. Accountability using safety goals based on incident rates and loss ratios does not take into account changes in experience levels, employee turnover, hours worked, production problems, and other conditions that may not be under the control of the leadership team.

By focusing on activities within the control of specific roles, accountability becomes a valid leadership tool. For example, the use of activity-based safety as discussed in Chapter 11, "Developing an Activity-Based Safety System", provides a method of straightforward preloss activities that can be readily implemented.

Get in Agreement on Objectives

Before you can hold employees accountable for their actions you must ensure that they know your expectations for their position. To accomplish this, each position must have established goals set for its personal performance that are related to the overall organizational goals.

When objectives are unclear, the ball can easily get dropped, and it will be hard to determine whose performance is lacking. To get started, use the S.M.A.R.T. concepts for developing goals and objectives as discussed in Chapter 4, "Setting the Direction for the Safety Culture". By using the S.M.A.R.T. concept, you can check assignments of responsibility to ensure that they specify who does what and that goals are reasonably attainable.

The key to success is to routinely refer to written objectives and expectations in meetings, conversations and performance discussions with all employees.

Writing Your Objectives for Each Job Position

Documenting objectives in writing clarifies the meaning and intent of what is to be accomplished. When you write each objective for the given position, ensure that you state your expectations in specific, measurable terms. Include what you want the position to achieve and to what degree you want the objectives to be accomplished in a specific time frame. Goals and objectives are not set in stone, as the conditions under which they were developed may change over time and must be reviewed.
The ongoing dialogue with leaders and employees should allow for constant adjustments.

Objectives must be based upon performance measurements, which are milestones that tell if the person did or did not perform his or her assigned safety-related task as defined.

Once you have defined your objectives and have put them in writing, if questions arise, then you will have a written document that outlines specific expectations. The written document shows that you are serious about the meeting of goals and objectives.

The following are examples that can be used to ensure that strong objectives have been developed:

- Hold a daily preshift review with all of employees at a specified time.
- Meet face-to-face with each employee and have a discussion about safety concerns.
- Conduct a weekly or monthly meeting as appropriate.
- Track all identified hazards until corrective actions are completed.
- Review the equipment/machine-specific checklist with your employees.
- Complete one job hazard analysis each month for the department.

Through these discussions, your message is directly communicated to the employees for each of these activities. Refer to Chapter 11, for more details on "Developing an Activity-Based Safety System".

Nonsupervisory Employees

When developing responsibilities for nonsupervisory employees, do not confuse specific responsibilities with work rules and/or work practices. A short statement about the employee's responsibility to report hazards, understand and follow rules, and work practices is applicable may include:

- Attend the daily preshift review at its specified time.
- Participate in weekly or monthly meeting as appropriate.
- Assist in correcting identified hazards.
- Report hazards and associated risk.
- Assist in loss-producing event investigations and weekly planned inspections.
- Use the equipment/machine-specific checklist.
- Assist in job hazard analysis.
- Use the risk guidance tool when conditions or hazardous conditions are identified.

Review Assigned Activities Regularly

If a managers or supervisors meet their written objectives, but their areas of responsibility continue to have too many risk-related issues such as close calls or no improvement in hazardous conditions, then their objectives may need to be updated to reflect the current situation.

Lesson Learned # 3

At a plant, a supervisor was assigned five activities that needed to be accomplished and he successfully completed four of the activities. During performance reviews, it was discussed why the supervisor could not accomplish the one activity. The supervisor presented the case that the activity was not in his control. The supervisor was asked what activity could he successfully be accomplish. An agreement was reached to replace the one unachievable activity with one that the supervisor had more control over. The new activity was successfully completed and safety conditions improved.

When you assign responsibilities to a role, it is essential that you delegate the necessary authority and/or commit sufficient resources to back those responsibilities. A conscientious employee can become demoralized when given an assignment without the necessary means to carry it out (Managing worker safety and health, n.d.).

> *Where individuals perceive their roles as irrelevant or demeaning, there is a potential for negative stress. Responsibility without authority, work which is too challenging or not sufficiently challenging, expectations of high performance without adequate training and feedback are examples of the issues under consideration here.*
>
> **(Petraeus & Amos, 2006)**

Elements of Delegation

Delegation is more than just "passing the buck" to a subordinate, telling them to "do what you need to do". Leadership is still responsible for the results and cannot totally blame the subordinate if things go wrong. The leadership must still be involved through periodic reports, performance reviews, and face-to-face discussions to assess the status of what has been delegated.

Network analysis can be used to determine where and how responsibilities are shifted by assigned accountabilities and authorities. The network may reveal how these are cascading up through the organization. Refer to Chapter 3, "Analyzing and Using Your Network". The triad of responsibility, authority, and accountability cannot be separated and must be considered as interrelated and together as a whole.

> *Trust but verify.*
>
> **Vladimir Lenin, Russian Proverb**

Several key leadership definitions would include:

- "Responsibility—is the duty of the person to complete the task assigned to him. A person who is given the responsibility should ensure that he accomplishes the tasks assigned to him. If the tasks for which he was held responsible are not completed, then he should not give explanations or excuses. Responsibility without adequate authority leads to discontent and dissatisfaction among the person. Responsibility flows from bottom to top. The middle level and lower level management holds more responsibility. The person held responsible for a job is answerable for it. If he performs the tasks assigned as expected, he is bound for praises. While if he doesn't accomplish tasks assigned as expected, then also he is answerable for that" (Delegation of Authority, n.d.).
- "Authority—in context of a business organization, authority can be defined as the power and right of a person to use and allocate the resources efficiently, to take decisions and to give orders so as to achieve the organizational objectives. Authority must be well-defined. All people who have the authority should know what is the scope of their authority and they shouldn't misutilize it. Authority is the right to give commands, orders and get the things done. The top level management has greatest authority. Authority always flows from top to bottom. It explains how a superior gets work done from his subordinate by clearly explaining what is expected of him and how he should go about it. Authority should be accompanied with an equal amount of responsibility. Delegating the authority to someone else doesn't imply escaping from accountability. Accountability still rest with the person having the utmost authority" (Delegation of Authority, n.d.).
- "Accountability—means giving explanations for any variance in the actual performance from the expectations set. Accountability cannot be delegated. For example, if 'A' is given a task with sufficient authority, and 'A' delegates this task to 'B' and asks him to ensure that task is done well, responsibility rest with 'B', but accountability still rest with 'A'. The top level management is most accountable. Accountability, in short, means being answerable for the end result. Accountability can't be escaped. It arises from responsibility" (Delegation of Authority, n.d.).
- "Often, the word responsibility is used in conjunction with the word accountability. When hearing the word accountability, many people immediately equate it with responsibility and see the two as being the same" (Artley, 2001).
- "Assignment of Duties—The delegator first tries to define the task and duties to the subordinate. He also has to define the result expected from the subordinates. Clarity of duty as well as result expected has to be the first step in delegation" (Delegation of Authority, n.d.).
- "Granting of authority—Subdivision of authority takes place when a superior divides and shares his authority with the subordinate. It is for this reason; every subordinate should be given enough independence to carry the task given to him by his superiors. The managers at all levels delegate authority and power which is attached to their job positions. The subdivision of powers is very important to get effective results" (Delegation of Authority, n.d.).

- "Creating Responsibility and Accountability—The delegation process does not end once powers are granted to the subordinates. They at the same time have to be obligatory towards the duties assigned to them. Responsibility is said to be the factor or obligation of an individual to carry out his duties in best of his ability as per the directions of superior. Responsibility is very important. Therefore, it is that which gives effectiveness to authority. At the same time, responsibility is absolute and cannot be shifted. Accountability, on the other hand, is the obligation of the individual to carry out his duties as per the standards of performance. Therefore, it is said that authority is delegated, responsibility is created and accountability is imposed. Accountability arises out of responsibility and responsibility arises out of authority. Therefore, it becomes important that with every authority position an equal and opposite responsibility should be attached" (Delegation of Authority, n.d.).
- "Therefore every manager, i.e. the delegator has to follow a system to finish up the delegation process. Equally important is the delegate's role which means his responsibility and accountability is attached with the authority … " (Delegation of Authority, n.d.).

Relationship between Responsibility, Authority, and Accountability

Refer to Table 7.2 for a brief overview of "Differences between Responsibility, Authority, and Accountability".

Assigning Authority

Built into managerial and supervisory job titles are the levels of authority that define the scope and limits of authority. The safety management system can aid in defining the flow of authority through the organization and support responsibility, authority, and accountability.

Table 7.2: Differences between Responsibility, Authority, and Accountability

Responsibility (Delegation of Authority, n.d.)	Authority (Delegation of Authority, n.d.)	Accountability (Delegation of Authority, n.d.)
"The duty of the person to complete the task assigned to him."	"It is the legal right of a person or a superior to command his subordinates."	"It is the obligation of the subordinate to perform the work assigned to him."
Leadership is still responsible for the results and cannot totally blame the subordinate if things go wrong.	"Authority is attached to the position of a superior in concern."	"Responsibility arises out of superior-subordinate relationship in which subordinate agrees to carry out duty given to him."
Overall responsibility remains with the leadership.	"Authority can be delegated by a superior to a subordinate."	"Responsibility cannot be shifted and is absolute."
The defined network of interrelations can show the shared relationships.	"It flows from top to bottom."	"It flows from bottom to top."

Source: Adapted from Management Study Guide, "Delegation of Authority", with permission (Delegation of Authority, n.d.).

A form of authority that must be understood by all employees is that they have, under defined criteria, the authority to stop or cease doing something that is an obvious hazard. We have seen too often in weak safety cultures the mindset that an operation has to continue regardless of any potential perceived hazards. Hazardous Conditions in many cases may exist because of problems that were built into a process and may not be resolved without significant changes in a number of operational factors.

> *Clear and unambiguous lines of authority and responsibility for ensuring safety shall be established and maintained at all organizational levels within the Department ...*
>
> **(Volume 1: Concepts and principles, human performance improvement handbook, 2009)**

According to Bob Frost, Measuring Performance, as cited in The Performance-Based Management Handbook, Volume 3, Establishing Accountability for Performance:

> *Authority is the right to act without prior approval from higher management and without challenge from managing peers. Authority is assigned. On the other hand, responsibility is delegated.*
>
> *Authority and Resources. Responsible personnel must have adequate authority and resources to perform the desired tasks. Commitment of necessary resources for workplace health and safety must be documented and must address staffing, space, equipment, training, and promotions. Budget and capital expenditures for health and safety improvements must also be included.*
>
> **(Voluntary Protection Program (DOE-VPP), n.d.)**

Defining Accountability

The final element of the triad is accountability, which activates responsibility and authority. Without accountability, even if authority is granted and responsibility delegated, nothing may happen, actions may or may not be taken, nobody may care about results, and the safety management system fails.

Accountability is

- "The obligation of an individual or organization to account for its activities, accept responsibility for them, and to disclose the results in a transparent manner" (Accountability, n.d.).

> *When managers and employees are held accountable for their safety responsibilities, they are more likely to press for solutions to safety problems than to present barriers. By implementing an accountability system, positive involvement in the safety management system is created.*
>
> **(Geigle, n.d.)**

One area of concern is to ensure that the assignment and the acceptance of accountability is linked. The person being held accountable must fully understand how and for what he or she is being held accountable. The assignment must be completed up front and made clear prior to establishing responsibility and authority. Accountability cannot be retroactive.

> *" ... most people view accountability as something that belittles them or happens when performance wanes, problems develop, or results fail to materialize"*
>
> **(Connors & Smith, 2004)**
>
> *"What gets measured, and rewarded, gets done"*
>
> **(Petersen, 1993)**
>
> *"The first step in the accountability process is to explain how one plans to achieve expected results"*
>
> **(Artley, 2001)**
>
> *" ... many think accountability only crops up when something goes wrong or when someone else wants to pinpoint the cause of the problem, all for the sake of pinning blame and pointing the finger"*
>
> **(Connors & Smith, 2004)**
>
> *"The buck stops here."*
>
> **Harry Truman**

Refer to Table 7.3 for some "Essential Components of an Effective Accountability System".

Table 7.3: Essential Components of an Effective Accountability System

- Established formal standards of behavior and performance, for example, policies, procedures, or rules that clearly convey standards of performance in safety to employees.
- Resources provided to meet those standards, for example, a safe workplace, effective training, and adequate oversight of work operations.
- An effective system of measurement.
- Appropriate application of consequences, both positive and negative.
- Consistent application throughout the organization.

Source: How to Effectively Assess and Improve your Safety and Health Program through Safety and Health Evaluation, OR-OSHA 116, Public Domain, Permission to Reprint, Modify, and/or Adapt as necessary, n.d.

Assigning Specific Responsibilities

Loss-producing events are triggered by initiating an action that may be due to human errors. These errors are not necessarily the result of lower level employees physically doing something that creates the event but can stem from latent errors built into the organization (Volume 1: Concepts and principles, human performance improvement handbook, 2009).

The safety management system establishes responsibilities, authorities, and accountabilities as well as the structure to guide the organization in reducing such errors in combination with hazards and associated risk.

Roles and responsibilities will define the types of leadership skills that are necessary to support the safety management system and avoid or reduce the potential for active and latent errors that lead to loss-producing events. The following are examples of some assigned safety roles for each category:

The Leadership Team

The leadership team establishes and provides the leadership, budget, and resources for carrying out the organization's safety policy, goals, and objectives.

The leadership team ensures that the stated organizational safety policy remains effective and is actively followed. Refer to Appendix J for a list of "Sample Responsibilities for the Leadership Team".

Managers

Managers maintain safe working conditions in their respective area of responsibilities. This position carries out the stated objectives of the leadership team, which provides the leadership, budget, and resources for carrying out the organization's safety policy, goals, and objectives. To recap, managers have the task of ensuring that the goals and objectives established by senior leadership get done. Refer to Appendix K for a list of "Sample Responsibilities for Plant/Site Superintendents/Managers".

Supervisors

Supervisors are the direct link between management and the line employees who carry out the work. They must directly ensure safe conditions are maintained in their respective area of responsibilities. Refer to Appendix L for a list of "Sample Responsibilities for Supervision".

Employees

Employees are directly exposed to operational hazards and associated risk. This is where the potential for a loss-producing event can be triggered by an initiating action.

Refer to Appendix M for a list of "Sample Responsibilities for Employees".

> Senders and Moray, *Human Error: Cause, Prediction, and Reduction*, 1991, p. 20. (Volume 1: Concepts and principles, human performance improvement handbook, public domain, 2009. U.S. Department of Energy) define "The initiating action is an action by an individual, either correct, in error, or in violation, that results in a facility event".

Establish Consequences for Performance

Consequences for performances can enhance or degrade the safety culture depending on how they are established. If consequences are primarily regarded as negative, then the overall culture of the organization will perceive safety from a negative point of view. A balance of both positive and negative consequences is essential, with an emphasis on positive feedback.

Newly assigned responsibilities require employees go through a learning process to develop new skills and develop new behavior patterns and habits. With any new system, a period of learning with minimal or no consequences for inadequate performance must be established. To ensure a high level of quality performance in safety-related activities, a period of extra coaching, immediate corrective action without penalty, and enhanced discussions on why and how things should be done must be given the necessary time, budget, and acquired knowledge.

Although the goal of any accountability system should be to develop acceptance for actions, employees must understand that there are positive and negative consequences for performance. Consequences can reinforce and encourage the changing of old habits and the importance of meeting agreed upon objectives. Consequences need to be appropriate to the situation.

Lesson Learned # 4

When viewing consequences though the lens of networking and how a message can spread through a network, overreacting positively or negatively may rapidly send the wrong signal through the organization. For example, a weak safety culture may discipline a supervisor after the first poorly conducted incident investigation. This is an obvious example of overreacting to a performance problem. An assessment of why the performance was poor should have been completed (Mager & Pipe, 1970). By disciplining the supervisor, valuable information can be lost about the depth, scope, and understanding by supervisors of the intent of an incident investigations. The desired message is to get supervisors to do a better job with investigations. In this case, a negative message was sent and actually may cause other supervisors to hide loss-producing incidents as much as possible.

Summary

When the word "leader" is heard, it immediately conveys an image of someone who is able to step to the front of a group, clearly communicate what he or she desires the group to do, inspire the group to do more than what is normally expected, and get the group to achieve what others did not think was possible.

The leadership team by its visible presence and behavior sets the tone that drives the development of the safety management system. Your personal leadership goal should be to increase and influence your authority, which in turn allows you to better advise the leadership team on solutions that weave the safety management system deep into the culture of the organization.

The safety management system and the safety culture are influenced and shaped by how the organization is managed and how it is led.

A strong central structure can have an organizational culture that readily reinforces the safety culture at all locations. The organization may have a general culture that espouses a participative style that is ideal for an in-depth safety culture when driven by a well-led and managed participatory safety management system.

Achieving safety excellence begins with leadership commitment. This statement tends to get overused. However, while this statement is overused, it is vital for the leadership team to show their commitment to the safety process.

Most organizational leaders and managers have a general and basic understanding of the value of safety. However, as the organization sets priorities for all of its other mandates, it will tend to see safety only as a priority on the list of agenda items, especially if loss-producing events are low.

Leaders must be visible and in constant contact with all levels of the organization. To be a leader, you must remain visible to all levels of the organization.

Most levels of management are routinely challenged with the completion of assigned tasks with limited direction. Management's objective is to complete their tasks as assigned using the resources they have been given.

Effective implementation of a safety management system is a leadership team's responsibility. As an in-depth system must reach all levels of the organization, the leadership team should actively involve all employees of the organization in the process. To reach the real potential of a system, it is important that roles, responsibility, delegation, and authority for safety-related assignments are clearly defined and coupled with a strong accountability structure.

To determine specific tasks leadership and employees will be assigned and how you want these roles to contribute, first conduct a review of your entire organizational structure. The review will help identify the key roles that you deem vital to your safety culture.

Delegation is more than just "passing the buck" to a subordinate and telling them to "do what you need to do". Leadership is still responsible for the results and cannot totally blame the subordinate if things go wrong.

The triad of responsibility, authority, and accountability cannot be separated and must be considered as interrelated and together as a whole.

Consequences for performance can enhance or degrade the safety culture depending on how they are established. If consequences are primarily regarded as negative, then the overall culture of the organization will perceive safety from a negative point of view.

Chapter Review Questions

1. When you hear the word "leader", describe in your own words the image that you see in a person.
2. List the five basic questions that define real leadership. Discuss a person who you know that possesses these skills.
3. Provide a list of what General Bradley considered leadership expectations. Refer to Table 7.1 for Bradley's thoughts on leadership expectations.
4. Discuss the impact that true leadership has on a safety management system.
5. Describe what is meant by management by walking around.
6. Discuss an open door policy and what effect it may have on the safety management system.
7. Why do you need to review your organization to determine safety-related tasks for each role?
8. What are some of the questions that can be asked to define leadership?
9. What is the value of developing job descriptions?
10. What methods would you use to write a basic job description?
11. Why do you have to define clear goals when assigning responsibilities?
12. Why do you need to obtain agreement on stated objectives?
13. Why do you have to take into consideration the difference between supervisory and nonsupervisory employees?

14. What are some of the safety activities that you can write into a basic job description?
15. Describe the elements of delegation.
16. Compare and contrast the relationship between responsibility, authority, and accountability. Refer to Table 7.2 in the textbook for a brief "overview between responsibility, authority, and accountability".
17. Why is it important to ensure that you assign the correct lines of authority?
18. Why is it important to define clear accountabilities?
19. List the various organizational roles that need to be held accountable for their actions.
20. Define the types of leadership skills that are necessary to support the safety management system.

Bibliography

A Reference for Developing a Basic Occupational Safety and Health Program for Small Businesses. (March 2008). State of Alaska Department of Labor and Workforce Development, Public Domain. Retrieved from http://1.usa.gov/YdsojM.

Accountability. (n.d.). BusinessDictionary.com. Retrieved from html#ixzz2BxQQ3AwI.

Army Leadership, ADP 6-22. (August 2012). Army Knowledge Online. Headquarter, Depart of the Army, Approved for Public Release; Distribution is Unlimited, Public Domain. Retrieved from http://bit.ly/Vs6YyK.

Artley, W. (September 2001). The performance-based management handbook. *Establishing accountability for performance* (Vol. 3). Oak Ridge: Institute for Science and Education. Retrieved from http://1.usa.gov/1118Hxc.

Connors, R., & Smith, T. (June 2004). *Create a culture of accountability*. DIETARY MANAGER. Retrieved from http://bit.ly/UBAmhT.

Delegation of Authority. (n.d.). *Management study guide, pay your way to success*, Permission to Reprint/Modify/Adapted for use. Retrieved from http://bit.ly/VQe5xg.

Duties, Responsibilities, and Authority explained. (n.d.). Army study guide. Retrieved from http://bit.ly/Xhwql3

Geigle, S. J. (n.d.). Safety Responsibility and Accountability: What's the difference? OSHAcademy™ Course 703 Study Guide, Permission to Reprint, Modify, and/or Adapt as necessary. Retrieved from http://bit.ly/VximGD.

How to Effectively Assess and Improve your Safety and Health Program through Safety and Health Evaluation, OR-OSHA 116 Public Domain, Permission to Reprint, Modify, and/or Adapt as necessary. (n.d.). Public Education Section, Oregon OSHA, Department of Consumer and Business Services.

Leadership Quotes. (n.d.). The Happy Manager, a Better Way to Manage. Retrieved from http://bit.ly/VLLvNE.

Mager, R. F., & Pipe, P. (1970). *Analyzing performance problems; or "you really oughta wanna"*, David S. Lake Publications, Blemont CA.

Managing worker safety and health. (n.d.). Illinois OSHA Onsite Safety & Health Consultation Program, Public Domain, Adapted for Use. Retrieved from http://bit.ly/WTsneh.

Managing Worker Safety and Health Appendix 5-2 Sample Assignment of Safety And Health Responsibilities. (n.d.). Missouri Safety & Health Consultation Program (OSHA), Public Domain, Adapted for Use. Retrieved from http://on.mo.gov/VQwB9z.

Peters, T. J., & Waterman, R. H. (2004). *In search of excellence: Lessons from America's best-run companies.* New York: HarperBusiness.

Petersen, D. (1993). *The challenge of change: Creating a new safety culture.* Safety Training Systems.

Petraeus, D., & Amos, J. F. (December 2006). *The US army/marine corps counterinsurgency field manual.* FM 3-24, MCWP 3-33.5, Approved for public release, Headquarters Department Of The Army, Public Domain. Retrieved from http://bit.ly/105MvMj.

Roughton, J. E., & Crutchfield, N. (2008). *Job hazard analysis: A guide for voluntary compliance and beyond. Chemical, petrochemical & process*. Elsevier/Butterworth-Heinemann. Retrieved from http://amzn.to/VrSAq5.

Roughton, J. E., & Mercurio, J. J. (2002). *Developing an effective safety culture: A leadership approach, adapted for use*. Butterworth-Heinemann. Retrieved from http://amzn.to/X8Gaz8.

Volume 1: Concepts and Principles, Human Performance Improvement Handbook, Public Domain. (2009). U.S. Department of Energy. Retrieved from http://1.usa.gov/WdIqoP.

Volume 2: Human Performance Tools For Individuals, Work Teams, And Management, Human Performance Improvement Handbook, Public Domain. (2009). U.S. Department of Energy. Retrieved from http://1.usa.gov/11Ex7vE.

Voluntary Protection Program (DOE-VPP). (n.d.). Department of Energy, Public Domain. Retrieved from http://1.usa.gov/14ob6Rt.

CHAPTER 8

Getting Your Employees Involved in the Safety Management System

Start with good people, lay out the rules, communicate with your employees, motivate them and reward them. If you do all those things effectively, you can't miss.
—Lee Iacocca

Introduction

For an effective safety culture, one core element is employee involvement in the safety management system. As part of the leadership team, you do not have to solve all of your safety culture issues alone. Your organization already has great "built-in" resources available that may not be fully utilized as effectively as they should. These resources are your employees! If fully engaged, your employees can be your best problem solvers as they are closest to the real and/or potential hazards and associated risks.

Employees know the hazards and risks better than anyone in the organization as they are exposed to these every day when they perform their jobs, its steps, and its tasks. While they may know the hazards and risks, and have developed informal ways of controls or avoiding loss-producing events, these controls may or may not be the best methods to follow. This is where employee involvement in the safety management system becomes vital.

The success of your safety management system depends on the interaction between all employees including the leadership team. The ultimate goal of this interaction is to develop systems and/or methods that will reduce exposure to hazards and associated risks. With open and positive interaction and communication, better solutions are defined, developed, and implemented.

To be effective, you must determine how to engage your employees at all levels of the organization. A review of how employees are networked as discussed in Chapter 3, "Analyzing and Using Your Network", and an understanding of the overall organizational cultural as discussed in Chapter 2, "Analyzing the Organizational Culture" begins the process of employee involvement assessment.

In this chapter, we outline how improved employee involvement strengthens your safety management system. Increased positive relationships will enhance the organization's safety culture as employees become part of the solution and not part of the problem.

After completing this chapter you will be able to:

- Discuss the reasons employees do not want to be involved in the safety management system.
- Discuss why all employees should be involved in the safety management system.
- Discuss the guidelines for employee involvement.
- Identify the best method to get employee insights about the safety management system.
- Discuss the central safety committee and its permanent subcommittees.
- Understand ad-hoc subcommittees and their purpose.

Reasons Employees Are Not Involved in the Safety Process

Before we start our discussion on how to get employees involved in the safety process, we need to step back and look at reasons why employees do not always want to get involved.

A complaint we have heard from managers is that they cannot get their employees to comply with basic safety requirements, whether it is wearing the appropriate personal protective equipment (PPE) or following safe work practices. Their questions are, "How do you change an employee's behavior? How can I get them to comply with company and/or department safety work practices?"

In many cases the perception about safety held by supervision and leadership is that it is only about tangible items such as PPE. PPE is easy to see as it is a tangible "artifact" that can be observed. However, PPE is the last stage of the hierarchy of controls (Identifying and Controlling Hazards, Online Course 104, n.d.) as advocated in safety management systems. In an effective safety culture where employees are involved in PPE management, the selection, distribution, wearing, inspection, and replacement of PPE does not have to be discussed. The use of PPE is an espoused value backed by deep-seated knowledge.

If the first thing that must be focused on when having safety discussions with employees is "Where are your safety glasses, gloves, and the like", then you have immediate feedback that employees are not actively involved and engaged in the safety management system where a safety culture does reach the deep levels of the organization's culture.

> As a point of reference, a situation was observed while researching employee involvement. James was watching a reality show called "Red Jacket Firearms—Sons of Guns". In this reality show, participants build, repair, and modify all types of guns for customers. In one situation, the manager of "Red Jacket Firearms" wanted a historic cannon used in World War I restored for a customer. He held a meeting with his gunsmiths to discuss his ideas and how they could best complete the restoration. The immediate response from his employees was, "The gun barrel is broken, what are you going to do with it?" The manager replied: "That is the difference between me and you, I do not look at problems, I only look at solutions". The employee's response was: "But who are you looking to solve the solution, the people with the problem!" (Sons of Guns, n.d.).

> *Managers often complain that they cannot get workers to comply fully with required safety measures, whether that means wearing appropriate personal protective equipment or following safe work procedures.*
> **(Managing worker safety and health, n.d.)**

Lesson Learned # 1

James remembers a case where an employee got dust in his eye. There was no recorded history in this plant regarding eye injuries. However, the leadership team decided to change its entire safety eye protection policy and program. They then discarded all currently used safety eye protection and replaced it with new types to solve this perceived problem, which was just a speck of dust in the eye.

In this case, the leadership team was focusing on the wrong issue. No risk assessment, no discussions, and no analysis of the work environment were conducted. They simply jumped to a solution for what may or may not have been a safety or risk problem.

Lesson Learned # 2

In another situation, an organization took action each time an injury occurred that involved a tool by banning the tool in question from use. Case in point, an employee was using a utility knife to cut material and the knife slipped, cutting the employee's leg. Almost immediately, without conducting an investigation to determine the root cause, all utility knives in the organization were simply banned.

You can take the position of banning things to improve safety; however, the question is, "Does banning resolve the immediate issue?" Banning is an easy solution to a problem, but where does it end?

Lesson Learned # 3

When manually feeding material into any equipment the risk of this material getting jammed can exist. If the equipment is not maintained properly, for example, employees may be constantly trying to unjam, modify or adjust the equipment.

In one situation, a general manager observed employees un-jamming a piece of equipment while it was running and did not question the underlying issues. The employees were just told to stop! Consequently, the employees had to keep shutting down equipment to unjam the product. And of course, when equipment is shut down, line management wants to know why.

Employees were sent different messages on what to do.

Using these examples, what do you believe the perception held by employees was of the safety efforts at these various organizations? Each scenario resulted in actions that made an employee tasks harder to complete without changing how the job was to be completed.

The common theme in each of these cases is that employees were not involved in the process nor asked for their insight. Instead, a mandated solution was forced upon them. This type of approach is authoritarian and not a participatory organizational culture. If you continue to respond to problems with limited discussions and analysis, why would you then expect employees to want to be involved in solving safety issues? Based on these lessons learned, take a look at your own organization to see if you can identify similar situations.

Why Should All Employees Be Involved?

The leadership team provides the vision/mission of the organization with the goals and objectives that they believe will guide the organization through a continuous improvement process. One thing that the leadership team can easily overlook as it deals with many obligations and general issues is that employees have developed a detailed knowledge of the overall steps and tasks needed to get their jobs done everyday.

The following are basic elements about why you should encourage your employees to get involved:

- Each employee can be a valuable problem solver based on his or her experience.
- Employees are in close contact with real and/or potential hazards and associated risks every day. However, they may have developed their own way of doing the job more efficiently that may or may not be adequately controlling the hazards and associated risks. Refer to Chapter 9, "Risk Perception—Defining How to Identify Personal Responsibility".
- Employees who participate in safety management system development are more likely to support and use the process.
- If employees are encouraged to offer their ideas/suggestions and their ideas/suggestions are taken seriously, they will be more productive and more motivated to become involved in the safety process (Managing worker safety and health, n.d.).

Listen to Your Employees

The most efficient approach in developing any safety management system is to "just listen" to what employees have to say. This is the most direct way you can learn about the effectiveness of your safety management system. However, this approach can be difficult for various levels of management who have not had the opportunity to experience the true meaning of employee involvement. The culture of the organization may have a value system that reduces or prevents direct involvement of employees. The leadership team may have a basic understanding of the operational aspects of the organization but have a limited understanding of the leadership concepts that focus on getting employees involved in the process.

Lesson Learned # 4

James has learned one rule of thumb: “When interviewing employees during an investigation of an injury, I have learned to listen 80% of the time and only talk 20% of the time. In this way, I am able to usually get to the potential root cause of the loss-producing incident quicker by listening to all of the facts before I ask questions”.

Often we do an “Information Dump” on employees and others. An information dump is where we unfortunately spend most of the time letting people know how much we know as opposed to listening and being aware of what their questions and concerns may be. As a result, we leave the discussion feeling like we have accomplished our task when in reality, nothing of value was learned or communicated. An information dump stops learning and prevents employee involvement. We have learned that if you listen more than you talk, you can learn far more about the employee and your organization.

> Daniel Pink suggests listening for employee use of “we” and “they” in discussions. He gives credit for this to Robert Reich, former U.S. Labor Secretary. If employees use “we”, it indicates they “feel they are part of something significant and meaningful”. If “they” is used, it signals that “some amount of disengagement or perhaps even alienation” may be present (Pink, 2011).

Getting Employees Involved in the Safety Process

It is human nature to resist change. In general, taking the approach of “my way or the highway” can result in hidden resistance that prevents any constructive dialog about concerns or problems.

As a point of beginning, determining if employees have a level of trust that the leadership team is serious about asking for their involvement in the safety process is crucial. In your evaluation of the organizational culture, an assessment of the level of trust between all employees provides an indication of the potential success of the safety management system.

Employee perception may be based on cases similar to the lessons learned discussed above. To overcome this perception, it is important that you try to mend any mistrust and miscommunication between the leadership team and all employees.

The leadership team can accomplish this by showing visible support with positive action by:

- Communicating their intentions clearly with all employees in the organization that they will provide a safe work environment.
- Discussing with employees the expectations about maintaining of a safe work environment.
- Assigning all employees specific responsibilities as they relate to safety efforts.
- Providing adequate resources for all assigned tasks.
- Taking employees' suggestions and reports of safety-related issues seriously.
- Implementing all safety suggestions in a timely manner or taking time to explain safe alternatives that can be used.
- (Managing worker safety and health, n.d.)

Refer to Chapter 11, "Developing an Activity-Based Safety System" on methods on communicating with all employees about safety suggestions when they are successful.

Guidelines for Employee Involvement

Employees who are asked to participate in the development and implementation of safety rules and procedures have more of a personal stake in the safety process. We have a tendency to support ideas that we help to develop and implement (Employee Involvement, eTools, Public Domain, n.d.).

Basic key elements that can be used as a guide to achieve involvement at all levels of the organization may include the following:

- Communicating regularly with all employees in regard to workplace safety-related issues. Refer to Chapter 11, "Developing an Activity-Based Safety System".
- Providing all employees with complete access to relevant information of specific programs.
- Allowing employees to help identify methods to become familiar with assessing and identifying hazards and prioritizing results of assessments, training, and program evaluations. Refer to Chapter 9, "Risk Perception—Defining How to Identify Personal Responsibility".
- Establishing a way to report injuries, near misses, loss-producing events, and hazards found.
- Providing prompt responses to all reports and suggested recommendations.
- Allowing all employees to be involved in defining methods to identify specific hazards and assist in fixing identified hazards under their direct control. This can be accomplished by utilizing a team to develop a job hazard analysis (JHA) to detail safety-related issues and developing solutions to work practices. Refer to Chapter 12, "Developing the Job Hazard Analysis".
- Allowing employees to develop and revise workplace safety rules.
- Providing training for both current and newly hired/transferred and refresher training for seasoned employees on site-specific safety issues.
- Assisting in developing and presenting safety-related information at safety meetings.
- Supporting coworkers by providing feedback on risks and assisting them in eliminating hazards.

- Participating in site inspections and safety walk-throughs.
- Performing a preloss review to identify hazards and associated risk. Refer to Chapter 11, "Developing an Activity-Based Safety System".
- Providing the time and resources for employees to participate with the leadership team and other advisory or special-purpose safety committees.
- Participating in safety decision-making throughout the organization's operations (Draft Proposed Safety And Health Program Rule: 29 CFR 1900.1, Docket No. S&H-0027, n.d.; Employee Involvement, eTools, Public Domain, n.d.; Managing worker safety and health, n.d.).

As part of your assessment of employee involvement, review the current methods used by the leadership team to encourage employees to participate in other organizational activities. Does leadership have internal activities that encourage employees to become involved? How successful and what methods are used to communicate and gain employee involvement? Who are the key players that make that involvement happen? To the degree possible, use a similar approach in working with leadership to emphasize involvement in the safety management system. Use existing successful involvement models that are already in use to design the approach to employee safety management system involvement.

"Just Ask" Your Employees to Get Involved

Lesson Learned # 5

James once discussed with a general manager why his employees would not get involved in the safety management system. The general manager said he had done everything that he knew to get his employees involved and could not understand why no one wanted to be involved. After discussing this issue for a few minutes, James asked a question that shocked him. "Have you just asked them?" His response was "NO!" James and the general manager then began to walk through the plant talking to as many employees as possible and received many positive responses.

In this lesson learned, James found that the plant was also multicultural, with a majority of the employees being non-English speaking. The general manager had been reluctant to ask the employees for their involvement because he did not speak their language. As he and the general manager began walking through the plant, they asked employees (using interpreters where needed) if they would like to participate in resolving safety issues. The overwhelming response was "YES".

So the moral of the story is all you need to do is "Just ask the employees" if they want to become involved in the safety management system. You then have volunteers instead of employees dealing with mandated involvement. Refer to Chapter 7, "Leadership and the Effective Safety Culture" for more details.

Simple Beginnings Can Generate Major Impact

The leadership team and employees represent "nodes" in the network of communications as discussed in Chapter 3,"Analyzing and Using Your Network". When involvement is viewed as a network, leadership can better understand that employee involvement provides the vital link to more effective safety management systems. As a network, influence begins to spread even if only a relatively small number of involved employees are active. As safety issues are resolved and communicated, a positive reputation and trust develops. As a result, other employees recognize that leadership is serious and that there is a benefit to being involved.

Employees' opinions and suggestions must be reviewed with an open mind. We return to the importance of listening to the employee's side of the problems and solutions so that an informed decision can be made as a cooperative effort between the leadership team and employees.

Safety Committees

Safety committees have traditionally been used as a focal point of employee involvement. An effective safety committee can readily spread a positive influence for safety-related activities and efforts rapidly through an organization. To develop a positive safety committee, a charter of its mandate and structure, training on the basics of holding a meeting, and core training on the safety management system are needed to sustain its effectiveness.

An issue we have noted, is that safety committees are established without much thought regarding their purposes and goals. Safety committees are established on the basis that organizations are supposed to have one. We agree that they can serve a purpose. However, they can have the potential for becoming misguided gripe sessions that degrade instead of enhance the safety culture. Prior to a committee being established, thought must first be given about what the committee would be able to accomplish.

Before developing or restructuring a safety committee, review how other committees, if any, are being used in the organization. If successful meeting formats and charters exist in other parts of the organization and have been effectively used over time, model the safety committee after those committees. For example, if a successful quality committee has been in operation, its meeting structure, charter, and administration may already be familiar to leadership and employees. The safety committee could make minor changes to an existing charter and not have to reinvent the wheel as to how to organize and manage itself.

Group decisions benefit from their many points of view and varied experiences of the safety committee members (Managing worker safety and health, n.d.). A committee may not be the best and fastest way to reach a decision, but group decisions are often the best choice for getting employees to buy in on safety-related matters. If the safety committee has a good cross-section of employees drawn from throughout the organization, the diversity of the group can better communicate the committee's efforts throughout the organization as well as increase the potential for improving the committee's problem solving skills.

A safety committee is not the best approach to determine situations that are immediately dangerous to life and health situations. In these cases, immediate corrective actions are taken and the committee can be used to determine why the situation was allowed to develop, to further improve the decisions made, and to correct gaps in the safety management system. The safety committee does not replace the responsibility and accountability of the leadership team to ensure that a safe work environment is maintained.

The safety committee is advisory in nature but greatly magnifies the potential for focusing the creative energy and expertise of employees on hazards and associated risks. As it becomes embedded in the overall social network of the organization, it acts to expedite information rapidly to the leadership team and employees that may otherwise be blocked for whatever reason. A safety committee can transcend boundaries of the organization that exist.

Establishing the Team Charter

A charter is recommended to clearly define the scope and structure of the safety committee. A charter is the document that will provide guidance over the life of the committee and gives the reason for the establishment of the committee.

> *In project management, a project charter or project definition (sometimes called the terms of reference) is a statement of the scope, objectives and participants in a project. It provides a preliminary delineation of roles and responsibilities, outlines the project objectives, identifies the main stakeholders, and defines the authority of the project manager. It serves as a reference of authority for the future of the project.*
>
> **(Charter, n.d.)**

The charter is developed in a joint effort between the leadership team and the members of the initial safety committee. The charter is a living document that must be periodically amended as conditions warrant. However, its core purpose and structure should be kept as permanent as possible. The establishment of a safety committee is essentially the same as that prescribed for the development of a project team.

In order to develop a committee or team charter effectively, a series of questions need to be answered to assess as many aspects of the committee/team's potential duties, accountabilities, and authority as possible. These questions are answered prior to the initial meeting formalizing or revitalizing the safety committee. The questions develop the basic content for the safety committee charter.

> *Each organization will want to fine-tune the questions to fit their particular situation.*
>
> **(Turner & Turner, 1998, chap. 2)**

1. What is the purpose of this safety committee?
2. What will be the authorities, accountabilities, and responsibilities of this safety committee?

3. How will the safety committee measure its success?
4. Who is the safety committee sponsor or champion?
5. Who are the members? How long will each member serve on the committee? Create as diverse a committee as feasible.
6. How will safety committee leadership be determined and rotated?
7. What specific deadlines must the safety committee maintain?
8. How often is the safety committee expected to meet? How much time will members be allowed to participate?
9. When is the best time for the safety committee to meet?
10. Where will the safety committee meet?
11. What is the budget for the safety committee?
12. What resources and advisory expertise will be needed?
13. How will reports and communications of the safety committee be communicated to leadership and employees? What media will be used? (Adapted from Turner & Turner, 1998, chap. 2)

Once a core charter has been structured, it is then shared with the safety committee at its first meeting. The meeting is critical, as it will set the tone for all future meetings. Members are asked to review, discuss, and question the content of the charter at this first meeting and offer suggestions for changes. The final decision regarding the charter is mutually agreed upon by the leadership team and safety committee members.

> The first step is the most important step in growing a team. The Team Charter specifies the purpose of the team, the boundaries of its scope and authority, and team membership. One of the major reasons that teams fail is that the original charter was too vague. This leads to 'mission creep' in which teams spill into areas that were never intended, or teams become confused about how much authority they have and stumble into conflict with supervisors and other teams (Turner & Turner, 1998, chap. 2).
>
> "Mission creep" is the equivalent to "scope drift" we discussed in Chapter 4, "Setting the Direction for the Safety Culture".

The committee must continue to show its value to the organization. If this cannot be maintained, neither leadership nor employees will want to remain involved because they will feel that the safety committee is not a productive use of their time.

Refer to Appendix N for an Example of a "Safety Committee Team Charter".

Choosing Your Safety Committee Members

Use your knowledge and the employee networks to select safety committee members, always keeping in mind the need for a diverse group.

Lesson Learned # 6

After an initial call for committee volunteers, a method James has found useful in establishing a safety committee is to let employees recommend other employees that they feel would make good members of the safety committee. The person recommended is then asked to recommend someone who is qualified to be a safety committee member. This method uses the employees' personal network and increases the spread of influence of the committee as the selection process message travels through the organization.

The Central Safety Committee

Depending on the size of your organization, to provide better support for the safety management system, a central safety committee can be established. This committee has also been called the executive committee or joint health and safety committee (Joint Health and Safety Committees: A Practical Guide for Single Employer WorkHealth, n.d.). The central safety committee (Successful Health and Safety Management, hsg65, 1997) provides oversight of the activities that ensure all areas of the organization are brought into the activities that are developed by the committee (Managing worker safety and health, n.d.).

For the central safety committee, the leadership team retains overall responsibility for the committee. Committee leadership could be jointly shared with a responsible committee member and/or a safety manager or coordinator. The "Chair" for the committee can be elected by members with the various roles rotating through the committee membership.

Several key functions or roles must be filled for a committee to function include:

- A committee "Chair" that leads and moderates the meetings to ensure the agenda is followed and assignments are delegated.
- Secretary or note keeper—someone is assigned each meeting to keep notes on all discussions and develop the meeting minutes that will keep track of all projects and discussions.
- A timekeeper is assigned to ensure the meetings begin and end on time. It is critical that members keep the obligation to stay within the defined time constraints for meetings.

This format of having a chair, note taker, and timekeeper should be followed by all permanent subcommittees and ad-hoc project committees (Coble, Taylors, & Jones, n.d.; Joint Health and Safety Committees: A Practical Guide for Single Employer WorkHealth, n.d.).

Refer to Figure 8.1 for an "Overview of a Proposed Central Safety Committee" as outlined in this chapter.

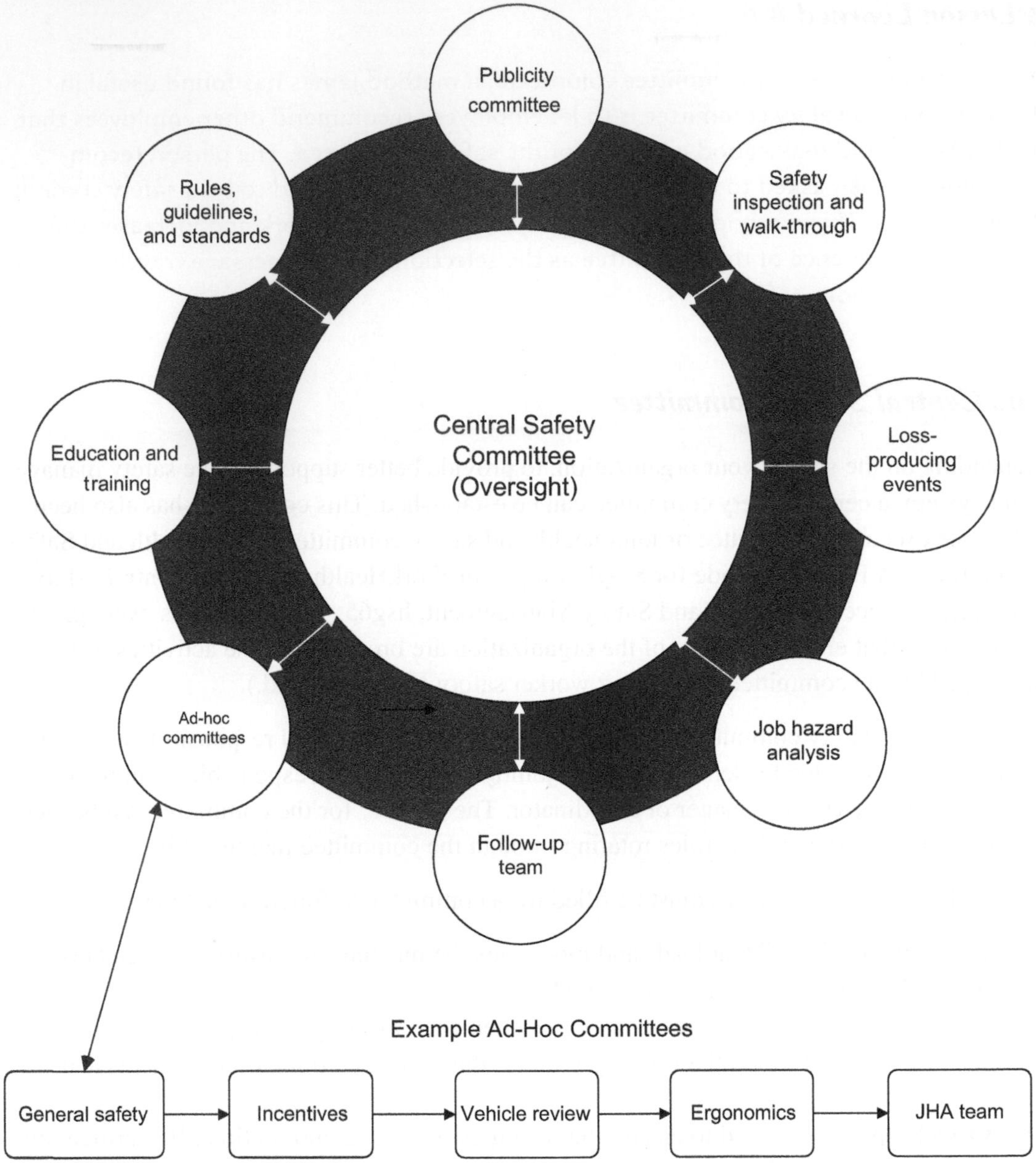

Figure 8.1

Overview of a Proposed Central Safety Committee. *Source: Adapted from central safety and health management system, reproduced with permission. CTJ safety associates, LLC. Retrieved from* http://www.ctjsafety.com/serv_cshc.htm.

Permanent Subcommittees

A wide array of activities can be assumed by the central committee. To reduce the time requirements and demands on the members, various tasks can be delegated to members who in turn will lead and manage the specific subcommittees. These subcommittees coordinate and share their information to provide overall consistency and ensure that no gaps between the subcommittees develop. Permanent subcommittees could include, but are not limited to, the following:

- Publicity committee,
- Inspections and walk-throughs,
- Loss-producing incident reports,
- Job hazard analysis,
- Rules and procedures,
- Education and training,
- Housekeeping,
- Follow-up team.

Publicity Committee

The publicity committee's main duty is to communicate and promote the safety management system. The committee establishes the style and media to be used and defines the network of communication that gets the safety message to all employees. It ensures that all employees are kept updated on all activities of the various subcommittee projects and activities. This committee is critical to ensuring that a positive and clear message is communicated about the desired goals and objectives designed to enhance the safety culture.

> *Safety publicity, communications, bulletins, signs, contests, awards, etc., are activities the Publicity Committee would coordinate.*
>
> **(Coble et al., n.d.)**

Inspections and Walk-Throughs

Inspections, observations, and walk-throughs are a core feature of a safety management system. This subcommittee assists in coordinating the various types of surveys, reviews, and inspections of facilities and operations. It would review checklist and inspection formats to ensure that these remain current, are kept up to date and consistent and are routinely scheduled. The subcommittee assists in training on each of the various formats used (Coble et al., n.d.).

This committee works with the follow-up team to ensure that hazardous conditions and safety issues identified are corrected or alternatives developed until full corrective action can be taken.

Loss-Producing Incident Reports

This subcommittee assists in developing and using a process to complete a review of loss-producing events. Depending on the complexity of the event, an ad-hoc committee specific to the incident can be established to work with the leadership team. The committee would also provide a review on all reports for consistency and completeness. It will determine if the basic causes of an event were determined and if effective corrective measures have been taken to the degree possible (Coble et al., n.d.).

This subcommittee must work closely with leadership as well as risk management, insurance, and legal advisors wherever applicable to ensure all liability and workers' compensation criteria are properly followed.

Job Hazard Analysis

This subcommittee can be used to assist and review JHAs to ensure that it is kept updated and continues to meet current requirements, as detailed by the task. This is a critical committee in that it can assist in ensuring that an ongoing process of JHA is maintained and that hazards, associated risk, and residual risk are consistently controlled and maintained.

For more details, refer to Chapter 12, "Developing the Job Hazard Analysis".

Rules and Procedures

This subcommittee assists in developing, updating, and ensuring consistency in the rules, guidelines, and standards necessary for the safety management systems and maintaining safe operations in the work environment.

Lesson Learned # 7

In another organization, each time an employee was injured, a process of recording the injury-related causal factors became incorporated into a "Book of Safety Rules". This book had grown to 35 pages of mandated rules over the years, along with the many documented causes of injuries. Each time a new injury occurred, the book of rules would be reviewed to find out what "violation" may have transpired. If a rule was found in the book based on a past injury, then that rule would be applied to the current injury and disciplinary action would be taken against the injured employee. This ever-growing book took on a life of its own. The book became a bible of past occurrences and used as a disciplinary tool, with employees who were expected to know the 35 or so pages of rules.

Education and Training

The education and training subcommittee assist in coordinating and supporting safety-related training with new training ideas, materials, references, and other resources. It ensures that safety training remains consistent, is kept current, and meets the needs of the organization. It would also shed light on gaps in training and provide oversight of the quality and delivery of safety training.

Housekeeping

This subcommittee reviews housekeeping quality, schedules, and issues. It assists in resolving underlying obstacles to maintaining a safe work environment based on controls in place for physical conditions, hazards and associated risks. It reports findings to the central safety committee.

Follow-up Team

The follow-up subcommittee provides oversight of safety-related maintenance and work order requests. This subcommittee requires special expertise and should consist of maintenance technicians, employees, and a supervisor providing authority to take corrective action. It provides an unbiased review of each safety-related maintenance and work order created, schedules, and completed work within a designated time period or still outstanding. Its objective is to assist the maintenance staff in setting risk-related priorities and ensure that corrective actions are fully implemented.

The subcommittee reviews each maintenance and work order with the employee who turned the order in and uses the Risk Guidance Tool to establish a priority for corrective action, moving the request to the next level of decision-makers. When the maintenance or work order is logged in as completed, a member of this team will "follow-up" with the employee reporting the condition to determine if it was corrected. The team member and the employee then sign off on the correction.

For more details, refer to Chapter 11, "Developing an Activity-Based Safety System".

Ad-hoc Committees

Ad-hoc committees are formed to complete specific projects and are limited to a specific time frame. One of the concerns that has been expressed about safety committees is that a number of employees do not want to become involved in committees that are long term. However, as projects are developed by the central safety committee, short-term projects appeal to different employees having different expertise and experience. Some employees may like the central committee, where they can use their skills in administration and organizing. Other employees may join an ad-hoc committee on a topic of interest.

Ad-hoc safety committees meet as often as needed to complete the assigned projects. Ad-hoc committee chairpersons meet with the central safety committee as directed and provide regular status reports on group activities and recommendations (Coble et al., n.d.).

Lesson Learned # 8

Nathan once had a client that wanted to re-establish its safety committee. A message was sent to all employees about the new committee, asking for volunteers who might be good members for the new effort. It was learned that interest was high and that in-depth expertise already existed in the form of retired military safety personnel who had joined the organization. As a result, the new committee had an immediate wealth of expertise that was already experienced in safety. Expertise had been hidden until the action of "asking the employees" was used!

Summary

To be effective, you must determine how to engage your employees at all levels of the organization.

A complaint we have heard from managers is that they cannot get their employees to comply with basic safety requirements, whether it is wearing the appropriate PPE or following safe work practices.

In many cases, the perception about safety held by supervision and leadership is that it is only about tangible items such as PPE, which is easy to see as it is a tangible "artifact" that can be observed. However, PPE is the last stage of the hierarchy of controls.

The leadership team provides the vision/mission of the organization with the goals and objectives that it believes will guide the organization toward continuous improvement.

The most efficient approach in developing any safety management system is to "just listen" to what employees have to say. This is the most direct way you can learn about the effectiveness of your safety management system. However, this approach can be difficult for various levels of management who have not had the opportunity to experience the true meaning of employee involvement. The culture of the organization may have a value system that reduces or prevents direct involvement of employees.

Initially, it is crucial to determine whether employees possess a level of trust that the leadership team is serious about asking for their involvement in the safety process. In your evaluation of the organizational culture, an assessment of the level of trust between all employees provides an indication of the potential success of the safety management system. Trust can be developed by providing employees with the opportunity to get involved in activities such as establishing, implementing, and evaluating your safety management system.

When involvement is viewed as a network, leadership can better understand that employee involvement provides the vital link to more effective safety management systems. Influence begins to spread even if only a relatively small number of involved employees are active.

An effective safety committee can readily spread a positive influence for safety-related activities and efforts rapidly through an organization. To develop a positive safety committee, a charter of its mandate and structure, training on the basics of holding a meeting, and core training on the safety management system are needed to sustain its effectiveness.

The safety committee is advisory in nature but greatly magnifies the potential for focusing the creative energy and expertise of employees on hazards and associated risks. As it becomes embedded in the overall social network of the organization, it acts as a way to expedite information rapidly to the leadership team and employees that may otherwise be blocked for unknown reasons. A safety committee can short circuit across boundaries of the organization that exist.

A charter is recommended to clearly define the scope and structure of the safety committee. A charter is the document that will provide guidance over the life of the committee and gives reason for the establishment of the committee.

Depending on the size of your organization, to provide better support for the safety management system, a central safety committee can be established. To reduce the time requirements and demands on the members, various tasks can be delegated to members who in turn lead and manage the specific subcommittees. These subcommittees coordinate and share their information to provide overall consistency and ensure that no gaps between the subcommittees develop. Ad-hoc committees are formed to complete specific projects and are limited to a specific time frame.

In Chapter 9, we will continue our discussion on risks. Its control must permeate your organization if an in-depth safety culture is to be sustained. A safety management system using only inspections and observations to identify hazards will not provide a full appreciation of the potential for injury and damage without linking to the potential risk.

Chapter Review Questions

1. As part of the leadership team you have valuable resources. Who are those valuable resources? Please describe these resources in detail.
2. Discuss the reasons employees do not want to be involved in the safety management system.
3. Discuss why all employees should be involved in the safety management system.
4. Discuss the guidelines for employee involvement.
5. Identify the best method to get employee insights about the safety management system.
6. Discuss the central safety committee and its permanent subcommittees.
7. Discuss the use of ad-hoc subcommittees and their purpose.

8. Why is it important to listen to your employees? Please describe in detail your thoughts and ideas.
9. List the elements outlined to demonstrate the guidelines for employee involvement.
10. Explain the "Just Ask" concept discussed in this chapter.
11. Provide a list of the permanent subcommittees that may be used.
12. Define the roles of each permanent subcommittee. How does this list compare to your own safety committee efforts?

Bibliography

Charter. (n.d.). Wikipedia. Retrieved from http://bit.ly/VRaByi.

Coble, D., Taylors, B., & Jones, J. (n.d.). Central Safety and Health Management System, Adapted with Permission. CTJ Safety Associates, LLC, Adpated with Permission. Retrieved from http://bit.ly/14Eu85r.

Draft Proposed Safety And Health Program Rule: 29 CFR 1900.1, Docket No. S&H-0027. (n.d.). Occupational Safety and Health Administration (OSHA), Public Domain, Adapted for Use. Retrieved from http://1.usa.gov/TLkITh.

Employee Involvement, eTools, Public Domain. (n.d.). Occupational Safety and Health Administration (OSHA), Adapted for Use. Retrieved from http://1.usa.gov/VIFfdd.

Identifying and Controlling Hazards, Online Course 104. (n.d.). Oregon Occupational Safety and Health Division (Oregon OSHA), Public Domain, Permission to Reprint, Modify, and/or Adapt as necessary. Retrieved from http://bit.ly/12rYEfK.

Joint Health and Safety Committees: A Practical Guide for Single Employer Work Health. (n.d.). The Canadian Centre for Occupational Health and Safety. Retrieved from http://bit.ly/X4VEqh.

Managing worker safety and health. (n.d.). Illinois OSHA Onsite Safety & Health Consultation Program, Public Domain, Adapted for Use. Retrieved from http://bit.ly/WTsneh

Pink, D. H. (2011). *Drive: The surprising truth about what motivates us*. USA: Penguin Group. Retrieved from http://bit.ly/VaQStm.

Roughton, J., & Crutchfield, N. (2008). *Job hazard analysis: A guide for voluntary compliance and beyond, chemical, petrochemical & process*. Elsevier/Butterworth-Heinemann.

Sons of Guns. (n.d.). Retrieved from http://bit.ly/WuVrND.

Successful Health and Safety Management hsg65. (1997). Health and Safety Executive, Crown Publishing, Permission to Reprint. Retrieved from http://bit.ly/YKDzRk.

Turner, L., & Turner, R. (1998). *How to grow effective teams and run meetings that aren't a waste of time*. Ends of the Earth Learning Group. Adpated for Use. Retrieved from http://bit.ly/UYqQqP.

PART 3

How to Handle the Perception of Risk

CHAPTER 9

Risk Perception—Defining How to Identify Personal Responsibility

You only see what you know.
–Albert Einstein

Dice have no memory; they change all of the time.
–CSI TV

Introduction

An understanding of risk and its control must permeate your organization if an in-depth safety culture is to be sustained. A safety management system using only inspections and observations to identify hazards will not provide a full appreciation of the potential for injury and damage without linking the results to the potential risk.

As safety is an emergent property of all aspects of an organization, without constant focus on the potential changes in the organization, the potential for loss-producing events may have a way of slipping out of control. Too often the true scope of hazards and associated risk are only identified after a loss-producing event has occurred.

By training and educating all employees on the concepts of risk and providing a method for use in risk assessment, two issues are addressed:

1. The perception of risk is changed from the belief that no risk exists if no loss-producing event has occurred.
2. When controls are not considered effective or have degraded, the communication of risk can be greatly improved at the point of exposure.

This chapter provides a risk assessment method that will initiate discussions intended to reach a mutual agreement on hazard-related and associated risk concerns. This method creates the opportunity for employees to alert leadership that they believe they are at risk in some aspect of their job. The method provides a format for discussion and agreement on the level of risk. The intent of the method is to prevent employees from knowingly putting themselves in harm's way or continuing to complete a task believing they are safe when they are not safe.

When you complete this chapter, you should be able to:

1. Discuss why we take unnecessary risk.
2. Discuss why a shift is needed from loss to risk.
3. Discuss why injury rates are not a true measure of risk.
4. Discuss the concepts of a risk guidance tool.
5. Define hazard and risk.
6. Discuss how the risk guidance tool aids in communicating risk.
7. Discuss risk tolerance and its issues.
8. Discuss how risk assessment can improve the safety culture.

> *Judgments also alter with time, what was acceptable yesterday may not be acceptable today and tomorrow.*
>
> **(Mullai, 2006)**

Why Do We Take Unnecessary Risk?

Risk is intangible and hidden in the fabric of the work environment. When we deal with any type of hazard-related condition and its associated risk, we have a built-in affinity to judge our environment based on our past experiences. We see a situation, mentally scan our memory for similar situations, and act based on what consequences we experienced or saw in the past. A problem develops when what we think we see is not what the reality is at that point in time. Having never had a hand caught when reaching into a moving piece of equipment, driving while texting, smoking around gasoline and so on, and having nothing happen, mask the true potential for injury and severity.

The safety management system must include a process to not just identify hazards but include the assessment of associated risk. The structure of such a process should include a continuous proactive identification of existing, new, and potential hazards.

Many hazards may be considered routine and almost a permanent part of an organization and controlled as such. However, as an organization undergoes change, introducing new services, new equipment, new employees, etc., the control of hazards and associated risk cannot wait for the formal inspection traditionally used but must be assessed at the point of exposure by the employee who may be at potential harm.

The organizational culture may celebrate or embrace taking risk as part of doing business, as part of the tradition as a way of life, or as a way of proving one's self. We may observe potential risks but to remain considered part of the group, we follow the behavior of others and continue to take risk knowing a potential for injury is present.

For example, if we see an employee putting his or her hands into a piece of running equipment to unjam it without getting injured, we might think that it is natural and mimic the behavior over and over. The "meme" that is activated is that getting the job done takes precedence over following a safe course of action that may slow a process down. The risk taking behavior creates a sphere of influence in the immediate work environment extending into the minds of future employees.

Refer to Chapter 3, "Analyzing and Using Your Network" on to how define "meme" and risk perception.

> *People's views on risks and their value judgments are not static, but change according to circumstances.*
>
> **(Mullai, 2006)**

Once something risky is done one time and nothing happens, we tend to do it again and again without thinking. Then one day, the many factors involved in our risky behavior come together resulting in a situation where injury, damage, or other harmful events occur. When it comes to taking risk, we perceive from normal experience, harm "cannot happen to me".

Lesson Learned # 1

A colleague told the story of stopping an employee from cleaning a printer roller by wrapping a rag around his hand and holding it against moving rollers with a nip point. The employee's response was, "I've done this for many years and never been caught!!" Several weeks later, he was finally caught and had a serious hand injury.

When James started out in the printing industry, he was told, "A good printer always has a missing finger". The mindset of accepting risk is still with us.

Each time a risk is taken and nothing happens, at-risk behavior gets reinforced. This behavior is not limited to the work environment but is reinforced through daily behavior. For example, driving distracted is something that many of us have done-eating, using the cell phone, texting on your cell phone, changing radio stations, reaching for something in the back seat, etc.

> To put what we have discussed in perspective, the following is a story about a man who loves to eat ham. Only after several generations was a key question asked: Why?
>
> A man loved ham so much that he had his wife cook ham very often. When she cooked the ham, he noticed that she always cuts both ends off of the ham before baking it. He never would say anything because he did not want to upset his wife but in his mind he was always questioning the reason that she cut the ends off the ham.
>
> One day he got the nerve to ask his wife: "Honey, why do you cut the ends off the ham before you place it in the pan?" She replied, "Because my mother always did it". He asked, "Can we call your mother and ask her why she cuts the ends off of the ham?" Her mother's response was "Because my mother always did it". This was very interesting, so he wanted to pursue this a little more. So they called the grandmother and asked, "Grandmother, I would like to know, when you bake a ham, why do you cut off the end of the ham before you put it in the pan?" He was shocked at her response: Grandmother stated "Because, in my day, the pans were too small and the ham would not fit my pan so I had to cut the ends off".
>
> This influence of the grandmother moved through several generations!

What is the moral of this story? Sometimes we see someone doing something that looks right and then we do it ourselves without asking questions. It may or may not be best way of performing a task. By developing and maintaining a "questioning attitude" at all times while doing a task, awareness can be developed about changes and a "Stop when Unsure" meme spread in the work environment by asking ourselves why are we doing this. Refer to Table 9.1 for a method for "Developing a Questioning Attitude" (Volume 1: Concepts and principles, human performance improvement handbook, 2009).

> *The important thing is not to stop questioning; curiosity has its own reason for existing.*
>
> **Albert Einstein**

Table 9.1: Developing a Questioning Attitude

Stop, look, and listen—Proactively search for work situations that flag uncertainty
• Periodically pause-timeout to check the work situation. • Pause when a flag is recognized. • Identify inconsistencies, confusion, uncertainties, and doubts. • State or verbalize the uneasiness or question in clear terms.
Ask questions—Gather relevant information.
• What are the "knowns" and "unknowns"? • Use independent, accurate, and reliable information sources, especially other knowledgeable persons. • Compare the current situation (known) with independent sources of information. • Consider "what if ... ?" and/or use a "devil's advocate" approach in a spirit of helpfulness. • Identify persistent inconsistencies, confusion, uncertainties, and doubts.
Proceed if sure—Continue the activity if the uncertainty has been resolved with facts.
• Otherwise, do not proceed in the face of uncertainty.
Stop when unsure—If inconsistencies, confusion, uncertainties, or doubts still exist, do the following.
• Stop the activity. • Place equipment and the job site in a safe condition. • Notify your immediate supervisor.

Source: Volume 2: Human performance tools for individuals, work teams, and management, human performance improvement handbook, 2009.

We sometimes see things we do not understand but continue doing them without asking questions—Is what I am doing right or is it wrong?

- Are you DOING something or NOT DOING something because the "ham will not fit in the pan" or
- Are you DOING something or NOT DOING something because you know from factual evidence it is the right thing to do or not to do?

> *You become what you think of most.*
>
> **Dale Carnegie**

A workplace without fear of retribution allows employees to question work procedures and raise issues concerning their perception of a hazard and associated risk, making the decision to proceed without a loss-producing event. Deming discusses removing fear from the workplace as necessary to ensure problems and issues are reported (Deming, 2000).

> *The notion that there is some level of risk that everyone will find acceptable is a difficult idea to reconcile and yet, without such a baseline, how can it ever be possible to set guideline values and standards, given that life can never be risk-free?*
>
> **(Fewtrell & Bartram, 2001)**

Shifting the Thought Process to Risk

The thought process for the need of developing and implementing a risk perception method for use by employees came after a black belt research project that James conducted. The project was designed to develop an understanding of why hand and finger injuries were so high at an organization. The project looked into injury data collection, used direct observations of hand usage, and training methods (Roughton, 2005).

A design of experiment (DOE) was used to determine the effectiveness of four training methods. It incorporated a series of pictures with four scenarios of a training method.

Training Method	Type of Discussion	Type of Visual Aids
1	Verbal discussion	No visuals provided
2	No verbal discussion	Visual aids provided with text only
3	No verbal discussion	Visual aids provided without text
4	Verbal discussion	Visual aids provided with text

Source: Roughton, unpublished.

The DOE led to questions on why the safety message was not being retained by employees. Different concepts were tested for communicating the desired safe behavior. It was found that the best actions that improved information retention included verbal discussion and pictures with text coupled with preshift reviews, supervisor/employee one-on-one discussions to verify by demonstration that information was retained, and use of a machine/equipment-specific checklist. Supervisors were also asked to describe the at-risk behaviors they wanted to change. It was found that when risk identification was aimed at carefully selected areas, it brought into view peripheral issues of high risk that must be addressed.

Based on the project requirements, James had to evaluate and use of a risk matrix and the failure mode and effects analysis. While he was familiar with these tools, the black belt project gave him new insights into a different line of thinking with regard to risk. When combined with the findings about one-on-one discussions and use of visual aids with pictures and text, the potential benefits of directly bringing the concepts of risk to the employee became apparent.

> *Any risk assessment is subject to the 'perception of risk' in the eyes of the individual undertaking the risk assessment. While the information available to different individuals is the same, those individuals can perceive surprisingly different levels of risk.*
>
> **(Improving Health, Improving Services (Risk Form Assessment), 2012; Mishap Prevention Program, 2011)**

Building the Foundation for Risk Perception

A risk perception tool has been used to develop an understanding of operational risk for a number of years. The US military has adopted the concept to address risk in its daily operations and activities. The model recommends that personal as well as operational risk as identified by leadership be part of the overall training in risk concepts (Operational Risk Management (ORM) Fundamentals, 2010; Operational Risk Management, Marine, 2002).

However, many risk assessment tools used in industry approach the concept as a management tool and not one that immediately involves the employee. A personal risk assessment tool can be an effective process to implement at the "grass roots level" for determining appropriate actions based on the potential associated risk.

> *When undertaking a risk assessment, the consequence or 'how bad' the risk being assessed is must be measured. In this context, consequence is defined as: the outcome or the potential outcome of an event. Clearly, there may be more than one consequence of a single event.*
>
> **(A Risk Matrix for Risk Managers, 2008)**

To implement an effective risk assessment method, a combination of employees and leadership is the best approach to address risk perception. In turn, this enhances the education of employees in the identifying, recognizing, evaluating, and control of associated risk. Use of a standardized personal risk assessment tool builds a bridge between the differences and variations in individual risk perception.

Hazard Recognition Tools

A number of tools can be used to identify hazards (e.g., job hazard analyses, standard operating procedures, hazard hunting, task analyses, etc.). Based on the frequency of exposure and potential severity of loss, each hazard has a range of potential risk.

While these tools can identify operational hazards, they do not directly approach the assessment of risk. They have been used by most organizations based on loss assessments (total case incident rates (TCIR), workers' compensation, etc.). Hazard recognition programs should be reviewed to understand work environment real-life situations and develop a foundation for the use of a personal risk assessment tool.

> *Many acceptable risk decisions have to be made on the basis of incomplete information even by professionals specializing in the issues of concern.*
>
> **(Klapp, 1992)**

Risk Assessment Tool Defined

A risk assessment tool can be used by employees to conduct a self-assessment before performing a task that may have inherent hazards. The risk assessment tool is in the form of a small card carried by all employees. We will discuss the use of the tool in detail later in this chapter. The intent is to have all employees (including the leadership team) be able to:

- Recognize hazards specific to their work environment;
- Analyze the potential associated risk—its potential severity and the exposure to the hazard;
- Develop with leadership a mutually agreed upon risk control for the hazard.

The use of a risk assessment tool may require a shift in the perceptions held by the leadership team on what its role is in risk management. It requires allowing employees to raise issues without fear of reprisal and with an understanding that disagreement is okay.

The intent of using a tool is to reach agreement on the risk, and if none is reached, the discussion is taken to a higher level. The overall intent is to ensure that communication of concerns about risk is allowed to move through the organization quickly without obstruction or delay. The main format that defines the tool includes:

- A simple risk matrix that is easy for all employees to understand;
- A process that can be used at a moment's notice or as a routine part of task assessment;
- Consistent use by all levels of the organization;
- Being able to assess a broad range of risks.

> *"When undertaking a risk assessment, the consequence or 'how bad' the risk being assessed is must be measured. In this context, consequence is defined as: the outcome or the potential outcome of an event. Clearly, there may be more than one consequence of a single event"*
>
> *"When assessing likelihood, it is important to take into consideration the controls already in place. The likelihood score is a reflection of how likely it is that the adverse consequence described will occur"*
>
> **(A Risk Matrix for Risk Managers, 2008)**

> *"Likelihood can be scored by considering: frequency (how many times will the adverse consequence being assessed actually be realized?) or probability (what is the chance the adverse consequence will occur in a given reference period?)"*
>
> **(A Risk Matrix for Risk Managers, 2008)**

Changing Perceptions

The shift from a loss-based process using TCIR or other post-loss ratios to a risk-based process has been long advocated. TCIR can be a valuable measure but cannot be the only factor used to assess organizational safety performance. In order to develop a basic measuring system, the elements of the activity-based system as discussed in Chapter 11, "Developing an Activity-Based System" can be used.

> *A risk is acceptable when it is acceptable to the general public.*
>
> **(Fewtrell & Bartram, 2001)**

However, what is acceptable in the general public may not be acceptable in an organization and may be detrimental to its safety culture. What is allowed outside the organization may result in workers' compensation claims, third-party liability claims and lawsuits, or damage to operations if allowed within the organization.

Uncontrolled risk is unacceptable within an organization as the rules and regulations it must legally follow are different. For example, employees may on their own time take extreme risk, such as sky diving, rock climbing, driving a vehicle on weekends without using a seat belt, etc. The concern is that this mindset of risk acceptance may get into the workplace unless a mechanism is in place to prevent unacceptable risk.

Meeting and Getting to Sustainability

Using a personal risk assessment tool reduces the gap between various perceptions of risk. We each perceive risk differently and we need an objective tool rather than subjective personal opinion to analyze risk levels.

The value of the risk assessment tool must be clearly communicated to the leadership team who can support and champion its use. The process starts with getting your immediate leadership to see the value of the tool through demonstration and going through the risk perception process. By using the tool to evaluate existing known hazards for inherent associated risk, the value of the tool is demonstrated. Knowledge of the tool will spread through the network in multiple directions, making its full use throughout the organization potentially more successful.

The next level of introducing the tool begins with small group meetings where it is explained and its value demonstrated. As buy-in from these core groups is achieved, the word will again informally spread through the organization. Refer to Table 9.2 for an introduction to "Identifying Risk Perception".

As we discussed in a previous chapter, "rollout" has the connotation of being just another program or flavor of the month. The term rollout carries a negative connotation to many people. How we frame a discussion or action goes a long way toward getting it accepted. To "introduce" something has a different connotation, implying a more mutual approach.

Lesson Learned # 2

Nathan was once in a meeting that was discussing current workloads and ongoing "initiatives". One of the participants spoke out saying, "I have 20 to 30 initiatives from a number of corporate and regional departments, all with short time frames and listed as high importance! These are in addition to my normal workload. Can we have a moratorium on initiatives so that I can get some work done around here?" We discussed the planning process in the Chapter 4, "Setting the Direction for the Safety Culture".

Table 9.2: Identifying Risk Perception

To further increase the potential for successful introduction of the risk perception concept and tool, review the following:

- What are the espoused values that drive the use of other forms and guidelines currently in use?
- How have other tools, rules, guidelines, etc. been "rolled out"? (See discussion.)
- How many other such items have recently been released from various groups?
- How were they accepted? Were they accepted with enthusiasm, subdued patience, or with benign neglect?
- How many other endeavors or "flavors of the month" have been rolled out?
- Have they tainted the implementation of the risk assessment tool?
- What approach can be taken to avoid becoming the next "flavor of the month"?
- Who in your network can best assist with implementation of the risk assessment tool?
- A goal is to have all levels of the organization trained and using the risk assessment tool—What are the primary obstacles to getting this level of acceptance?
- What introduction materials and communications are needed for implementation of the tool?
- Has its value been clearly determined and part of the communication?
- Is a policy needed to mandate utilizing the tool or would this not be beneficial to its introduction?
- Is the use of the tool consistent with other elements of the safety management system?
- What is the risk tolerance of the organization?
- What is the current understanding and perception of risks within the organization?
- What organizational factors might influence the normal perception of risk-making decisions?

Answering these questions can set the stage for better planning for introduction of the risk assessment tool and establishing its goals and objectives.

Implementing the Risk Assessment Tool

> *Risk is inherent in all tasks, training, missions, operations, and in personal activities no matter how routine.*
>
> **(Operational Risk Management (ORM) Fundamentals, 2010; Operational Risk Management, Marine, 2002)**

The underlying goal is not to add to the burden of the leadership team and employees who will be using the tool. Its introduction should enhance and be integrated into the current safety management system. Review your system and determine if it can replace other program elements that are not showing value or improvement to the system. The introduction must be kept simple, show that the tool is of value, and be quickly implemented. The preparation for the introduction is crucial and should include step-by-step instructions on its use.

The introduction of the tool should include at a minimum the following elements:

1. An introduction to the risk assessment tool;
2. A brief explanation of hazard and associated risk;
3. An explanation on the use of the tool and the risk assessment matrix;
4. A discussion of methods to use to communicate the hazards and perception of associated risk to establish the ground rules for discussion of personal risk perceptions and when leadership team and/or safety department intervention is required.

In order for the introduction and ongoing process to be effective, they must meet the following criteria:

1. The use of the tool should be incorporated into the training calendar and be a part of the orientation process for new hires and/or relocated individuals.
2. The introduction should be jointly completed by at least one member of the leadership team and a safety manager or coordinator.
3. A diverse combination of leadership, administrative, operations, and maintenance employees should attend the introduction session to reduce miscommunication and the silo effect that keeps groups separate from each other. The key is to ensure that all employees acquire the same knowledge in the use of the tool as this diversity increases the influence of safety activities.
4. The introduction session should begin with an endorsement from the senior leadership of the organization.
5. Leadership must talk about the tool and ask questions about its use and findings when visiting facilities and operations to emphasize its support.

Individuals must feel free to question any operation or action felt to be an unacceptable risk without fear. The question that should be maintained in the minds of all employees is "What are we doing that we have always done and that is no longer an acceptable risk?" And "What routine and nonroutine tasks have risk that must be addressed in more detail?"

> *Judgments also alter with time—what was acceptable yesterday may not be acceptable today and tomorrow.*
>
> **(Mullai, 2006)**

Hazard and Risk Defined

Hazard is defined as "a condition, set of circumstances, or inherent property that can cause injury, illness, or death" (Occupational Health and Safety Systems, 2012). In addition, we add the potential to cause damage or create a loss-producing event of any type.

Risk is "an estimate of the combination of the likelihood of an occurrence of a hazardous event or exposure(s), and the severity of injury or illness that may be caused by the event or exposures" (Occupational Health and Safety Systems, 2012). Again, we expand this to include damage or any other type of loss.

Risk assessment is the "process(es) used to evaluate the level of risk associated with hazards and systems issues" (Occupational Health and Safety Systems, 2012).

Why Are Risk Analysis and Risk Reduction Important?

The assessment of risk goes beyond simply looking for areas that cause injury. Hazards in all probability also have the potential of causing damage to operations, harm to third parties, and even harm to the brand of the organization. Impact on the brand of the organization (its overall reputation, customer trust, and public perception) can result in loss of market share, increased regulatory fines, lawsuits, as well as public mistrust.

> What are the beliefs of the culture regarding these impacts? Are they considered a "cost of doing business", a price that must be paid to get the job done? That loss is simply fate and cannot be avoided? Beliefs such as these will not be visible in any written publications but may be verbally expressed in post-loss conversations. The shift to an understanding of risk increases the scope and depth of the safety culture.

Personal Risk Tolerance—How Do We Decide What Is a Risk?

The perception of risk may be very different from person to person. Without a baseline method to assess risk, risk acceptance, or tolerance, your assessment and overall concerns may be challenged. Use of the risk assessment tool reduces reliance on one person's assessment or opinion of what a "risk" is.

We are risk takers by nature. This is magnified through activities and media constructs that have unconsciously trained us to take risks and ignore the consequences. Consider the perception communicated in many movies and reality shows where an extreme risk is taken in direct defiance of common sense, defies the laws of physics, or is shown as necessary to save the world. A lifetime of watching actors jump from high buildings and landing without harm, ignoring the "geeky" safety guy who just does not understand the situation, and surviving all of the explosions and risk events "prove" that taking risk was essential. This form of entertainment is what is funneled into our subconscious daily. Our entertainment has trained us that taking a physical risk is essential and the end results are always positive. This attitude affects all of us—from senior leadership through all levels of employees.

We have already discussed how the perception of safety must be addressed to improve the safety culture. You must begin with assessing your own personal risk tolerance. Is it coloring your own perception of risk and, in turn, affecting your approach and communication of risk-related issues? By understanding your own risk tolerance and personal/organizational factors, we can start to understand how to develop better strategies that cut through perceptions about the acceptance of risk.

> We routinely see actors jumping or falling from high buildings and slamming into the pavement, stunned but immediately able to get up and continue the chase.
>
> According to White and Bowen (As cited in Willie Hammer's *Occupational Safety Management and Engineering*, 2001), their research found that "50% of all persons impacting against a hard surface with a velocity of 18 miles per hour (27 feet per second) will be killed. This is equivalent to a free fall of 11 feet" (Hammer & Price, 2001). The entertainment does not show the mechanisms used in the scene to prevent harm yet may leave your employees the risk perception that they can fall from high places and recover without a scratch!

The Risk Assessment Tool Process—The Risk Guidance Card

If a safety management system is in place, a hazard control program should have already been implemented and have addressed general safety-related criteria. A "Risk Guidance Card" would be designed and distributed to all employees. The card is used to begin a dialog when employees perceive that a risk could affect them as they perform their assigned tasks. An employee or a work team can use the card as part of a "pre-job assessment" (Volume 2: Human performance tools for individuals, work teams, and management, human performance improvement handbook, 2009) and when concerns are expressed that additional or existing hazard controls are needed.

A risk assessment matrix is used to estimate the potential consequences of an adverse outcome based on a combination of what is perceived as the potential severity and the probability of exposure to the hazard being evaluated.

> *A risk assessment matrix is a useful tool to identify the level of risk and the levels of management approval required for any Risk Management Plan. There are various forms of this matrix, but they all have a common objective to define the potential consequences or severity of the hazard versus the probability or likelihood of the hazard.*
>
> **(Safety Management System Toolkit, 2007)**

Assessing risk must be done with full knowledge that risk perceptions can differ. The intent behind the use of a risk guidance card is to allow employees to immediately assess levels of risk and also get guidance on where the authority to accept risk resides. The objective is to reduce the probability and severity of a high risk being accepted without informing and discussing with immediate leadership. The card allows a range of decisions about risk, from acceptable to unacceptable as well as communication of risk and necessary controls to be made.

The matrix used below is a variation on the one used in our book, *Job Hazard Analysis* (Roughton & Crutchfield, 2008). The risk guidance card has been abbreviated as it is to be used to develop an awareness of immediate risk. Refer to Table 9.3 for way for "Identifying Probability and Severity".

Table 9.3: Identifying Probability and Severity

<table>
<tr><th colspan="2">Probability</th><th colspan="2">Severity</th></tr>
<tr><td colspan="2">Unlikely—Rare chance of occurrence, some possibility injury exists
Likely—An event or an injury has happened in the past
Very likely—Harm is certain—An obvious condition</td><td colspan="2">Marginal—Minor injury, no first aid
Critical—Minor injury, potential first aid
Severe—High change of an injury or permanent disability</td></tr>
<tr><td colspan="4">The objective is to identify conditions(s) and/or behavior(s) before performing a task.
STOP—THINK—AND ASK questions
Stop when the hazard and risk are not understood!
Use a questioning attitude to determine the likelihood of an injury.</td></tr>
<tr><td>What is the frequency of this task?</td><td>Seldom—Less than one time per shift/day</td><td>Occasionally—At least daily, no more than three times per shift/day</td><td>Frequently—More than three times per shift/day</td></tr>
<tr><td colspan="4">As the frequency of a task increases, so does the probability of a loss-producing event.</td></tr>
<tr><td colspan="4">The questions to ask before conducting any task
What is involved in this task that can cause harm or damage?
How can we keep harming myself or others while performing this task?
If I get hurt and damage occurs, how serious could it be?
How likely is it to happen?
What should be done to reduce the risk?</td></tr>
</table>

To use the risk guidance card (Figures 9.1–9.4) and information in Table 9.3, the user(s) first ask the following questions using the definitions from Table 9.4 and 9.5:

- Given what I see and believe about the situation or condition, what are the consequences of exposure if something goes wrong?
- What are the odds or probability that something harmful or damaging can happen? Will an event be marginal, critical, or severe?

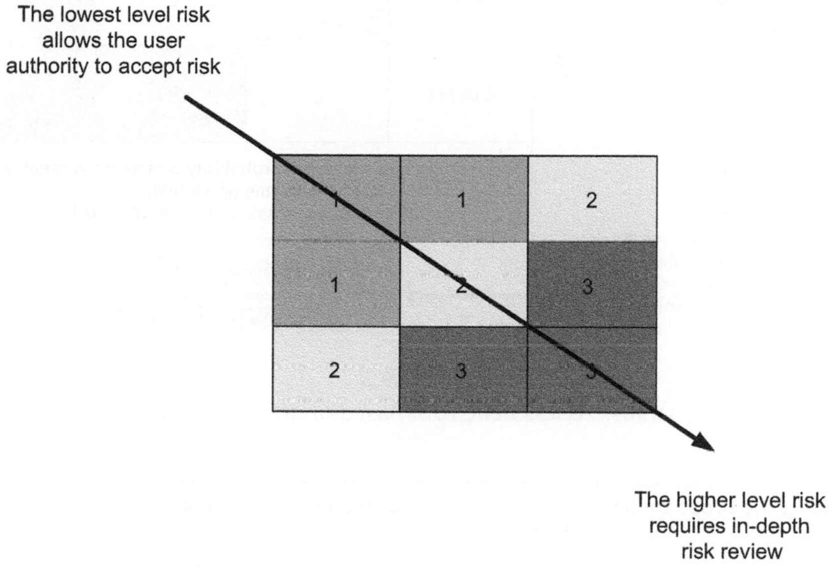

Figure 9.1

Provides an Overview of Assigning Authority Level to Identified Hazard and Associated Risk. *Source: Safety Management System Toolkit, 2007; System Safety Process Steps, 2005.*

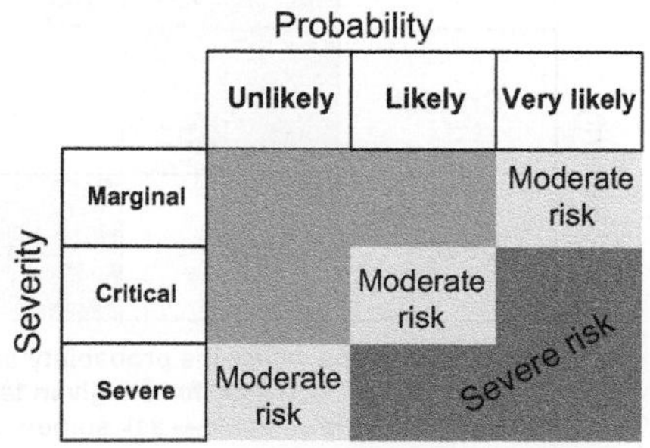

Figure 9.2

Provides an Overview of Risk Tolerance as Related to Identified Hazard and Associated Risk. *Source: Risk Management System Guidance Pack, The Residual Risk Matrix, 2011; System Safety Process Steps, 2005.*

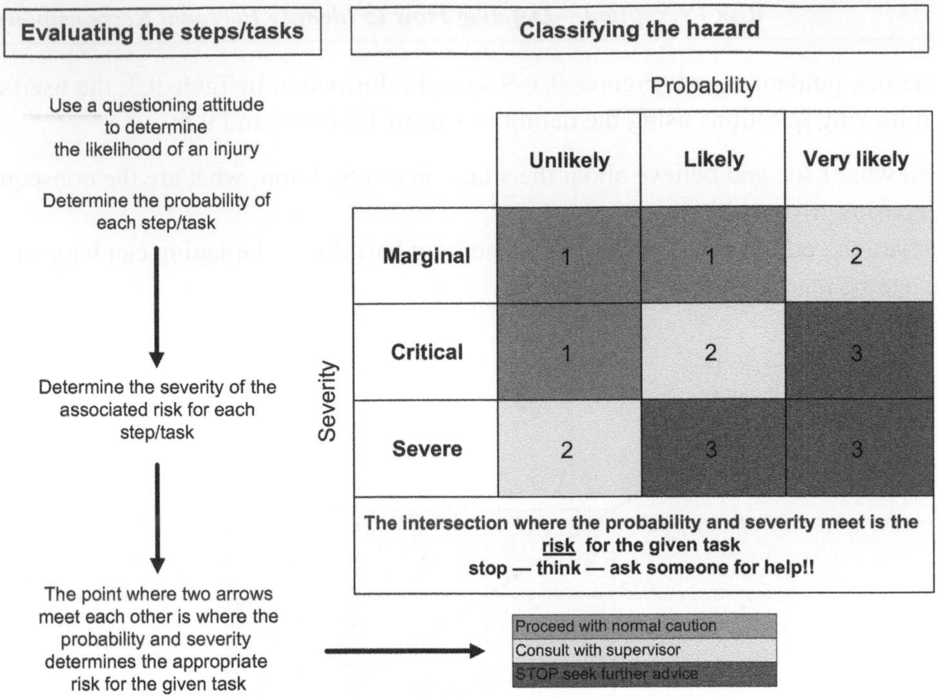

Figure 9.3

Provides a Model Risk Assessment Tool Based on the Probability and Severity.

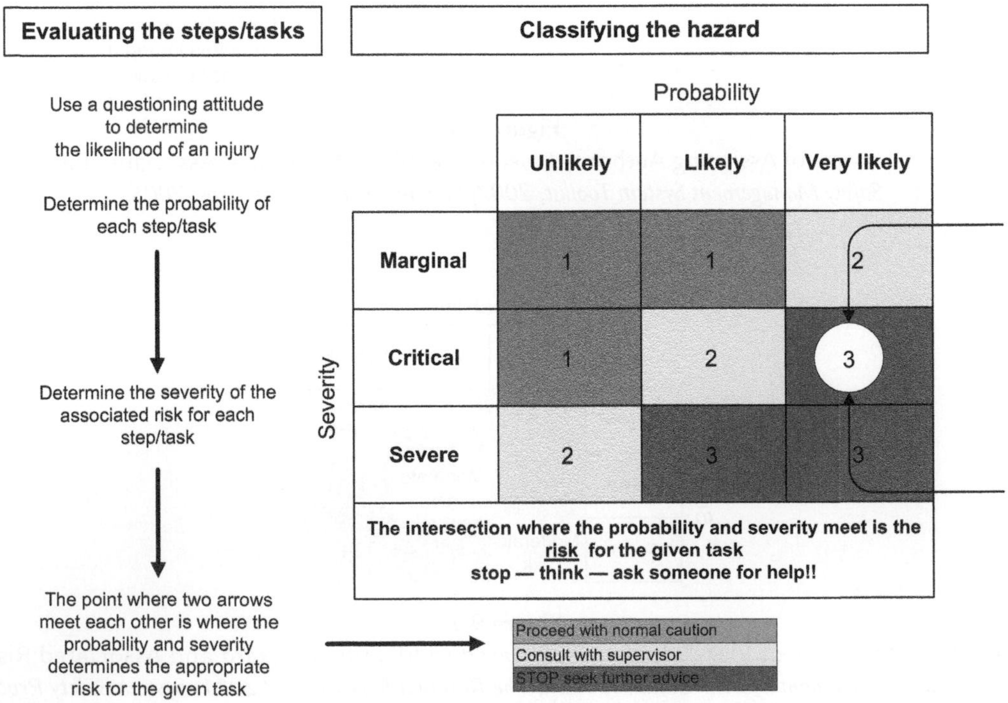

Figure 9.4

Example Risk Assessment Analysis.

- What is included in this task that can cause a loss producing event?
- What should be done about this situation or condition?

The objective is to explicitly define the potential harm in terms of the adverse consequence(s) that might arise from the hazard and its level of risk. Brief simple definitions are used to define probability and severity ratings to begin the overall discussion. Refer to Table 9.4 "Probability Rating" and Table 9.5 "Severity Rating".

When assessing likelihood, it is important to take into consideration the controls already in place. The "likelihood" score is a reflection of how probable an adverse consequence as described will occur.
(A Risk Matrix for Risk Managers, 2008)

Risk Scoring

Probability and severity are scored on a scale of 1–3. Where the two scores intersect on the matrix is the "Risk Rating". Refer to Table 9.6 for "Risk Guidance Card–Risk Rating Color Codes".

Table 9.4: Probability Rating

Unlikely	Rare occurrence—the possibility of an injury or damage exists
Likely	Reasonable—a strong possibility for damage or injury
Very likely	High possibility of a loss-producing event

Table 9.5: Severity Rating

Marginal	Minor injury or damage—first aid
Critical	Injury requiring medical attention or moderate damage
Severe	High possibility of a serious injury with permanent disability or major damage

Source: Army System Safety Management Guide, 2008; Improving Health, Improving Services (Risk Form Assessment), 2012; Mishap Prevention Program, 2011.

Table 9.6: Risk Guidance Card-Risk Rating Color Code

Risk Rating	Color	Action
1	Green	"Proceed with caution"
2	Yellow	Suggest "consult with supervisor, discuss risk, and identify a possible solution"
3	Red	Suggest "Based on your assessment, stop work, discuss with appropriate levels of leadership. Discuss risk and identify solution"

Source: Adapted from Operational Risk Management (ORM) Fundamentals (Operational Risk Management (ORM) Fundamentals, 2010; Operational Risk Management, Marine, 2002).

An objective of the risk guidance tool is to develop a situational awareness of the potential for a loss-producing event. For a visual representation of the risk assessment, the risk guidance tool uses a color code. An example of the use of color is found in *The US Navy's Operational Risk Management (ORM) Fundamentals*:

- "Green indicates that errors may occur, but errors will be caught by the individual.
- Yellow indicates that the potential for consequential errors has increased.
- Red indicates that errors may occur that cannot be caught and, therefore, become consequential to the task or mission.

The target is used during team communication to focus one or more individuals into an understanding of situational conditions, or an individual can use it as a self-assessment tool to increase SA (Situational Awareness).

(Operational Risk Management (ORM) Fundamentals, 2010)

If you have a risk rating of 3, then the task is stopped, the supervisor is contacted, and the perceived hazard or associated risk situation or condition is discussed. The employee shares the perception of risk with the supervisor and how the assessment was made using the risk guidance card. The supervisor cannot override the concerns of the employee without providing a solution or clearly showing how the hazard and its associated risk can be controlled. Both supervisor and employee must agree with the analysis.

The risk assessment discussion moves to the next level of leadership if:

- The supervisor and employee cannot agree on the risk rating and a solution. The supervisor cannot make a decision to proceed with the task without the agreement of the employee.
- A risk rating 3 has conditions or is a situation that cannot be resolved at this level of the organization. All level 3s must move to the next leadership or higher level.

Refer to Figure 9.4 for an "Example Risk Assessment Analysis".

If the organization has a viable safety culture, the use of the risk assessment will make its strength clear and apparent. Given the need for an organization to meet specific production or service goals, the pressure to maintain a schedule of activity may tempt to make employees see risk as acceptable even when conditions or the situation indicates otherwise.

Because the potential for error or surprise exists in many organizational activities... information that could constitute error signals must be widely available through communication nets that can cut across departments and hierarchical levels. Communication barriers or blockages can pose a threat to feedback, evidence accumulation, and the sharing of cautionary concerns.

(Boin & Schulman, 2008)

Summary

An understanding of risk and its control must permeate your organization if an in-depth safety culture is to be sustained. A safety management system using only inspections and observations to identify hazards will not provide a full appreciation of the potential for injury and damage or loss producing events without linking the results to the potential risk.

When we deal with any type of hazard-related condition and its associated risk, we have a built-in affinity to judge our environment based on our past experiences. We see a situation, mentally scan our memory for similar situations, and act based on what consequences we experienced or saw in the past. A problem develops when what we think we see is not what the reality is at that point in time.

Each time a risk is taken and nothing happens, at-risk behavior gets reinforced. This behavior is not limited to the work environment but is reinforced through daily behavior.

A risk assessment tool has been used to develop an understanding of operational risk for a number of years. However, many risk assessment tools used in industry approach the concept as a management tool and not one that immediately involves the employee. A personal risk assessment tool can be an effective process to implement at the "grass roots level" for determining appropriate actions based on the potential associated risk.

The risk guidance card is used by employees to conduct a self-assessment before performing a task that may have inherent hazards. The risk guidance card is in the form of a small card carried by all employees.

The risk guidance card may require a shift in the perceptions held by the leadership team on what its role is in risk management. It requires allowing employees to raise issues without fear of reprisal and with an understanding that disagreement is okay.

The main format that defines the tool includes:

- A simple risk matrix that is easy for all employees to understand;
- A process that can be used at a moment's notice or as a routine part of task assessment;
- Consistent use by all levels of the organization;
- Being able to assess a broad range of risks.

The underlying goal is not to add to the burden of the leadership team and employees who will be using the tool. Its introduction should enhance and be integrated into the current safety management system.

The assessment of risk goes beyond simply looking for areas that cause injury. Hazards in all probability also have the potential of causing damage to operations, harm to third parties, and even harm to the brand of the organization.

The perception of risk may be very different from person to person. Without a baseline method to assess risk, risk acceptance or tolerance may challenge your assessment and overall concerns.

If the organization has a viable safety culture, the use of the risk assessment will make its strength clear and apparent.

In our next chapter we will discuss risk management as being an essential element of a strong safety culture. The concepts of risk management should be considered an essential part of the leadership team's decision-making. All employees need a basic understanding of the terms risk, risk control, and risk management if the organization's safety culture is to be sustained.

Organizations continue to use post-loss data (injuries and damage) as the primary guide to how well the organization is doing with regard to control of loss-producing events. As we discussed in Chapter 4, "Setting the Direction for the Safety Culture", with the potential for loss information to be influenced by business and organizational changes, an inaccurate picture of the true state of the safety culture and the safety management system can exist.

The shift from a loss-based safety management system to one that is balanced between a focus on loss with a stronger emphasis on risk identification and control requires a fundamental change in perception held about safety. It requires letting go of long-held opinions and exclusive focus on post-loss events that are tangible and can be measured and recognize that risk can be intangible and require a more detailed study of the organization. Detailed reviews and analysis of loss-producing events remain essential, but the focus shifts from blame to understanding the overall risk that was allowed to develop and was left uncontrolled.

Chapter Review Questions

1. Explain safety as an emergent property. Provide sufficient details to support your answer. Part of the answer may be in another chapter we covered.
2. Discuss why we take unnecessary risk.
3. Discuss in detail why injury rates (TCIR) are not a true measure of risk. Provide some examples.
4. Compare and contract concepts of a risk guidance card.
5. Define the terms of hazard and risk. Provide some examples to support your response.
6. Discuss how the risk guidance tool aids in communicating risk.
7. Discuss personal risk tolerance and some related issues. Provide some examples as provided in this chapter. In addition, provide some additional examples from your experience.
8. Discuss why it is important that risk assessment is used and how it can improve the safety culture.

9. Refer to Table 9.3. Discuss part of the method that was outlined in "Identify Probability and Severity" of a risk. Use Table 9.4, "Probability Rating" and Table 9.5, "Severity Rating" to discuss your response.

Bibliography

A Risk Matrix for Risk Managers. (January 2008). 4-8 Maple Street, London: The National Patient Safety Agency. Retrieved from http://bit.ly/12hoCH4.

Army System Safety Management Guide. (November 2008). Department of the Army, Pamphlet 385–16, Public Domain. Retrieved from http://bit.ly/11f0gib.

Boin, A., & Schulman, P. (2008). Assessing NASA's safety culture: the limits and possibilities of high-reliability theory. *Public Administration Review*, *68*, 1050–1062.

Deming, W. E. (2000). *Out of the crisis*. MIT Press. Retrieved from http://amzn.to/14o8bbI.

Fewtrell, L., & Bartram, J. (2001). *Water quality: Guidelines, standards, and health. Assessment of risk and risk management for water-related infectious disease*. IWA Publishing.

Hammer, W., & Price, D. (2001). *Occupational safety management and engineering* (Vol. 240). Prentice Hall.

Improving Health, Improving Services (Risk Form Assessment). (March 2012). Lincoln, LN4 2HN: NHS Lincolnshire: NHS Lincolnshire, Cross O'Cliff, Bracebridge Heath, Public Domain. Retrieved from http://bit.ly/XOLXZo.

Klapp, M. G. (1992). *Bargaining with uncertainty: Decision-making in public health, technological safety, and environmental quality*. Auburn House Publishing Company.

Mishap Prevention Program. (August 2011). United States Air Force, Instruction 91-202. Retrieved from http://1.usa.gov/XMhhrU.

Mullai, A. (2006). *Risk management system: Risk assessment frameworks and techniques*. DaGoB Project Office, Turku School of Economics.

Occupational Health and Safety Systems. (2012). The American Industrial Hygiene Association. ANSI/AIHA Z10.

Operational Risk Management (ORM) Fundamentals. (July 2010). Department of The Navy Office of The Chief of Naval Operations, Pubic Domain. Retrieved from http://bit.ly/VuvB8D.

Operational Risk Management, Marine. (2002). Marine Corps Institute, ORM 1-0, Pubic Domain. Retrieved from http://1.usa.gov/14wk9z4.

Risk Management System Guidance Pack, The Residual Risk Matrix. (November 2011). University of Derby. Retrieved from http://bit.ly/Y2JM4p.

Roughton, J. (August 2005). Safety Black Belt Projects, Certified by BMG. Breakthrought Management Group.

Roughton, J., & Crutchfield, N. (2008). *Job hazard analysis: A guide for voluntary compliance and beyond. Chemical, petrochemical & process*. Elsevier/Butterworth-Heinemann. Retrieved from http://amzn.to/VrSAq5.

Safety Management System Toolkit. (2007). Developed by the Joint Helicopter Safety Implementation Team of the International Helicopter Safety Team, *The international helicopter safety symposium, Montréal, Québec, Canada*. Retrieved from http://bit.ly/YaSCOg.

System Safety Process Steps. (January 2005). FAA's Safety Risk Management Order, 8040.4, Public Domain. Retrieved from http://1.usa.gov/13m7wsA.

Volume 1: Concepts and principles, human performance improvement handbook. (2009). U.S. Department of Energy, Public Domain. Retrieved from http://1.usa.gov/WdIqoP.

Volume 2: Human performance tools for individuals, work teams, and management, human performance improvement handbook. (2009). U.S. Department of Energy, Public Domain. Retrieved from http://1.usa.gov/11Ex7vE.

CHAPTER 10

Risk Management Principles

The essence of risk management lies in maximizing the areas where we have some control over the outcome while minimizing the areas where we have absolutely no control over the outcome and the linkage between effect and cause is hidden from us.

—Peter L. Bernstein, "Against the Gods"

Introduction

Risk management is an essential element of a strong safety culture. Safety management systems such as ANSI Z10-2012 (Occupational health and safety systems, 2012) have criteria for a risk assessment to be completed as part of the overall analysis of an organization. The concepts of risk management should be considered an essential part of the leadership team's decision-making. All employees need a basic understanding of the terms risk, risk control, and risk management if the organization's safety culture is to be sustained.

Organizations continue to use post-loss data (injuries and damage) as the primary guide to how well the organization is doing with regard to the control of loss-producing events. As we discussed in Chapter 4, "Setting the Direction for the Safety Culture", with the potential for loss information to be influenced by business and organizational changes not directly associated with the cause of injuries or loss, an inaccurate picture of the true state of the safety culture and the safety management system can exist.

In a July 19, 2010, letter to OSHA staff on "OSHA at 40", OSHA administrator David Michaels wrote, "Ensuring that American workplaces are safe will require a paradigm shift, with employers going beyond simply attempting to meet OSHA standards, to implementing *risk-based* workplace injury and illness prevention programs" (our italics) (Linhard, 2010).

The shift from a loss-based safety management system to one that is balanced between a focus on loss with a stronger emphasis on risk identification and control requires a fundamental change in perception held about safety. It requires letting go of long-held opinions and exclusive focus on post-loss events that are tangible and can be measured and recognize that risk can be intangible and require more detailed study of the organization. Detailed reviews and analysis of loss-producing events remain essential, but the focus shifts from blame to understanding the overall risk that was allowed to develop and was left uncontrolled.

In this chapter we will discuss risk, its definition, and concepts for the communication of risk to leadership and to all employees.

After completing this chapter you will be able to:

- Define risk.
- Discuss how definitions of risk may be different in the organization.
- Discuss the obstacles to risk management.
- Discuss the eight steps of a risk assessment.
- Discuss acceptable risk and its potential impact.
- Discuss the risk spectrum.

What Is Risk?

Risk management establishes the criteria that are necessary for the identification of hazards, assessing the risk inherent within those hazards and selecting the treatments or methods for the avoidance, control, transfer, or acceptance of risk. Before managing risk, several definitions are needed.

The concept of risk is relatively recent and was developed in parallel with the advance of statistics and probability. The concept of risk began to evolve around the sixteenth century and has gradually found its way into all aspects of organizations. Before that time period, the general consensus was that nothing could be done to prevent many loss-producing events as they may have been preordained or simply fate. We still find, however, that many people still believe that fate plays a strong role in loss-producing events and hold that "things just happen".

This belief is understandable, as the underlying cause of the loss-producing event may be from well-hidden latent human error incorporated in the overall design of the job, its steps, and task, as noted in literature on Human Performance Improvement (Volume 1: Concepts and principles, human performance, improvement handbook, 2009). From a risk standpoint, since a separation between the hazard, associated risk, and the time when a loss-producing event occurs, most of us simply cannot make the connection. We have referred to this mindset as "No loss = no risk".

Infrequent losses do not necessarily mean low or controlled risk, as evidenced by catastrophic events such as the space shuttle losses, the 2010 BP Gulf Deep Water Horizon oil spill (Deep Water, The Gulf Oil Disaster and the Future of Offshore Drilling, 2011), refinery explosions, and (Blast at BP Texas Refinery, 2005) and even disasters such as the sinking of the Titanic (Haverin, 2012).

New ventures, implementing new technologies, modifying a process, and starting a new production line are untaken after leadership has been convinced that the new "whatever it is" is good for the organization. Once the decision is made to start the endeavor, enthusiasm and an optimistic attitude take over. This attitude becomes rigid and defends itself against any criticism or even logical well-designed risk presentations. Those individuals who made the case for the endeavor can become blind to risk of any type.

For example, in our view, the commonalities in these events appear to be that communications about risk concerns were not relayed, risks were not fully assessed, potential loss indicators were ignored, standards were not followed, or inadequate regulations were followed even though conditions had changed.

Lesson Learned # 1

Nathan was part of a team charged with the development of a risk assessment process that could be used with client organizations. Having primarily worked on programs that were focused on "loss control", he found the study of risk a professionally life-changing experience. Once he moved through the "doorway" to a risk-based outlook, looking at risk and not strictly at loss, his perspective on developing a safety management system shifted from just preventing loss-producing events to the search for methods to control the various components of risk that generate the conditions for loss-producing events. He found that much effort had been going into reducing events of low risk and missing the high severity potential conditions that can remain hidden in operations.

Risk has been widely discussed and researchers have long advocated that risk assessments be completed. William Lowrance defined the concept of determining acceptable risk that related probability and severity for harm (Lowrance, 1976). Recent standards have defined risk and built assessments into their guidelines. For example:

ISO 31000 (A structured approach to Enterprise Risk Management (ERM) & the requirements of ISO 31000, 2010; ISO 31000, risk management—principles and guidelines, 2009) defines risk as the "effect of uncertainty on objectives". "Organizations using it can compare their risk management practices with an internationally recognized benchmark, providing sound principles for effective management and corporate governance."

ANSI Z10-2012 (Occupational health and safety Systems, 2012) defines risk as "an estimate of the combination the likelihood of occurrence of a hazardous event or exposure(s), and the severity of injury or illness that may be caused by the event or exposures". In ANSI Z10, a risk matrix is used to develop an estimate of the specific risk and is based on a combination of experience, industry history, science, and understanding of hazards. The risk matrix provides on what the potential severity may be in a worst case scenario based on the type of exposure.

> *Risk is the probability that exposure to a hazard will lead to a negative consequence.*
>
> **(Ropeik & Gray, 2002)**

Confusion over Definitions of Risk

A major issue develops when we try to discuss "risk" within an organization. Risk is defined differently by each department of an organization. Organizations face strategic risk that is related to economic needs, sales, and marketing decisions. Risk taking is considered inherent and a valuable trait in the general business environment as organizations see risks from the perspective of the probability of increasing market share and revenue stream. This definition can create situations that blind leadership to the consequences created by "risk" as defined and identified from a safety perspective (Quadrant of Risk: Hazard, Operational, Financial, and Strategic, n.d.).

Risk as understood by safety professions is hazard or operational risk—conditions that are related to operational physical hazards and have the potential to cause injury or damage. When discussing risk, it is important to ensure that your audience is using the same definitions.

Obstacles to Risk Management

One of the cultural obstacles that we have is that loss-producing events that cause injury in any form or damage are still viewed as having an element of fate. Both of us have experienced conversations with managers or supervisors that implied that nothing could be done to have prevented an "accident".

A major obstacle we face is that if no loss or major issue has recently occurred, then the mental model forms that the risk is low or maybe nonexistent. Being able to define the various types of risk faced by your organization requires use of many of the tools we are discussing throughout this book. An essential skill is being able to define or estimate how probable and severe a loss-producing event might be.

Lesson Learned # 2

Nathan once visited a poultry processing plant to do a safety program needs assessment. On the night shift, it was noted that an employee was shoveling out a drainage trench that conveyed offal (by-products) to what was supposed to be an automated system. When an employee was asked why he had to do this task as the system was supposed to be automatic, he responded that the design of the system of conveying offal was the idea of a member of senior management. Each night a team of maintenance and cleaning personnel had to work to get the system running for the next day as no one would tell the senior management that the expensive system was really not functional and clogged up every day. A series of questions that started out as a basic question concerned with the ergonomics of a task ended up finding multiple issues and more complex concerns.

Lesson Learned 2 clearly points out how risk communications can be easily stopped within an organization. The need for risk assessment built into overall decision-making is one of the overall goals of risk management. However, given the politics and egos that may be present, the approach must consider multiple approaches when presenting issues.

Another obstacle we face is the continued use of the term "accident", which is firmly embedded in the overall society culture. We have used the term "loss-producing event" throughout this book as it does not carry the connotation that is implied by the term accident. Accident still carries the image that something is not preventable or has an element of fate. A dictionary definition is an "unexpected happening causing loss or injury that is not due to any fault or misconduct on the part of the person injured…" (accident, n.d.). This definition does not tell us anything about the risk encountered and implies that not much could be done to have prevented the particular loss-producing event.

Ethan and Little, in their article on risk discuss the confusion between the concepts of hazard and risk, where there is tendency to use the terms interchangeably. As they noted "a hazard is a potential source of harm". However, "Risk is the chance that the adverse effects of unidentified hazards will occur" (Ethan & Little, 2011). The source of the hazard (chemical, electrical, radiation, etc.) establishes the physical condition that creates the potential for injury or damage. However, the risk is a combination of the potential severity of the incident that might occur and its probability of occurrence or on the frequency of exposure to the hazard.

ANSI Z10 does not use the term accident but defines "Incident" as "an event in which a work related injury, illness (regardless of severity), or fatality could have occurred (commonly referred to as a "close call" or "near miss") (Occupational health and safety systems, 2012).

Risk Assessment

Per HSG 65, a three-part approach is used to establish the foundation for the management of risk. While the process can be simply stated, it does require an in-depth and methodical approach in the attempt to bring out hidden hazards and associated risk. The three parts consist of risk identification, risk assessment, and risk control (Successful health and safety management, hsg65, 1997). Another approach is defined as "the process of evaluating the risk to safety and health arising from hazards at work" (Guidelines for Hazard Identification, Risk Assessment and Risk Control (HIRARC), 2008).

Risk Identification is a methodical review of all of the activities and conditions that exist in the organization. As risk is inherent in all activities, the point of beginning a risk identification process is to map the organization to determine what hazards exist and where they are located. Basically, this is asking the question, "What do we do and how do we do it?" Job hazard analysis, systems and process reviews, data analysis, and any other available tools to describe the organization are used. Once the hazards are located and described, risk assessment begins.

A very simple definition of risk assessment is "Assessing the risk that may arise from hazards" (Successful health and safety management, hsg65, 1997). The identification of priorities based on the severity of loss-producing events is established.

Figure 10.1 and Table 10.1 provide an eight step process to accomplish a risk assessment beginning with defining objectives and ending with modifying the safety management system based on the assessment.

Acceptable Risk

Risk is always present in every action, endeavor, and function we undertake. It may not be visible but it is there, from simply walking across a floor, picking up a newspaper, putting gasoline in a car, etc. Various sports represent the voluntary taking of risk—rock climbing, snowboarding down a treacherous mountain, Formula One racing, etc. in which we occasionally see tragedy occur. It is the understanding and control of risk that must be present.

A concept that is internationally used for risk management decision-making is defined "as low as reasonably practicable risk". Once the risk of the operation or item in question has been assessed, a determination must be made of how effective any controls are that are put in place. Acceptable risk is defined as that "risk for which the probability of a hazard-related incident or exposure occurring and severity of harm or damage that may result are as low as

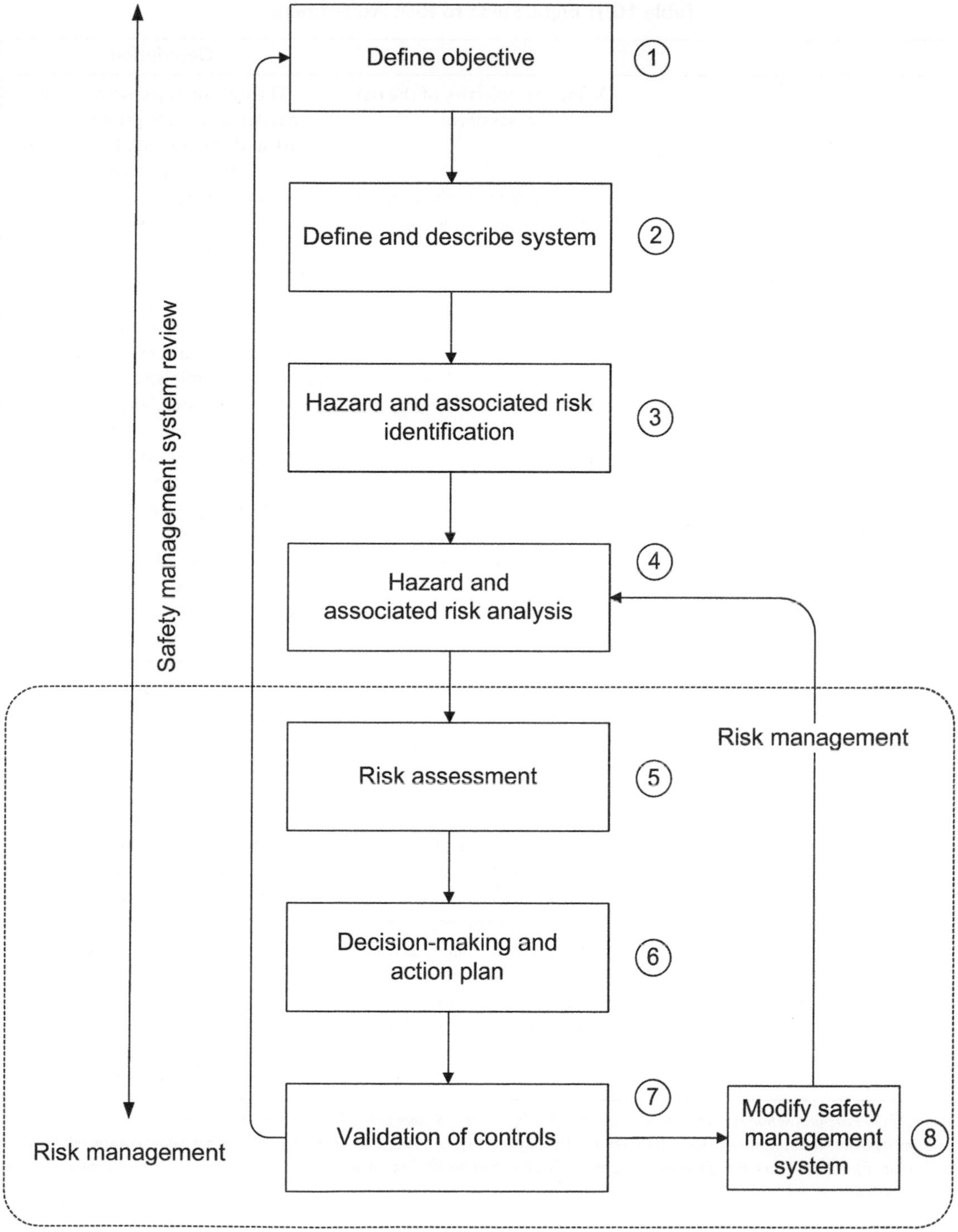

Figure 10.1
Eight Steps to Risk Assessment.

Table 10.1: Eight Steps to Risk Assessment

	Action	Description
1	Define the objective of the risk assessment.	The overall objective of a risk assessment is the prioritizing of risk and the potential impact on the organization.
2	Define and describe the organization or system under review.	An in-depth review of the organization/system elements is developed describing how they interact to create hazards and associated risk: people, tools/equipment/materials, environment, and policies/procedures/administrative.
3	Identify hazards in the organization/system.	Hazards are identified and analyzed. Asking "what do we do and how do we do it" is determined by a comprehensive review of the organization/system.
4	Assess and analyze the risk associated with the hazards.	Both quantitative and qualitative metrics are used to describe the scope of risk associated with specific hazards. An estimate of the probability of a loss-producing event and its potential severity is developed.
5	Prioritize hazards and associated risks by potential impact.	The risk assessment combines the impact of all hazards and risk, comparing them against defined criteria as to what is acceptable.
6	Develop action plans for management of risk.	After being prioritized, review each risk beginning with the highest priority or most severe risk.
7	Evaluate implementation of management methods and validate controls.	Evaluate the effectiveness of the methods used to manage risk.
8	Modify organization/system processes as applicable.	The risk management process continues throughout the life cycle of the organization/system, mission, or activity.

Adapted from Federal Aviation Administration , "Chapter 3—Principles of System Safety", 2000, and "Chapter 4—Safety Assessments Before Investment Decision", 2000 (Principles of System Safety, 2000, chap. 3, Safety assessments before investment decision, 2000, chap. 6). Adapted from Roughton and Crutchfield, 2008 (Roughton & Crutchfield, 2008); Acquisition Risk Management Probability Definitions, n.d.

reasonably practicable and tolerable in the situation being considered" (On Acceptable Risk, 2002). This leaves the acceptance of risk up to the individuals involved. Police, fire, military, mining, deep diving, flying an aircraft, and other occupations are very high risk but have various protections and controls designed to reduce the risk to a level considered acceptable for the task that must be completed.

Management of Risk

Your objective is to become a part of the current organizational approach to overall risk management. You may not be able to become a part of the process initially because of past perceptions and organizational design. However, having a working knowledge of the nature and type of your organizational risk through your own risk assessment may provide opportunities that show that you understand the organization's reasoning behind its risk acceptance.

Example of Risk Acceptance

The ongoing discussion on the use of cell phones while driving is a good example of risk acceptance. Data shows that cell phone use may increase the risk of auto accidents (CDC Adds Cell Phone Data to Annual Behavior Survey, 2011). It has been advocated that cell phone use while driving be illegal. The vast use of cell phones and the overall mindset that they are beneficial if certain controls are used has generated a resistance to a total ban. On-going technology changes may result in an acceptable control of the risk over time.

Many organizations have a risk management department that has a level of understanding about risk and may be developing or have risk assessment procedures. The risk management department, due to its involvement in placing insurance and financing risk, must have a procedure identifying exposures and loss potential.

In our experience, many safety professionals do not have a contact with their risk management department. James was not involved with the risk management departments of the organizations he worked with as safety and risk management were kept separate. Nathan had experiences where the safety and risk management departments were adversaries and caused conflict with different approaches to the control of risk. By developing lines of communication and rapport with the risk manager, the potential for overlap and conflict can be reduced.

Refer to Table 10.2, which offers steps to determine where and how risk is managed in other areas of your organization and actions to take to increase understanding risk.

Table 10.2: Checklist for a Risk Management Strategy

- Determine if your organization has a risk management department and if it follows or uses an enterprise risk management approach (ERM, ISO 31000) ("ISO 31000, risk management–principles and guidelines", 2009).
- Check your network to find the path to the leaders of the risk management program.
- Establish a rapport to the degree possible with the risk management personnel.
- Determine if safety issues are included in the current risk management process.
- Develop an understanding of how the risk management department assesses your organization's risk and the metrics used by that department.
- Develop an understanding of the resources that may be available from the risk management department.
- Develop an understanding of what the risk management lines of communication are and what they have been mandated to do.
- Determine what risk management knows about safety and what value you can bring to their process with the safety management system.
- Develop a mutual sharing of information about risk, issues, and concerns.

Consider a Risk Spectrum

To get a better idea of the nature and types of risk from a high level approach, the following may aid in determining the potential severity of a risk. The term "spectrum" is used depending on the situation as a range of possibilities that can cascade into higher levels of risk.

Risk to the organization's brand—The highest level risk impacts the "brand" of the organization. This is the organizational "killer" that drives customers away or causes a loss in organizational value with regulatory issues, legal liabilities, lawsuits, and general negative media commentaries brought against the organization. Depending on how resilient the organization is, the potential for not surviving a loss of trust in the brand is a possibility.

Risk to operations—This type of risk impacts an operation or facility that is providing a service or product(s). Losses due to this type of risk may stop production or services. It may not harm the brand to any high degree, and the organization should survive if insurance and alternative methods are available to meet customer demands. Any loss does not have to be total but may degrade the ability for that particular location to meet its services or production goals. The organization will probably survive if it has the ability to shift operations to other facilities. If not, then unless insurance or other funds can sustain the organization by buying or quickly rebuilding an operation, the organization will not survive. This potential requires extensive business continuity planning.

The development of a risk map that shows the interrelationships between inflows and outflows of products and services would aid in reviewing whether bottlenecks exist that could harm the entire organization. A risk map is essentially the same as a network map as described in Chapter 3, "Analyzing and Using Your Network". Refer to Figure 10.2 for a "Simplified Risk Map".

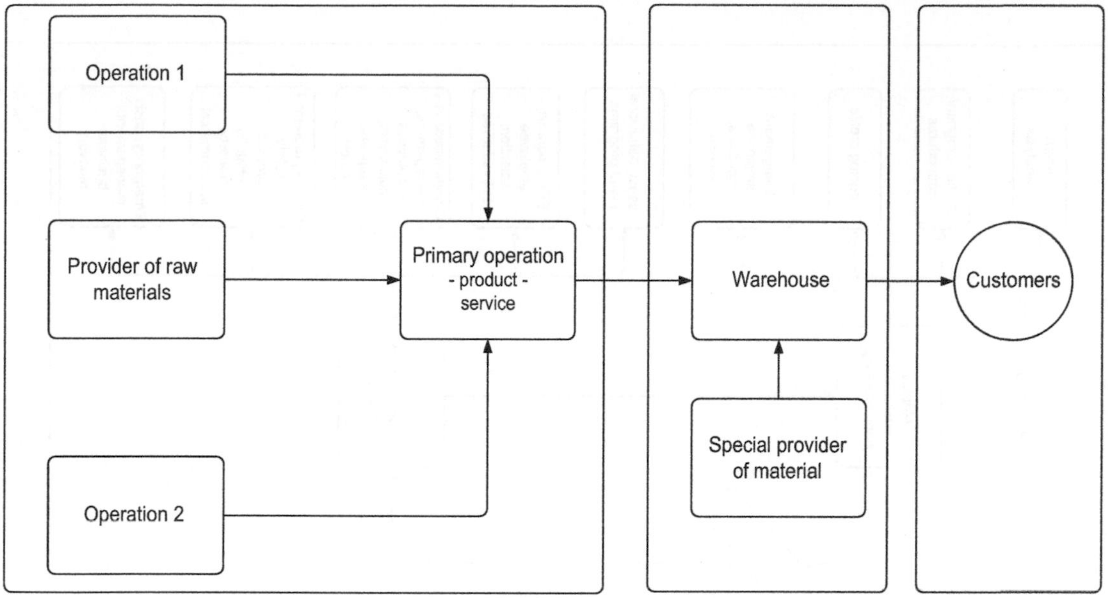

Figure 10.2
Simplified Risk Map.

Risk to nonorganizational entities or third parties—Risk of this nature may not cause damage to facilities and production continues. However, a substantial liability lawsuit can occur if the organization in some fashion causes harm through negligence, poor product design, vehicle incidents environmental spills or releases, etc. Such cases can have a financial impact. Liability insurance provides a level of coverage and normally minimizes the effects on the organization.

Employee loss events—Employee injury is a critical area of risk management and directly impacts the brand of the organization when frequency is high or when tragic severe events occur. Normally these injuries do not shut down operations for any length of time and are handled as medical cases subject to workers' compensation regulations. Generally, operations continue as normal.

Regulatory issues—In most cases, regulatory citations and impact do not shut down operations and do not result in any form of loss unless a fine is issued. Production continues even as regulatory visits are made. In extreme cases, an impact may be made if the organization has failed to reach compliance requirements and appears negligent in meeting standards.

The above order of risk is not a rigid hierarchy as the scope and size of a loss may shift the ranking. A plant explosion may cause employee injuries, harm nearby neighborhoods, cause environmental damage, and involve legal and regulatory claims or citations. Any combination

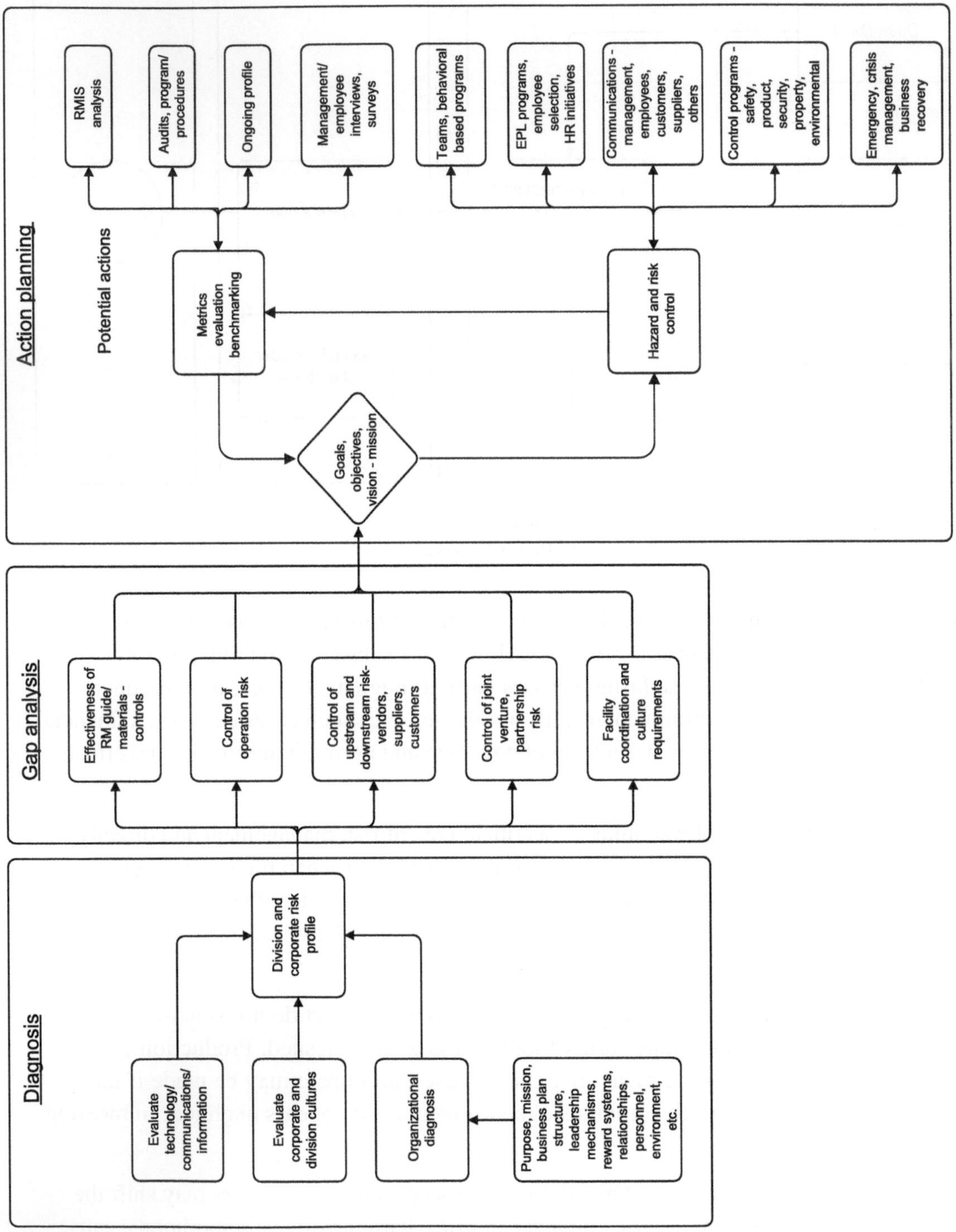

Figure 10.3
An Example Overview of a Risk Management Strategy and Assessment.

of these areas of risk may result in serious impacts and must be considered in the risk assessment and management process. Refer to Figure 10.3 for an "Example Overview of a Risk Management Strategy and Assessment".

The management of risk must be based on a strong safety management system that has been incorporated into the organization as a part of its real value system. Organizations may only know of the risk management from the perspective of ensuring that the organization has the appropriate insurance and finances to cover losses that might occur.

In Chapter 11 we will discuss the activity-based safety system (ABSS) that establishes basic activities that are performed at a specific time and place. The objective of ABSS is to quickly extend the influence of the safety management system throughout the entire organization to enhance and in turn strengthen the safety culture.

Summary

Risk management is an essential element of a strong safety culture.

The shift from a loss-based safety management system to one that is balanced between a focus on loss with a stronger emphasis on risk identification and control requires a fundamental change in perception held about safety.

Risk management establishes the criteria that are necessary for the identification of hazards, assessing the risk inherent within those hazards, and selecting the treatments or methods for the avoidance, control, transfer, or acceptance of risk.

A major issue develops when we try to discuss "risk" within an organization. Risk is defined differently by difference areas of an organization.

A major obstacle we face is that if no loss or major issue has recently occurred, then the mental model forms (a Meme) that the risk is low or maybe nonexistent.

An eight-part approach is used to establish the foundation for the management of risk. While the process can be simply stated, it does require an in-depth and methodical approach in the attempt to bring out hidden hazards and associated risk. The three core parts consist of risk identification, risk assessment, and risk control.

Many organizations have a risk management department that has a level of understanding about risk and may be developing or have risk assessment procedures. The risk management department, due to its involvement in placing insurance and financial risk, must have a procedure identifying exposures and loss potential.

The management of risk must be based on a strong safety management system that has been incorporated into the organization as a part of its real value system.

Chapter Review Questions

1. Define risk.
2. Discuss how definitions of risk may be different in the organization.
3. Discuss the obstacles to risk management.
4. Discuss the eight steps of a risk assessment.
5. Discuss acceptable risk and its potential impact.
6. Discuss the risk spectrum.

Bibliography

A structured approach to Enterprise Risk Management (ERM) & the requirements of ISO 31000. (2010). AIRMIC. Retrieved from http://bit.ly/Zg17Oa.

Accident. (n.d.). Merriam-Webster, Incorporated. Retrieved from http://bit.ly/XTAklU.

Acquisition Risk Management Probability Definitions. (n.d.). Risk Management Toolkit. Retrieved from http://bit.ly/VAZN7L.

Blast at BP Texas Refinery. (March 2005). Donate ProPublica, Journalism in the Public Interest in '05 Foreshadowed Gulf Disaster. Retrieved from http://1.usa.gov/13tv7Y8.

CDC Adds Cell Phone Data to Annual Behavior Survey. (2011). Centers for Disease Control and Prevention. Retrieved from http://1.usa.gov/Z6vqSU.

Deep Water, The Gulf Oil Disaster and the Future of Offshore Drilling. (January 2011). National Commission on the BP Deepwater Horizon Oil Spill and Offshore Drilling. Retrieved from http://1.usa.gov/XeVOw8.

Ethan, G., & Little, D. (July 2011). *Risk—assessing and mitigating to deliver sustainable safety performance*. Professional Safety.

Guidelines for Hazard Identification, Risk Assessment and Risk Control (HIRARC). (2008). Department of Occupational Safety and Health Ministry of Human Resources Malaysia. Retrieved from http://bit.ly/bBXZYW.

Haverin, C. B. (2012). *The short life and tragic end of RMS Titanic*. U.S. Coast. Retrieved from http://bit.ly/133fOAO.

ISO 31000, risk management—principles and guidelines. (2009). International Organization or Standardization. Retrieved from http://bit.ly/Uobg8J.

Linhard, J. (July 2010). OSHA At Forty: Assistant Secretary Michaels Outlines New Challenges and Directions. Mercer, ORC Network. Retrieved from http://bit.ly/VhCr5A.

Lowrance, W. W. (1976). Of Acceptable Risk: Science and the Determination of Safety.

Occupational health and safety systems. (2012). The American Industrial Hygiene Association. ANSI/AIHA Z10.

On Acceptable Risk. (January 2002). EHS Today. Retrieved from http://bit.ly/Z36Yma.

Principles of system safety. (2000). Federal Aviation Administration (FAA). Public Domain. Retrieved from http://1.usa.gov/YiEZvJ.

Quadrant of Risk: Hazard, Operational, Financial, and Strategic. (n.d.). ERM Strategies, LLC. Retrieved from http://bit.ly/YE0Spv.

Ropeik, D., & Gray, G. M. (2002). *Risk: A practical guide for deciding what's really safe and what's really dangerous in the world around you*. Boston, Mass: Houghton Mifflin Company.

Roughton, J., & Crutchfield, N. (2008). Job hazard analysis: a guide for voluntary compliance and beyond. In *Chemical, petrochemical & process*. Elsevier/Butterworth-Heinemann. Retrieved from http://amzn.to/VrSAq5.

Safety assessments before investment decision. (2000). Federal Aviation Administration (FAA). Public Domain. Retrieved from http://1.usa.gov/WSjtCr.

Successful health and safety management, hsg65. (1997). Health and Safety Executive, Crown Publishing: Permission to Reprint. Retrieved from http://bit.ly/YKDzRk.

Volume 1: Concepts and Principles, Human Performance Improvement Handbook, Public Domain. (2009). U.S. Department of Energy. Retrieved from http://1.usa.gov/WdIqoP.

CHAPTER 11

Developing an Activity-Based Safety System

Progress is impossible without change, and those who cannot change their minds cannot change anything.
—George Bernard Shaw

Introduction

In the ideal safety culture, the leadership team is constantly communicating and emphasizing the vision, goals, and objectives it believes are required for the organization to be successful. Refer to Chapter 4, "Setting the Direction for the Safety Culture". Therefore, a safety process is best implemented using a systematic approach that focuses efforts on key essential activities to drive improvement of communication, rapid feedback of issues and concerns, and improved coordination of safety activities. Refer to Figure 11.4, "Tailoring Safety and Health Metrics to Your Organization" and Figure 11.5, "Continuous Improvement".

If implemented correctly, the activity-based safety system (ABSS) will not be the "program of the month" but rather a continuous routine that is consistently followed.

The objective of this chapter is to introduce you to a basic ABSS that uses group and individual communication to help create awareness in a short period of time. This awareness is created through specific activities that are consistently and routinely performed on a daily, weekly, and monthly basis with no exceptions.

ABSS provides a framework that can help to quickly involve all levels of an organization by using the strengths of your internal network.

On completion of this chapter you will be able to:

- Discuss ways to communicate with specific levels of the organization.
- Define daily shift reviews and weekly and monthly meetings.
- Discuss the importance of one-on-one discussions with employees.
- Discuss the benefits of safety walk-through tours.
- Discuss the implementation of a machine/equipment-specific safety checklist.
- Define how to use a follow-up team to correct safety-related issues.
- Define the roles of the various levels of the leadership team.

Activity-Based Safety System

The ABSS approach leverages your internal network by establishing basic activities that must be consistently performed at a specific time and place. As activities are performed, they are tracked and reviewed to provide the ongoing development of real-time safety-related performance measurements. The objective of ABSS is to extend the influence of the safety process quickly throughout the entire organization to enhance safety awareness and in turn strengthen the safety culture. ABSS is intended to establish communication links that are consistently applied so that concerns and issues can be rapidly communicated between levels of your organization. This network increases the probability that hazards and risk-related issues are addressed without delay. Refer Chapter 3, "Analyzing and Using Your Network" for more details on how to increase and improve your network. Refer to Figure 11.1 for an "Activity-Based Management System Overview".

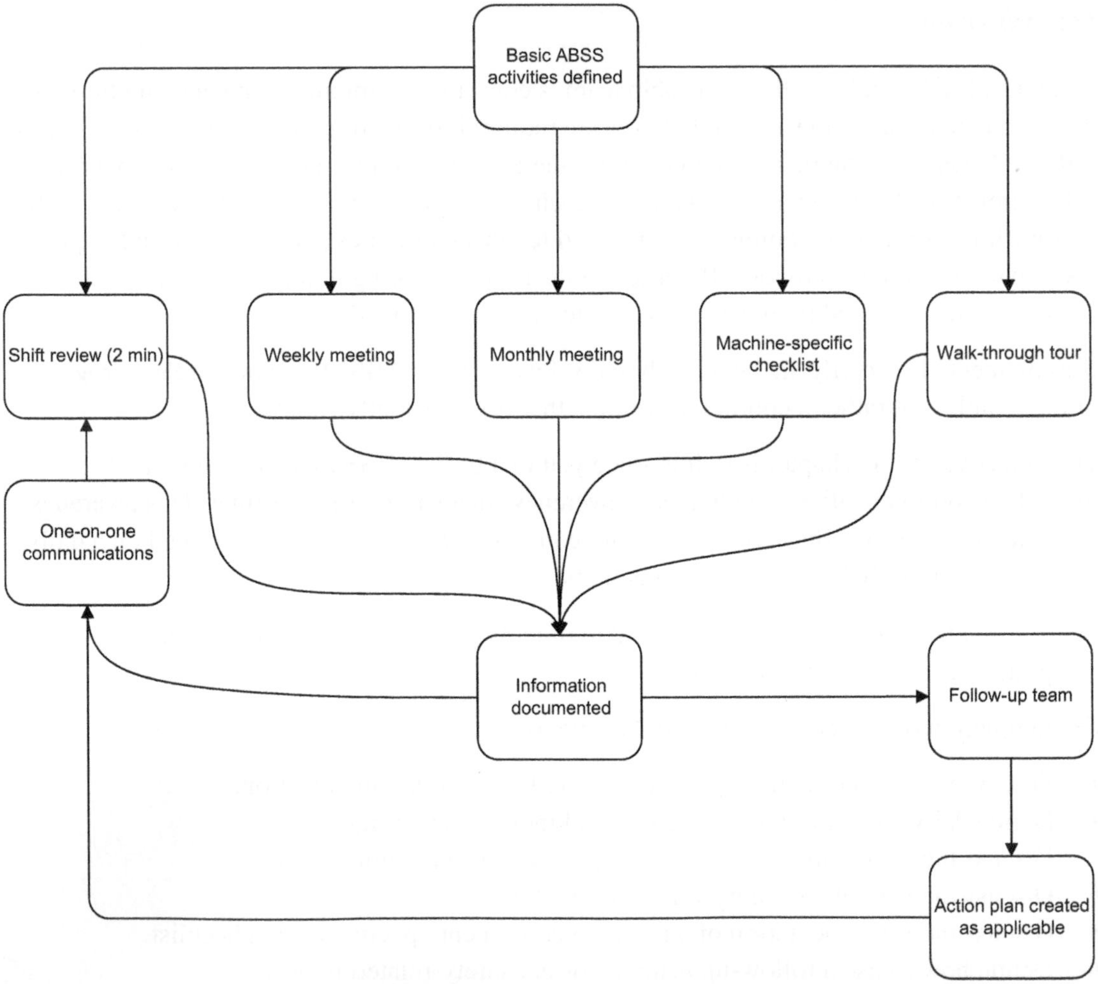

Figure 11.1
Activity-Based Management System Overview.

To get started, what you will need to do is take a look at existing safety management system-related activities and decide which activities add value to the process and which activities currently do not add value. Once all existing safety activities are identified, the decision can be made to remove, modify, and/or replace the nonvalued activities with new activities that will add value to the safety management system. Refer to Appendix O, Sample Activities and Results Measurements.

> When introducing a new concept such as ABSS, expect initial push back on the implementation. Begin by mapping out the current state of what is being required of leadership and determine what is really of value and what is being done based on traditional add-ons that have lost significance or benefit. If an activity does not add value (i.e., reduces and/or controls hazards or associated risk and improves safety awareness), eliminate that activity and replace it with an activity that brings value to the safety culture.

The key intent when implementing ABSS is not to create additional work but to complement specific safety-related efforts that may already be in place. All safety activities must be integrated with existing program elements to systematically formalize the communication process throughout the organization.

Advantage of Using ABSS

The advantage of the ABSS format is that it uses each supervisor's experience and areas of responsibilities to build safety-related information for discussion with employees. ABSS provides a method that allows supervision to go beyond potential and visible issues such as personal protective equipment (PPE). When fully implemented, supervisors will be able to develop their own area-specific topics to which their employees can relate. The key is that the supervisor is not alone, as each topic is developed with guidance from the safety manager or other resources.

> PPE is easy to see and control. By focusing on PPE, other more important hazards and associated risk may be overlooked. We are not saying that PPE is not an issue but that there are many other issues one must deal with prior to requiring PPE.

How ABSS Works

Many organizations already use some form of meetings and other various methods to convey their safety message. Unfortunately, some supervisors are not always knowledgeable about how to develop and present their own safety topics. So, to save time, many organizations will use a canned approach by using predetermined topics that are to be used at safety meetings. While these types of programs may be beneficial, this canned approach restricts supervision from using its own creativity and expertise. The problem is that a standardized script does not always take into account the nature of the work activity and is sometimes used as a "one size fits all approach".

> *Training Must Match the Application:*
>
> *Training packages that include a written script, and sometimes a video, that are designed to solve safety problems are plentiful. In many cases, the trainer only conveys what is provided, reading a prepared script that may or may not apply to the specific working conditions. As a result, trainees recognize the information is not relevant and imparting the information becomes more difficult than it needs to be.*
>
> **(Safety culture in Nuclear Installations, 2002)**

The ABSS system and its basic program elements move past the canned approach allowing supervision to start to use its experience and creativity. ABSS consists of the following basic elements:

- Employee meetings (daily preshift reviews, weekly meetings, and monthly meetings).
- One-on-one discussions with employees.
- Area walk-through tours (housekeeping, hunting for hazards, etc.).
- Machine/equipment-specific checklists.
- Follow-up teams.
- Performance metrics.

This is a communication system that begins the process of discussing and identifying immediate hazards and associated risk concerning area-specific safety issues that can be discussed on a daily basis. Refer to Chapter 9, "Risk Perception—Defining How to Identify Personal Responsibility".

> *...activities that comprise people's jobs must then be consistently in alignment with the targeted results. This mindset can become part of the culture only if people understand the results they are supposed to achieve in the job they are to perform.*
>
> **(Connors & Smith, 2004)**

Safety Meetings

A true safety culture provides a clear message that communication with all employees has a defined role to play in the safety management system. All employees should have an understanding of the requirements and importance of safety-related issues in their organization. For an organization to be successful in sustaining a safety culture, all areas of the organization should conduct specific types of safety-related meetings. The perception must be removed that safety meetings and information applies to "others but not me".

As discussed in the Chapter 3, "Analyzing and Using Your Network", your organization is essentially a network of individuals. Each work team or area of responsibility is more than just a box on an organizational chart. It is a network of individuals whose behavior is influenced by person-to-person and face-to-face interactions that can be influenced by the behavior and communications of everyone in the organization. The nature of this complex network and how it moves information through all levels of the organization is critical to the safety management system.

In using the "Three Degrees of Influence Rule" defined by Christakis and Fowler in their book *Connected: The Surprising Power of Our Social Networks and How They Shape Our Lives*, they found the following:

> *Everything we do or say ripples through our network, having impact on our friends (one degree), our friend's friend (two degrees), and the friend's friend's friend (three degrees). This rule applies to a broad range of attitudes, feelings, and behaviors.*
>
> **(Christakis & Fowler, 2009)**

Safety meetings can be used to affect the "three degrees of influence". The three types of safety meetings that have been found to be the most effective are daily preshift reviews, weekly meetings, and monthly meetings. These meetings can range from a free-form format to agenda driven and are used for maintaining general safety awareness directly with each work group, presenting specific topics for discussion, and providing safety-specific information and required regulatory compliance training.

Daily Preshift Review

The daily preshift review is a short, simple meeting that is held with each work group by its direct supervisor. As the name implies, this meeting is to be short, only long enough to ensure that all those involved receive the required safety-related topic of the day. This meeting can be called a number of things, two-minute drills, tool box talks, desktop, rally point, focus point, stand-down, etc. Whatever you choose to name these meetings, the name must be consistently used so that it becomes of symbolic importance to all employees.

It is important that this meeting be held preshift or prework assignment before the work day starts, be held every day, and becomes a daily habit, held at the same time and in the same area without fail. This daily preshift meeting sets the stage for success before the day begins. Everyone begins on the same page with the same message. It is a critical part of the safety process as it directly works to extend the safety influence of the supervisor through his or her immediate network.

The key here is that you want all employees to have a fresh mind and be engaged in listening to the message. Other meetings held later in the day are not as effective, as everyone has other work issues on their minds and tends not to retain the message.

These meetings become more important during a crisis or when you are under production pressures, as they provide a balance to a mindset that may desire to drop "safety" during the crisis. It may help in overcoming the overwhelming desire to take chances in the attempt to get back to a normal situation.

Not only is the supervisor communicating with those in the immediate meeting, but given the nature of networks, a magnifying effect occurs as each of the employees in turn influences and communicates through his or her own personal network. As employees talk to each other in break rooms, at lunch, friends off the job, etc. the message continues to travel throughout the organization.

> At the end of the preshift review, you can simply ask one question to get employees thinking: "Is there anything that will prevent us from working safely today?"
>
> Caution: All problems cannot be solved in the brief time of the preshift review. This can be accomplished in the one-on-one discussions presented later.

Lesson Learned # 1

One of the authors was attending safety meetings where safety scripts were being used. A supervisor was presenting to the group the topic of driving a fork truck and all of the rules that applied. However, the employees in the group were not fork truck drivers, they were machine operators. In another meeting, a supervisor was presenting a canned topic on snow chain usage and how to drive in the snow. The problem was that the plant was in a southern climate where snow is rare. In both of these meetings the employees, as a result, did not get much useful content about specific safety-related information and generally found the safety message to be a waste of time. In these cases, the safety culture is driven in the wrong direction as the perception becomes that safety meetings and content are not of value.

As discussed, the daily preshift meeting must become a habit to be conducted daily at a specific time and place, with no exceptions. In *Great by Choice*, Collins and Hansen call this the "20 Mile March". In their words, "It's about having a concrete, clear, intelligent and rigorously pursued performance mechanism that keeps you on track" (Collins & Hansen, 2011).

Multishift Operation

One thing that is often forgotten is the multishift operation. It is important that the transfer of information and knowledge between shifts is conducted. Prior to the preshift review, the supervisor leaving the previous shift should ensure that safety or operational issues that may have occurred during the previous shift are provided to the oncoming shift. This may be verbal or a brief report, email, or other communication. This should include concerns expressed by operators, staff-related issues, status of equipment, reasons equipment has been "red-tagged" or taken off line, and any abnormal conditions, problems, or issues that remain for the next shift.

One key point here is that these preshift reviews are conducted every work day, including the day that you conduct the weekly meetings and monthly meetings that will be discussed in the next sections.

Weekly Meetings

Regular weekly meetings must also be conducted at a specific time and place. This meeting is more formal than the preshift reviews and sets the tone for the week's safety agenda and brings closure to the past week's efforts.

A typical weekly safety meeting could last about 30 minutes or until all of the safety-related discussions have been addressed and completed. Due to the nature of the meeting, it may require more time to discuss safety-related topics, new issues, injuries in other areas, facilities with similar exposures, or other changes in the organization. Other concerns can be put on the table such as squelching rumors about injuries that have occurred, status of hazard correction, etc.

In addition, this weekly meeting can be used to recap and summarize the discussions from the daily preshift reviews in more detail to ensure that no obstacles or miscommunications are in the way of moving toward and enhancing a positive safety culture.

Based on the weekly safety meeting, you may identify problems or issues. The leadership team must prioritize each issue and develop an action plan. The action plan will outline the specific actions that will be taken, the person(s) responsible, and the completion date.

> By using the previous month's insights drawn from the ABSS, the accuracy and completeness of safety action plans can be assessed and reinforced.

One important point is that there are three weekly meetings. On the fourth week, you will conduct a monthly meeting as discuss in the next section.

Monthly Meetings

The regular monthly meeting provides a platform for in-depth training and/or refresher on regulatory compliance and/or organizational policies and procedures, protocols, guidelines, etc. This meeting is more formal than the weekly meetings and is usually one hour in length or until all of the safety activities have been completed.

The intent of this monthly meeting is to bring together a larger group, which provides a way of sharing a diverse set of ideas and perceptions. The important point is that these monthly meetings should not be held "just to have a meeting". There must be a purpose for each meeting and an agenda developed. This meeting could include outside speakers, members of the leadership team to provide an overview of the company performance, etc.

One-on-One Discussions with Employees

Supervision can easily drift into a traditional adversarial mode as supervisors typically only look for errors or production and service-related issues. In some cases, supervision may believe requesting or asking for feedback is a threat to its authority.

> According to George Robotham (as cited in *Health and Safety Risk Management*), "Research by Harvard professor T.J. Larkin suggests when communicating change with the workforce use the supervisor not senior management, use face to face communications and frame communications relevant to the immediate work area and processes"
>
> **(Robotham, n.d.)**

To address this type of issue, the one-on-one discussion is designed to be used as a direct face-to-face discussion with real-time feedback between the supervisor and each employee in the organization. Many employees will not speak up in meetings due to shyness, fear, mistrust, or other reasoning. By using a direct approach, employees can openly discuss with their supervisor the contents of safety meetings and any other concerns that they would not bring up in front of their peers.

> **Face-to-Face Discussion**
>
> Opportunities for employees to have face-to-face discussion support other communication activities and enable them to make a more personal contribution.
>
> Tours and formal consultation meetings are options but others include:
>
> - Planned meetings (or team briefings) at which information can be cascaded. These can include targeting particular groups of workers for safety critical tasks;
> - Health and safety issues on the agenda at all routine management meetings (possibly as the first item); monthly or weekly 'tool-box' talks or 'tailgate' meetings at which supervisors can discuss health and safety issues with their teams, remind them of critical risks and precautions, and supplement the organization's training effort.
>
> These also provide opportunities for employees to make their own suggestions (perhaps by 'brainstorming') about improving health and safety arrangements.
>
> **(Successful Health and Safety Management, hsg65, 1997)**

These discussions are not intended to trigger an immediate negative reaction or response from the supervisor. They are routine short discussions centered on safety concerns. These discussions must be conducted in the employee's work environment where the employee feels more comfortable, not in the supervisor's office.

In many cases, "we hear what the employee is saying, but we are really not listening to what the employee has to say". Supervisor listening skills are critical.

Lesson Learned # 2

James remembers a discussion with one of his managers when he was trying to provide information to the manager. Every time that he would say something, the manager would state the exact opposite of what he was trying to convey.

He finally made a bold statement to his manager, trying to get both of them on the same page. "You are listening to what I am saying, but you are not hearing me." After this was stated, the manager looked at him and stated: "I had not thought about it this way and maybe you are right." After he made this statement, the conversation got back on the right track and the issues were resolved.

It is critical that you listen and respond to your employee based on what is being discussed. You can use the same concept as you do when investigating a loss-producing event by listening to the employee(s) 80% or more of the time, without interruption, and talking to the employee(s) 20% or less of the time.

Supervision must show a sincere interest about what the employee has to say. Supervision may initially find this hard to do; however, it may even be harder for the employee to speak his or her mind if there is no trust or fear of retribution.

Safety Walk-through Tour

Safety inspections, observations, and walk-throughs are a core feature of the safety management program elements. Unfortunately, these walk-through tours can evolve or be perceived as a negative, fault-finding activity that becomes adversarial between the leadership team and employees.

Worksite inspections have been stressed as a core element of management for many years as suggested by Peters and Waterman in their *In Search of Excellence*, as "management by wandering around". They stressed that management must routinely get away from their office and desks and wander around reconnecting with the workplace (Peters & Waterman, 2004).

This can sometime be referred as "Boots on the Ground".

Blue Blood, TV show

The safety walk-through is an essential component of ABSS and is conducted as a joint effort by a diverse team that can consist of supervision, senior leadership, and employees. The objective is to have a group that brings a variety of perceptions, backgrounds, expertise, and different points of view about the work environment. The participation of a member of the leadership team is to provide awareness and to provide a presence to all employees that this effort is considered important and a value to the organization.

By varying the time of the safety walk-through, the group can meet and talk to different employees and see the work environment under different conditions. In multishift operation, make certain that all shifts have a walk-through including weekends.

The safety walk-through is completed once per month and at designated times built into the planning function. The safety walk-through is completed during the normal workday or shift.

This walk-through is typically no more than one hour in length and is not to be treated as a formal inspection, rather more of an awareness and general observation tool that allows the group to interact with employees to identify specific concerns about hazards.

At the conclusion of this walk-through, any safety issues noted will require a risk assessment and a safety work order created for the issues to be resolved. Refer to Chapter 9, "Risk Perception—Defining How to Identify Personal Responsibility".

Machine/Equipment-Specific Checklist

The next component of ABSS is a machine/equipment-specific checklist. This is a checklist that is used to focus on specific identified safety issues for each piece of machinery and/or equipment in the organization. This checklist is developed and updated by the employees and supervisor who works with and around the specific equipment with the input from the safety staff. The key to this list is not to make it overly detailed. We suggest that no more than 15 of the most important items are listed on this checklist, which may cover specific inherent hazards of a machine or equipment. Refer to Appendix P, "Operator General Observations and Machine/Equipment-Specific Daily Inspection Checklist". For an overview, refer to Figure 11.2, "Sample General Observations and Machine/Equipment-Specific Daily Inspection Checklist".

The checklist is not intended to be a maintenance checklist; as it is to be a quick reference to ensure that the primary hazards are reviewed and controlled.

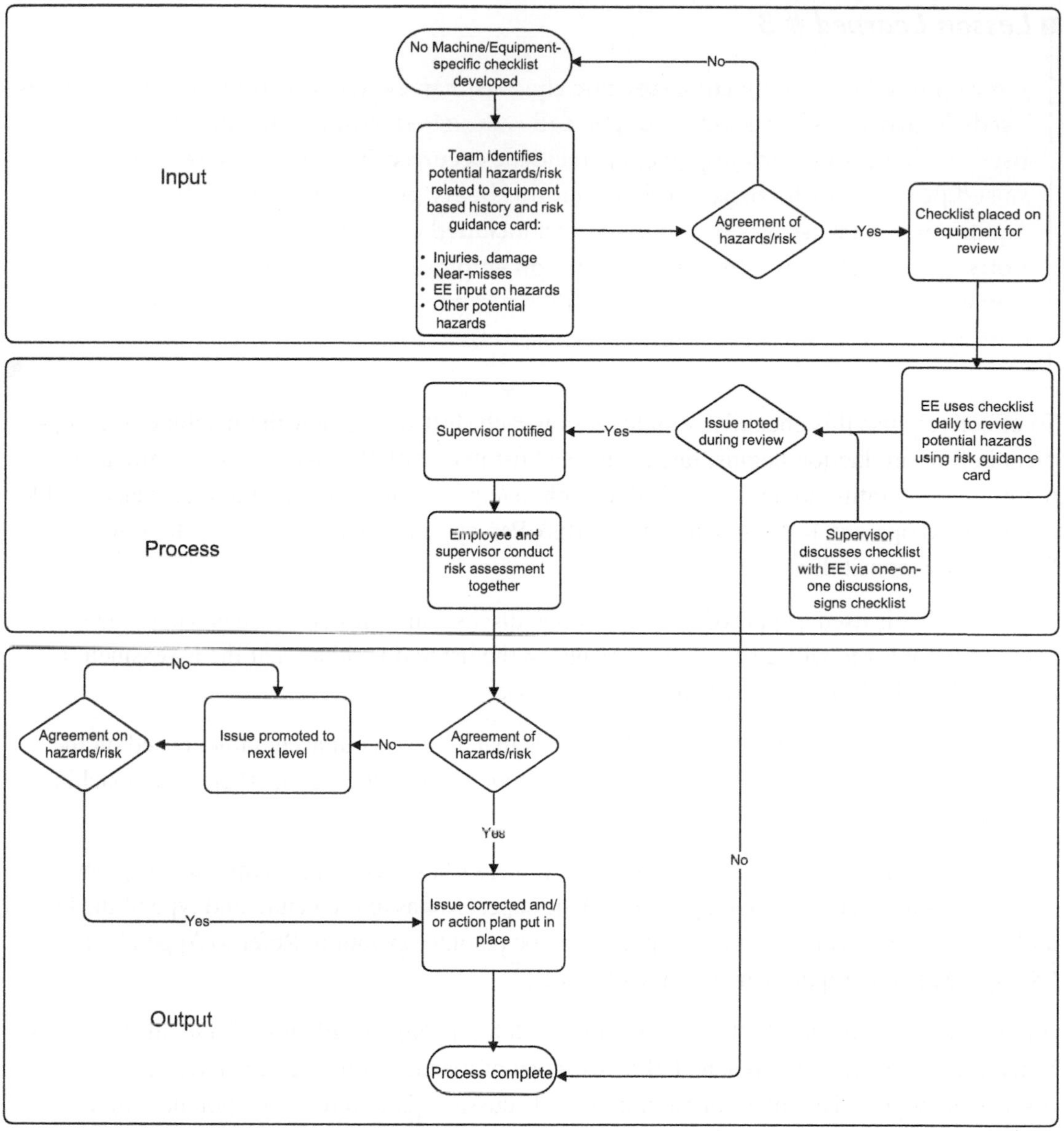

Figure 11.2
Sample General Observations and Machine/Equipment-Specific Daily Inspection Checklist.

Once the checklist format and content are completed and agreed upon, it is physically placed on the equipment so that it can be reviewed on a daily basis by the operator and used in the one-on-one discussion with the supervisor. The checklist is not static and is considered a living document as additional information is received.

Lesson Learned # 3

James noted how the machine-specific checklist can evolve over time. A load turner is used to turn a six-foot stack of paper and is designed with a 300-pound plate used to clamp the stack in place. An incident occurred in a plant where, while in a raised position, bolts broke and caused this heavy plate to fall. A machine/equipment checklist was put in place and included checking for missing or broken bolts. The checklist was developed through employee discussions, and uncovered a number of other safety concerns and unknown hazards that has never been addressed and were added to the checklist.

To effectively use this checklist, employees on each shift will review the machine or equipment before production begins, rate each checklist item with the risk guidance card, and record the risk rating to determine if the machine/equipment is in acceptable or unacceptable condition to operate. Refer to Chapter 9, "Risk Perception—Defining How to Identify Personal Responsibility".

This checklist is used as part of the one-on-one discussions with employees where the supervisor and employee will discuss all elements on the list and validate that the assessment is acceptable to both the employee and the supervisor.

If the employee who reviews the checklist finds that an item is unacceptable, based on the assigned risk the supervisor is contacted and a review is conducted together at the machine or equipment.

In reviewing the checklist as described above, if there is a safety issue identified, an agreement must be reached between the employee and the supervisor as to the nature and type of the hazard and the potential risk before the equipment can be put into operation. Refer to Appendix P, "Sample Machine/Equipment-Specific Checklist".

As discussed in Chapter 9, "Risk Perception—Defining How to Identify Personal Responsibility", if no agreement is reached, the next level of management is contacted and a second discussion begins. The intent of having these discussions is to reduce the potential and probability that employees will continue to operate machinery or equipment that is not functioning as intended and believe they have no alternative but to use the machinery or equipment in an unsafe condition.

> An effective safety culture reduces the acceptance of unnecessary risk. The machine/equipment checklist provides an immediate discussion and conduit for correction.

Follow-up Team

A fully functioning follow-up team is an important function that ensures that all safety-related work order requests are addressed and completed in a timely manner. This team should consist of a maintenance technician, several employees, and a supervisor, depending on the number of work orders that are under review. Refer to Chapter 8, "Getting Your Employees Involved in the Safety Management System".

Lesson Learned # 4

James remembers one situation that was used to motivate employees in assessing and correcting hazards. A team was created to assess and identify hazards in the workplace. This team (comprised of one manager, supervisor, maintenance technician, and production employees) was provided a budget of $10,000 to fix any safety-related issues that were identified. The objective was to talk to employees about potential safety hazards, establish a priority, and then correct as many issues as they could with their allotted funds.

The reporting employee was part of the discovery and kept in the loop until the hazards had been corrected based on the input from that employee. If the team did not feel that there was a hazard, it would be discussed with the employee and, as a team, a decision was made on what direction further activities would go.

The key to the process was to discuss the issue with the reporting employee as to whether it was a real hazard or associated risk. The reporting employee had the opportunity to either agree or disagree with the analysis. If there was an agreement, then the work order was closed and everyone would sign off on that agreement. If there was no agreement, the issue would be pursued until an agreement was reached. This program was a great success because only real hazards and associated risk were identified and corrected.

In the past, unsupported complaints about safety were used to have changes made even when there was not a true hazard or associated risk. After this process began, safety was no longer used as a crutch just to get something done.

The above lesson learned demonstrates that the key to employee involvement is to promote awareness, instill an understanding of the comprehensive nature of a safety management system, and allow employees to "own" the system. The key is ensuring to provide timely and appropriate feedback as to why something was either corrected or not corrected and what actions are to be taken if a valid issue exists.

The follow-up team provides an unbiased review of each safety-related work order created, scheduled, and completed within a designated time period or that remains outstanding. The team is charged with reviewing all safety-related work orders, using the risk guidance card to establish a priority for corrective action or moving the request to the next level of decision-makers. Refer to Chapter 9, "Risk Perception—Defining How to Identify Personal Responsibility".

The team is designed to break down any barriers that may exist between production, maintenance, employees, and the leadership team. To close this potential gap, the team uses a peer-to-peer approach to ensure that the safety-related hazard and associated risk is assessed. The intent is to involve the employee, ensure that the work order is valid, and that the hazard and associated risk is valid and needs to be corrected. When the work order is completed, a member of the follow-up team discusses the issue/conditions with the employee who requested a hazardous issue be corrected. If the employee agrees with the correction, then the work order is signed off and closed.

The follow-up team should have a preapproved budget as mentioned in Lesson Learned #4 that can be used for direct corrective actions without approval. This budget must be used responsibly and is subject to audit.

The follow-up team can present the status of safety-related work orders at weekly/monthly meetings. The status is posted both electronically and hard copy for all employees to review.

ABSS Roles and Responsibilities Defined

Refer to Figure 11.3 for the frequency and responsibility for safety-related activities.

The following is a basic overview of the roles and responsibilities of each level of the organization and can be modified as you see necessary in your organization.

Supervisor/Superintendent

The supervisor/superintendent's role is the key in presenting a common and consistent message to his or her area of responsibility. The supervisor documents the various activities of the ABSS process in an safety activity report. Refer to Appendix Q, "Sample Manger/ Supervisor Daily/Weekly/Monthly Safety Activity Report".

Supervisors participate each week in the following safety activities:

- Develop and present daily shift reviews.
- One-on-one discussions with employees.
- Reviewing machine/equipment-specific safety checklists.
- Coordinate with the follow-up team to correct safety-related issues.

They participate each month in the following safety activity:

- Safety walk-through tours.

Middle Management

Departmental management and other mid-level management should participate in safety-related activities for a minimum number of hours per month, as determined by the leadership team. This involvement will reinforce the safety culture and promote individual and department commitment to safety. These activities help create a positive safety environment and demonstrate leadership commitment to safety.

They participate each month in the following safety activities:

- Attend at least one daily shift review.
- Contact a defined percentage of employees for one-on-one discussions.

They participate quarterly in the following safety activities:

- Reviewing machine/equipment-specific safety checklists with operators.
- Safety walk-through tours.

Upper Management

General management should formally participate in or conduct safety walk-through/housekeeping tours at a minimum of one time per quarter. All areas within his or her safety responsibility should be covered in their entirety annually.

They participate each month in the following safety activities:

- Attend at least one daily shift review.
- Contact a percentage of employees to conduct one-on-one discussions.

They routinely participate in the following safety activities:

- Reviewing machine/equipment-specific safety checklists with operators.
- Safety walk-through tours.

Senior Management

Divisional management should periodically be visible by:

- Attendance at selected daily shift review.
- Conducting "roundtables" with selected employees.
- Periodic formal safety walk-through tours.

In summary, senior management should have an open door policy, conduct round table meetings with selected employees and make periodic unannounced visits to selected areas to meet one-on-one with employees.

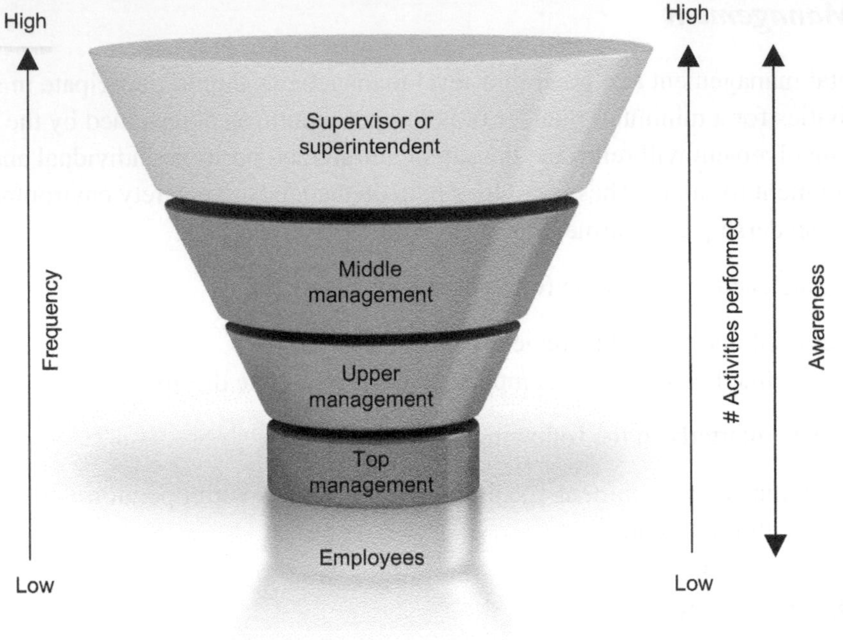

Figure 11.3
Frequency of Activities Performed by Level. *Source: Graphics provided with permission by* presentermedia.com.

Site Safety Professionals

Safety professionals are expected to participate each week in the following safety activities:

- Assist supervision in developing daily shift reviews.
- Conduct random one-on-one discussions with employees.
- Conduct random reviews of machine/equipment-specific safety checklists.
- Coordinate with the follow-up team to correct safety-related issues.

They are expected to participate each month in the following safety activity:

- Assist supervision in safety walk-through tours.

The site safety professional will provide support by providing safety-related materials, suggestions, and guidance to the leadership team tasked with achieving the desired results of selected activities. The safety professional in this role does not conduct meetings. The main responsibility is to be directly working with and supporting the leadership team, side-by-side assisting, advising, guiding, and supporting the team.

Refer to Figure 11.3, "Activities Performed by Level".

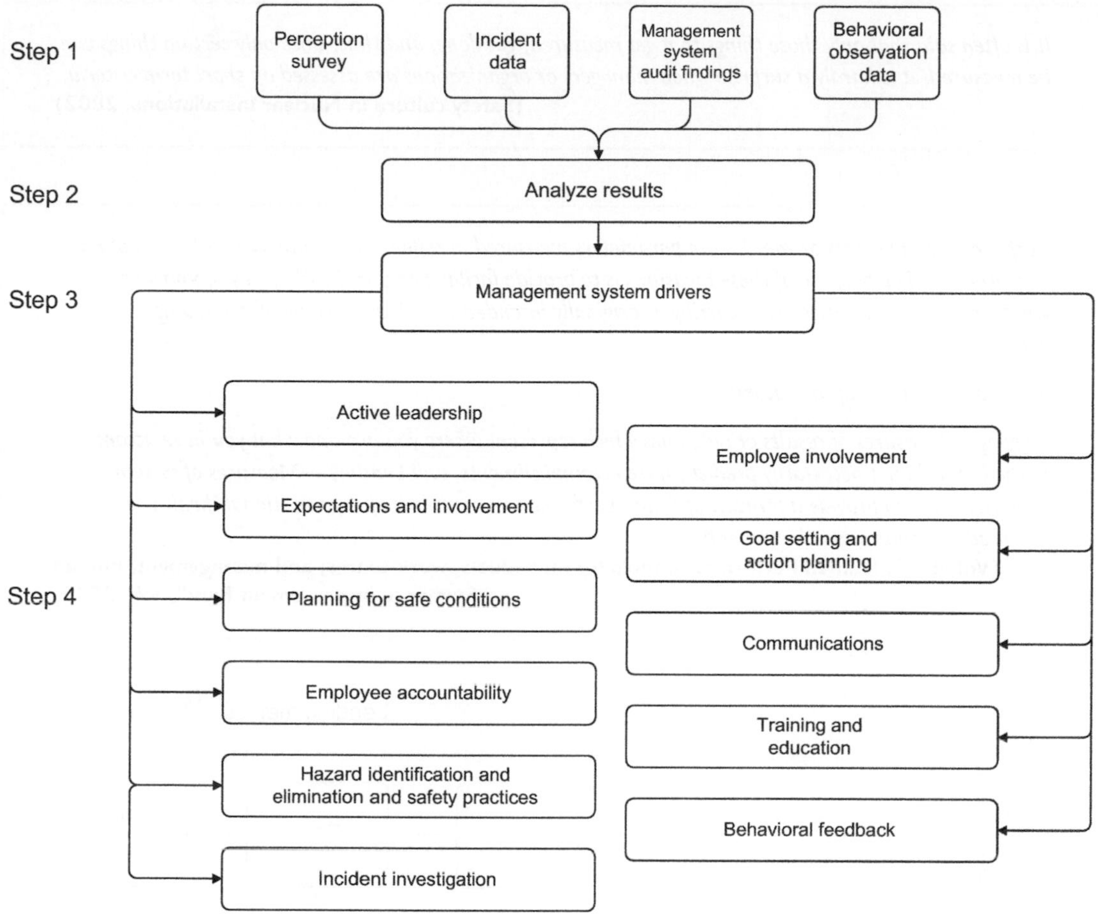

Figure 11.4
Tailoring Safety and Health Metrics to Your Organization. *Source: A New paradigm for safety and health metrics: safety and health metrics: framework, tools, applications, framework, tools, applications, and opportunities, Steve Newell, reproduced with permission.*

Measuring the Success of ABSS

Safety performance metrics and safety management system elements are of no benefit unless they are clearly and routinely communicated throughout the organization.

> *The majority of measures should be leading indicators that focus on proactive activities on the part of all employees—measures that track what people are doing daily to prevent accidents. With such measures in place, immediate and certain consequences can be engineered in to ensure those activities occur.*
>
> **(Judy Agnew & Aubrey Daniels, 2011)**

It is often said that only those things that get measured, get done; and that, since only certain things can be measured, it is hardly a surprise when managers or organizations are assessed on short term criteria.

(Safety culture in Nuclear Installations, 2002)

Performance indicators or metrics are parameters measured to reflect the critical success factors of an organization. The purpose of these measures is to provide facility personnel with a way of knowing whether planned activities are occurring as originally intended as well as warning of developing problems.

There are two types of indicators:

Lagging—Measures of results or outcomes which represent where you are and what you have accomplished, but do not necessarily predict future accomplishments, and Leading—Measures of system conditions, which provide a forecast of future performance; measures of organizational 'health', which can predict results and achievements.

(Volume 2: Human Performance tools for individuals, work teams, and management, human performance improvement handbook, 2009)

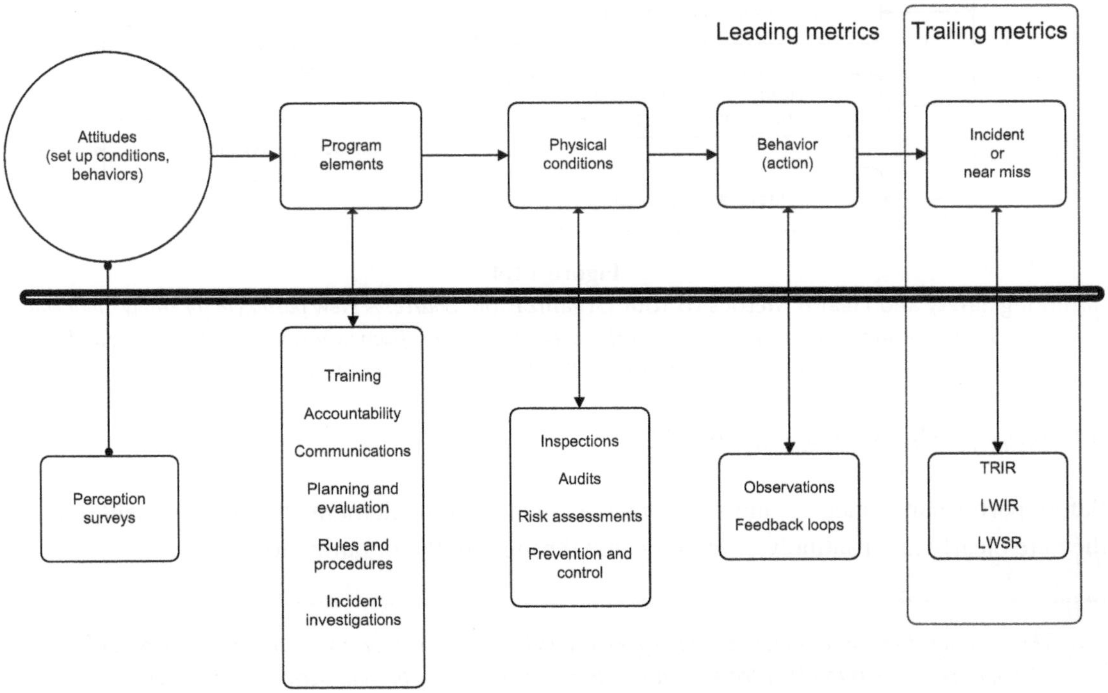

Figure 11.5

Continuous Improvement. *Source: A New paradigm for safety and health metrics: safety and health metrics: framework, tools, applications, framework, tools, applications, and opportunities, Steve Newell, reproduced with permission.*

Basic Tips for Using ABSS

- Be a good listener. Pay attention to employees. In many cases, we hear an employee talking but we do not listen to what he or she says, as many times we may knowingly or unknowingly have our own hidden agenda. One individual does not have all of the answers, and involving the person who directly performs the job can provide insights on what works and does not work.
- Make time for employees. Regular, one-on-one meetings with each employee are important. This is not to be a formal human resources process of personnel review. The goal is to reach out to those employees who best know the workplace and its ongoing potential for change.
- Give employees your full attention with no distractions (i.e., take the time to discuss safety, quality, etc. and nothing else). No phone calls, texting, etc.
- Keep the message consistent about the value of the safety process. Knowing who you are, and what you stand for, helps employees make better decisions on their own (or at least more informed decisions). If mixed messages have been sent concerning safety-related issues, a one-on-one can correct the communication.
- Give regular positive feedback. A tendency is to only approach an employee when something has gone wrong (i.e., "Why is the machine down?" "Why did you get hurt?" "Did you not know better?") Supervision may be hesitant to talk one-on-one to an employee about safety. Employees are valuable problem solvers because they are closest to the action (Roughton & Crutchfield, 2010).

Summary

In the ideal safety culture, the leadership team is constantly communicating and emphasizing the vision, goals, and objectives it believes are required for the organization to be successful.

The ABSS approach leverages your internal network by establishing basic activities that must be consistently performed at a specific time and place. As activities are performed, they are tracked and reviewed to provide the ongoing development of real-time safety-related performance measurements. The objective of ABSS is to extend the influence of the safety process quickly throughout the entire organization to enhance safety awareness and in turn strengthen the safety culture.

The advantage of the ABSS format is that it uses each supervisor's experience and areas of responsibilities to build safety-related information for discussion with employees.

ABSS consists of the following basic elements:

- Employee meetings (daily preshift reviews, weekly meetings, monthly meetings).
- One-on-one discussions with employees.
- Area walk-through tours (housekeeping, hunting for hazards, etc.).
- Machine/equipment-specific checklists.

- Follow-up teams.
- Performance metrics.

Be a good listener. Pay attention to employees. In many cases, we hear an employee talking but we do not listen to what he or she says, as many times we may knowingly or unknowingly have our own hidden agenda. One individual does not have all of the answers, and involving the person who directly performs the job can provide insights on what works and does not work.

Chapter Review Questions

1. Discuss the use of the activity-based safety system.
2. Discuss ways to communicate with specific levels of the organization.
3. Discuss daily shift reviews and weekly and monthly meetings.
4. Discuss the importance of one-on-one discussions with employees.
5. Discuss the benefits of safety walk-through tours.
6. Discuss the concept behind the implementation of a machine/equipment-specific safety checklist.
7. Discuss how to use a follow-up team to correct safety-related issues.
8. Discuss the roles of the various levels of the leadership team.
9. Explain the advantages of using ABSS.
10. Discuss how you would use the ABSS in your organization.
11. Compare and contrast the difference between the daily preshift review and the regular weekly safety meeting.
12. Why is it important that you communicate to all of a multishift operation?
13. How would you use the follow-up team in your organization based on what is presented in this chapter?
14. Discuss why different levels of the leadership have difference responsibilities defined in the ABSS system.
15. Discuss several ways that you can measure the success of ABSS.

Bibliography

Christakis, N. A., & Fowler, J. H. (2009). *Connected: The surprising power of our social networks and how they shape our lives*. Little, Brown.

Collins, J., & Hansen, M. T. (2011). *Great by choice: Uncertainty, chaos, and luck–why some thrive despite them all*. HarperCollins.

Connors, R., & Smith, T. (June 2004). *Create a culture of accountability*. Dietary Manager. Retrieved from http://bit.ly/UBAmhT.

Judy Agnew, P. D., & Aubrey Daniels, P. D. (November 2011). Developing high-impact leading indicators for safety. *Performance Management Magazine*. Retrieved from http://bit.ly/WTK66g.

Newell, S. A. (2001). A New Paradigm For A New Paradigm For Safety And Health Metrics: Safety And Health Metrics: Framework, Tools, Applications, Framework, Tools, Applications, And Opportunities, Reprinted with Permission. Mercer ORC.

Peters, T. J., & Waterman, R. H. (2004). *In search of excellence: Lessons from America's best-run companies.* HarperBusiness.

Robotham, G. (n.d.). More on 10 Sure Fire Ways To Stuff Up a Safety Management System. Health and Safety Risk Management. Retrieved from http://bit.ly/12UL6y8.

Roughton, J., & Crutchfield, N. (November 2010). Safety culture—six basic safety program elements. *Ezine Articles.* Retrieved from http://bit.ly/Z5oolV.

Safety culture in nuclear installations. (2002). International Atomic Energy Agency (IAEA). Guidance for use in the enhancement of safety culture, IAEA-TECDOC-1329. Retrieved from http://bit.ly/V1KsKt.

Successful Health and Safety Management, hsg65. (1997). Health and Safety Executive, Crown Publishing, Permission to Reprint. Retrieved from http://bit.ly/YKDzRk.

Volume 2: Human Performance Tools For Individuals, Work Teams, and Management, Human Performance Improvement Handbook, Public Domain. (2009). U.S. Department of Energy. Retrieved from http://1.usa.gov/11Ex7vE.

CHAPTER 12

Developing the Job Hazard Analysis

We do not want production and a safety program, or production and safety, or production with safety. But, rather, we want safe production.
—Dan Petersen

I have never met a man so ignorant that I couldn't learn something from him.
—Galileo Galilei

Introduction

The job hazard analysis (JHA) is the foundation for any successful safety management system. A safety culture can only exist when a full understanding of ongoing jobs, steps, and tasks is known and the various hazards and associated risk are managed and controlled. As such, the JHA is an essential element in assessing the depth and scope of risk within the organization.

By implementing a JHA process throughout the organization, the leadership team and employees develop an in-depth understanding of task-specific hazards and associated risk. As we have defined safety as an emerging property resulting from the interactions of all elements of an organization, the hazards and associated risk can only be tracked, identified, and controlled using an ongoing comprehensive process.

The primary objective of the JHA process is to ensure that a safety management system is designed to be self-sustaining and effective. After completing this chapter you will be able to:

- Discuss the process for developing a job inventory and its importance.
- Discuss the benefits of the JHA to the safety culture.
- Compare and contrast the benefits and drawbacks of a JHA process.
- Discuss why it is important to get employees involved in the JHA process.
- Discuss how to select a JHA ad hoc team and its benefits.
- Discuss how to build a case for having a JHA process.
- Discuss how to select a job for analysis.
- Discuss and define steps and tasks of a job.

Beginning the Job Hazard Analysis Process

Organizations are adaptive and complex structures that have varying levels of resilience. Many organizations have a tendency to operate in a reactive mode, making change only as conditions dictate or experience shows that changes are needed. Change is made after loss-producing conditions create a crisis or unacceptable conditions. The objective of a JHA process is to reduce the potential of finding hazards only after a job has begun and a loss-producing event has happened. The JHA process builds an increased resilience into the organization as hazard and associated risk are controlled before the job is begun.

The JHA process begins with a full assessment of the organization. This assessment develops an initial inventory and lists all the jobs being performed to accomplish the goals of the organization. The inventory is expanded to identify the potential inherent or built-in hazards and associated risk. The inventory is used to rank order or sort from highest to lowest the risk potential using the risk guidance card or similar risk matrix as outlined in Chapter 9, "Risk Perception—Defining How to Identify Personal Responsibility". The inventory categorizes the jobs to determine where to target efforts towards the jobs with the highest potential risk. The assessment takes into account the potential severity and frequency of exposure that exists and the potential impact on the organization. Refer to Chapter 10, "Risk Management Principles".

Use of the JHA shifts the perception about the controls or actions needed and eliminates the assumption that all jobs have been clearly designed with consistent steps and tasks.

Lesson Learned # 1

Nathan once conducted an incident investigation session with supervisors. The initial incident investigation had a recommendation for retraining the injured employee in a particular standard operating procedure. A JHA was completed to fully assess the job and what the employee was trying to accomplish. The procedure was requested as part of the analysis. It turned out that the procedure was nonexistent. Everyone had assumed one existed and it took the JHA to point out that gap.

To begin the JHA process consider the loss history of jobs to include severe injuries, damaging events, first aide, and near miss data that has been developed. This data will assist in identifying jobs that have a clear history of reoccurring loss or injury. When loss data is used in combination with the risk assessment, the picture may change and priorities may need to be reset. The emphasis should remain on high potential associated risk.

Review the job inventory and data collected with employees who perform the job. They will have additional insights that will bring out aspects of their experiences that will not otherwise be found. They may also point out jobs that may not be formally identified and where a gap exists in what is being done. High risk potential can exist in nonroutine jobs that are not identified, as, by definition, they are not done on a daily basis or are only rarely done.

The key point is to include employees in all phases of the JHA process and development. Employees are the active part of job performance. Their review of job steps and related tasks as well as the effectiveness of standard operating procedures and training is critical to creating an effective process (Roughton & Crutchfield, 2008).

Why Is a JHA Important?

The JHA provides a consistent method for developing an effective training program that will provide safer and more efficient work methods. A properly designed JHA is used as a comprehensive training aid that ensures all employees know the primary requirements of each assignment and how to make safe choices when performing specific tasks. Refer to Table 12.1 to review a list of "Why JHAs Are Important".

> *The JHA process provides a better method to identify opportunities for improvement in the safety management system and can help provide the foundation for a hazard and risk-based analysis. As this structured process is followed, issues may be uncovered that may require the leadership team to make some decisions if this task should be avoided modified.*
>
> *Once you understand the complexity and current design of a job, you can develop and/or update the job procedures to reflect the needed safety revisions and modifications.*
>
> **(Roughton & Crutchfield, 2008)**

By comprehensively and routinely using the JHA as a tool to identify hazards and associated risk, a performance measurement system can be established to determine if jobs are safely being performed.

Figure 12.1 provides an Overview of the "JHA Development Process".

Figure 12.2 provides an Overview of a "Typical JHA Implementation Process".

Benefits of Developing JHAs

When the JHA process is considered a primary component of the safety management system, it provides a method to bring reality to discussions about the nature of work being performed by employees. The completion of JHAs provides all levels of the organization not just a better understanding but a greater appreciation of the various tasks that are included in the job as a whole. As job understanding and the control of hazards and associated risk are communicated, the safety culture improves because of the following:

- Procedures, methods, and protocols incorporate safety-related criteria.
- Selection of materials, equipment, and tools are provided that make the job less hazardous.
- Selection and training of employees is improved to meet the skills and physical requirements of the job.
- Environmental issues can be addressed that affect the performance of the job.

Supervisors can use the findings of a JHA to eliminate and/or prevent hazards, thereby helping in reducing injuries; providing a safer and more effective work method for a given task; reduced workers' compensation costs, and in the end result help to increase productivity. The analysis can be a valuable tool for training new employees in the tasks required to perform their jobs safely.

(Roughton & Crutchfield, 2008)

The JHA provides a consistent reminder for all new, relocated, and/or seasoned employees. It can be applied to specific job steps or tasks that have been modified. It can be used as an awareness tool for those who need updated training and/or site-specific review of non-routine task.

(How to Identify Job Hazards, n.d.; Roughton, 1995)

Table 12.1: Why JHAs Are Important

- Helps to detect existing/or potential hazards and consequences of exposure.
- Helps to assess and develop specific training requirements. Refer to Chapter 13, "Education and Training—Assessing Safety Training Needs" for a discussion on training needs.
- Helps supervision understand what each employee should know and how they are to perform their job.
- Helps to recognize changes in procedures or machine/equipment that may have occurred.
- Improves employee involvement in the work design process.
- Helps to define specific at-risk events that might be occurring.
- Helps to outline preventive measures needed to modify or control associated at-risk events.

(Conducting a Job Hazard Analysis (JHA), n.d.)

Drawbacks of the JHA

The JHA process does require time and budget. As the common assumption is made that "our employees know how to do their jobs", resistance can be expected, especially if no visible problems are evident with the job. To move the process from just occasional use, its strength as part of any continuous improvement process must be made clear to leadership and employees as to its potential for improving the work environment (Conducting a Job Hazard Analysis (JHA), n.d.; Roughton, 1995).

Why Is It Important to Get Employees Involved in the Process?

As discussed in Chapter 8, "Getting Your Employees Involved in the Safety Management System", employee involvement is one of the key considerations in any successful safety management system. Employees are more apt to accept changes when provided an opportunity to become involved in the decision to design and/or develop changes in the jobs they perform. Benefits of this approach are that employees know the actual job as they must do it daily, know the obstacles and problems that may be inherent in the job. They provide to the JHA team, leadership, and safety management a better understanding of the "real way" jobs are completed. The initial impression of

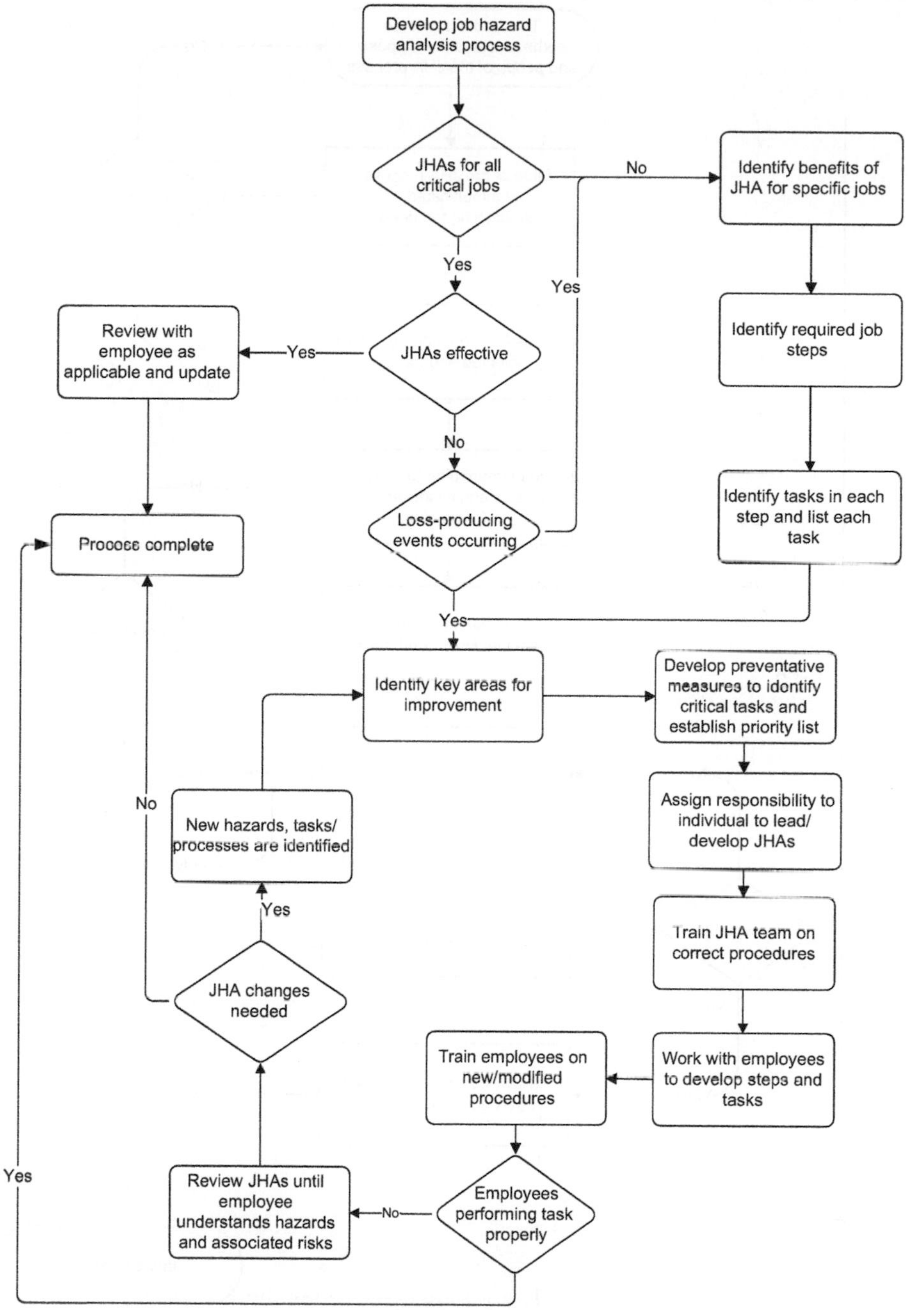

Figure 12.1

Overview of the JHA Development Process. *Source: Adapted from Roughton, J., Crutchfield, N. (2008). Job hazard analysis: a guide for voluntary compliance and beyond. Chemical, petrochemical & process. Elsevier/ Butterworth-Heinemann.*

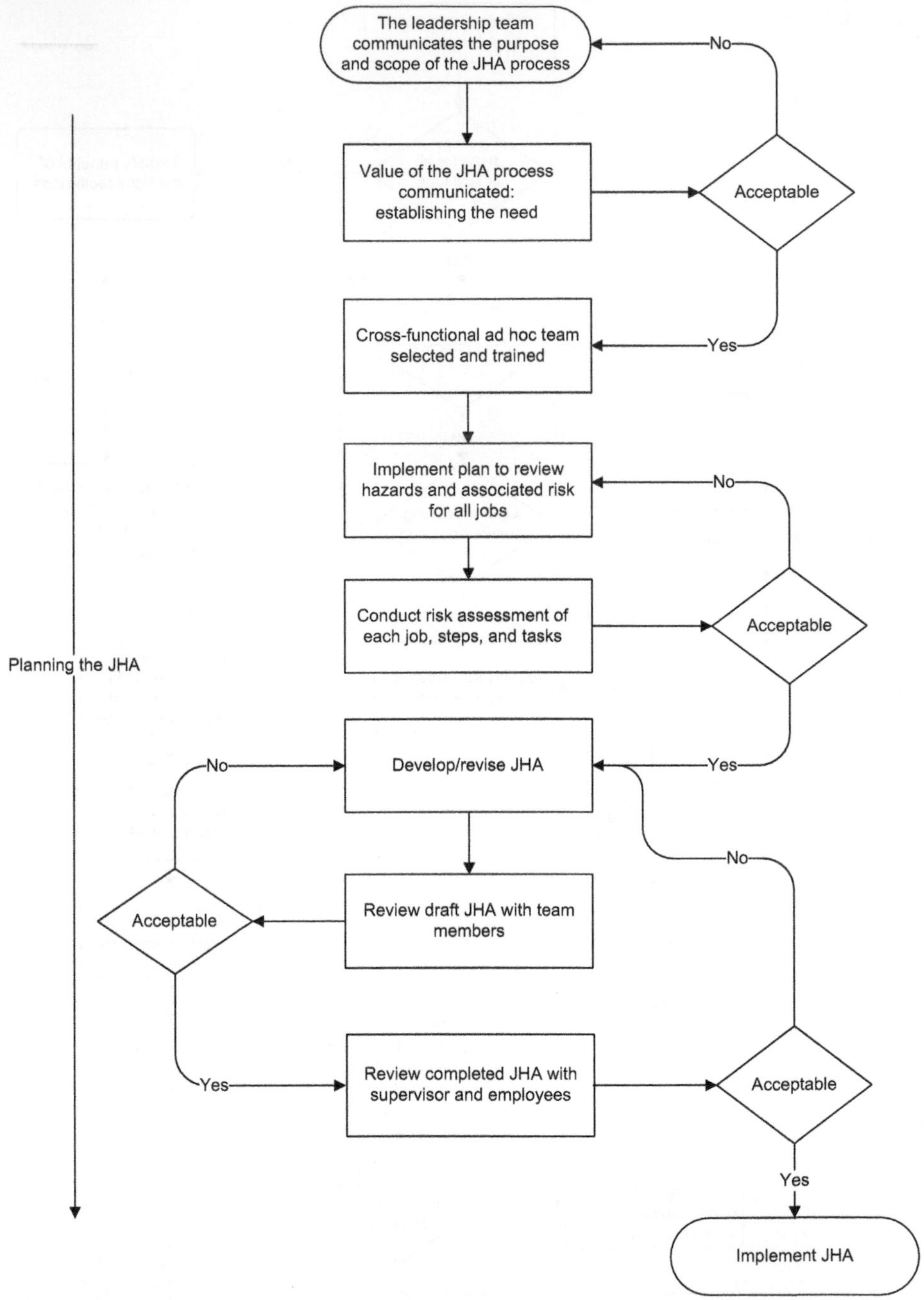

Figure 12.2

Overview of a Typical JHA Development Process. *Source: Adapted from Roughton, J., Crutchfield, N. (2008). Job hazard analysis: a guide for voluntary compliance and beyond. Chemical, petrochemical & process. Elsevier/ Butterworth-Heinemann.*

many jobs is that they are really simple. From our experience, even the simplest of jobs may require the coordination of many components when its steps and tasks, tools/equipment/materials, skills and abilities, general environment, and required administrative guidelines are considered. A simple job thought to be "an easy point of beginning" can be surprisingly complex (Roughton, 1995).

> *When employees are involved in the JHA development process they will be in a better position to understand the concepts and why it is important to develop the JHA.*
>
> **(Roughton & Crutchfield, 2008)**

It is important that employees understand the intended purpose of the JHA process. It is the job that is under review from a hazard and associated risk, safety-related perspective, and not an evaluation of their job performance or their productivity.

Selecting a Team

A member of the central safety committee should be designated to guide and administer the JHA process. This person would use cross-functional ad hoc teams to develop and maintain diversity and expertise for development of a portfolio of JHAs. The JHA ad hoc teams would develop specific JHAs based on the risk assessment list developed using the risk guidance card. They would draw on experience and expertise of employees and supervisors in the department or area where the job resides. Refer to Chapter 8, "Getting Your Employees Involved in the Safety Management System".

Use of ad hoc JHA teams can promote a more efficient process and ensure that a variety of perspectives and opinions are considered. Using multiple teams allows quicker development of JHAs. This approach spreads the time requirements through the organization and allows for more employee involvement. More involvement spreads job hazard information through the organization's networks. Refer to Chapter 3, "Analyzing and Using Your Network".

The individual selected to be the JHA process ad hoc team administrator should have a combination of the following traits:

- Be a good representative of the JHA process and be able to communicate verbally, in writing, and in presentations to all levels of the organization.
- Be able to communicate the value and benefits of the JHA process.
- Be respected by his or her peers as knowledgeable about operations and have an understanding of the various processes and methods used by the organization.
- Be able to apply risk assessment and problem solving techniques to uncover hidden hazards and associated risk.
- Understand the nature and relationship of hazards and associated risk with loss-producing events and how they apply to the process (Roughton & Florczak, 1999, p. 72).

This individual must be able to overcome barriers that may occur during implementation of JHAs and reinforce the leadership team's commitment to the process. A variety of methods and procedures would be used by the JHA administrator, such as:

- Provide input and advice to the ad hoc teams developing JHAs.
- Assist with evaluating the completeness and effectiveness of the JHAs based on job reviews, observations, and employee discussions.
- Assist in using the JHAs in loss-producing events, investigations, and reviews.
- Assist in the periodic evaluation of JHAs to ensure that they are using the current operational process, technology, task, and employee skill/training requirements.
- Assist in reviewing the JHAs with employees to ensure an understanding of what is expected of them and if they have any concerns with new procedures or actions required (Managing Worker Safety and Health, n.d.; Managing Worker Safety and Health, Appendix 9-4 Hazard Analysis Flow Charts, n.d.).

The Ad Hoc JHA team should:

- Use a charter to clearly define how they will operate and use the JHA process. The charter should be authorized by the leadership team as approved by the central safety committee. Refer to Chapter 8, "Getting Your Employees Involved in the Safety Management System" and Appendix O.
- Hold regular meetings and provide written documentation of meeting minutes to the JHA process administrator on the central safety committee.
- Encourage employee involvement by communicating the value and use of JHAs.
- Analyze job-related information and data concerning identified task and step hazards and associated risk to ensure controls are appropriate and comprehensively and effectively applied.
- Review employee suggestions, concerns, and/or complaints and bring these to the attention of the JHA process administrator.
- Ensure that insights provided by employees are accepted in a positive manner and ensure feedback is given without fear of reprisal.
- Immediately communicate to the central safety committee and the leadership team when high severity hazards and associated risks are identified and appropriate controls cannot be implemented (Roughton & Crutchfield, 2008).

Clear and concise expectations must be for the Ad Hoc JHA team. As each committee and subcommittee must have a charter, the JHA team charter must be developed to clearly define and outline the team's roles and responsibilities in the development of a comprehensive portfolio of JHAs (see Chapter 8, Getting Your Employees Involved in the Safety Management System" and Appendix N, Example of a Safety Committee Team Charter).

The JHA team charter will address such issues as:

- A clear statement that describes the improvement opportunity provided to the organization.
- The overall mission and objectives of the team and what it is established to do.
- The process and steps to be followed by the team.
- The scope of the process and the limits on the team to pilot the improvement or just make recommendations.
- The time frame of the charter and committee membership.
- The date and time the team will begin its activities.
- The process ownership of the JHAs completed.
- The methods of communication and documentation of the process (Conducting a Job Hazard Analysis (JHA), n.d.; Roughton, 1995).

Building the Case for a JHA Process

The JHA is a core process element of the safety management system and offers benefits in multiple areas. As mentioned above, if you have a JHA process, it becomes part of the network of communications that spreads job information and influences safe behaviors through the organization. It covers both the job design for potential latent, built-in errors that are hidden and for potential active errors by the employee (Job Safety Analysis: A Fundamental Tool for Safety, n.d.; Volume 1: Concepts and principles, human performance, improvement Handbook, 2009).

> Latent errors *result in hidden organization-related weaknesses or equipment flaws that lie dormant. Such errors go unnoticed at the time they occur and have no immediate apparent outcome to the facility or to personnel. Latent conditions include actions, directives, and decisions that either create the preconditions for error or fail to prevent, catch, or mitigate the effects of error on the physical facility … Managers, supervisors, and technical staff, as well as front-line workers, are capable of creating latent conditions. Inaccuracies become embedded in paper-based directives, such as procedures, policies, drawings, and design bases documentation.*
>
> Active errors *are those errors that have immediate, observable, undesirable outcomes and can be either acts of commission or omission. The majority of initiating actions are active errors. Therefore, a strategic approach to preventing events should include the anticipation and prevention of active errors.*
>
> **(Volume 1: Concepts and principles, human performance, improvement handbook, 2009)**

JHAs can be used as a problem-solving tool useful in assessing the potential for loss-producing events. Each job element has inherent hazards that have varying levels of risk (severity and probability). The JHA provides a structure that allows these elements to be analyzed and the desired proper controls better defined.

Possible issues that a JHA can provide further insights or explanations for would include:

- Job-related hazards and associated risk.
- Lack of knowledge of proper procedures.
- Lack of physical ability to complete specific steps and tasks.
- Improper use of "on-the-job training" where a new employee is trained by a "seasoned" (long-time) employee.
- Reducing or eliminating the perception that associated risk are acceptable by clearly identifying the potential severity of loss-producing events (Job Safety Analysis: A Fundamental Tool for Safety, n.d.).

> *Does an employee's perception of an existing and/or potential hazards and consequences of exposure differ from that of the employer? The employee sees a hazard and wants it fixed immediately. The employer may respond to this issue quickly and address the hazard but is often slowed down by internal structures, budget restraints, proper corrective actions, priorities, etc. Answers to critical questions must be clearly defined: Is the safety issue real? How big is the risk? What are the options? What is the best way to correct the identified risk? Who is going to correct the hazard? How long will it take to develop and implement preventative measures? How much will it cost? Is there need for additional training?*
>
> *One must remember that risk is based on probability even with a blatant hazard; no loss-producing events may have been developed.*
>
> **(Roughton & Crutchfield, 2008)**

> *JHA is a technique that can help an organization focus on specific tasks as a way to identify hazards before they occur. It focuses on the relationship between the employee, the task, specific tools, material, and equipment, and the work environment. Ideally, after uncontrolled hazards are identified, steps can be taken to eliminate or reduce the risk of hazards to an acceptable level.*
>
> **(Roughton & Crutchfield, 2008)**

> The Corps of Engineers use a form of the JHA process called Activity Hazard Analysis (AHA) on construction sites. The AHA is developed prior to performing any new task.
>
> *Before beginning each work activity involving a type of work presenting hazards not experienced in previous project operations or where a new work crew or sub-contractor is to perform the work, the Contractor(s) performing that work activity shall prepare an AHA.*
>
> **(Army System Safety Management Guide, 2008)**

Table 12.2 provides an overview of "JHA Basic Terms" that are used in developing the JHA.

Selecting the Jobs for Analysis

The following Table 12.3 provides an example list of "High Priority Jobs" that might be considered as having the potential for loss-producing events.

Table 12.2: JHA Basic Terms

Term	Discussion Points
Job	A job can be defined as a sequence of steps with specific tasks that are designed to accomplish a desired goal (Managing Worker Safety and Health, n.d.; Managing Worker Safety and Health, Appendix 9-4 Hazard Analysis Flow Charts, n.d.).
Steps	Steps are defined as a series of actions necessary to complete the job.
Tasks	Tasks are detailed actions taken to complete a step.
Analysis	Analysis is the art of breaking down a job into its basic steps and their tasks and evaluating each step/task for specific inherent hazards and associated risk. Each hazard or associated risk is evaluated for methods of control (avoidance, engineering controls, administrative practices, PPE, etc.) that are implemented as part of the standard operating procedures.

Source: Adapted from Roughton, J.E, & Crutchfield, N. (2008). Job hazard analysis: a guide for voluntary compliance and beyond. Chemical, petrochemical & process. Elsevier/Butterworth-Heinemann. Retrieved from http://amzn.to/VrSAq5.

Table 12.3: High Priority Jobs

- High frequency of injuries, illnesses, or damage.
- High degree of risk as found in industry history or from risk assessment.
- High duration of task.
- High physical forces.
  - Posture required of the person (i.e., ergonomics).
  - Point of operation requiring employee versus machine interface or exposure.
  - High pressure, mechanical, pneumatic, fluid, etc.
  - High and/or excessive vibration.
  - Environmental exposures.
- Nonroutine tasks involving high hazards and associated risk.
- High turnover or rotation of employees.
- Near misses or close calls; "almost" at-risk events.
- Recent process or operational changes or relocation of equipment.
- New jobs and/or tasks with little or no risk data.
- New equipment or process.

Source: Conducting a Job Hazard Analysis (JHA), n.d.; Roughton & Crutchfield, 2008.

Figure 12.3 shows the relationship of the JHA within a risk improvement model that incorporates an analysis of the job, behavioral consequences built into the job, and the structure of the administrative and management system within which the job is completed.

When conducting a JHA, make sure that you take into consideration applicable safety-related standards for your industry. Compliance with these standards is mandatory, and by incorporating their requirements in the JHA, you can ensure that your safety process meets known requirements.

(Roughton & Crutchfield, 2008)

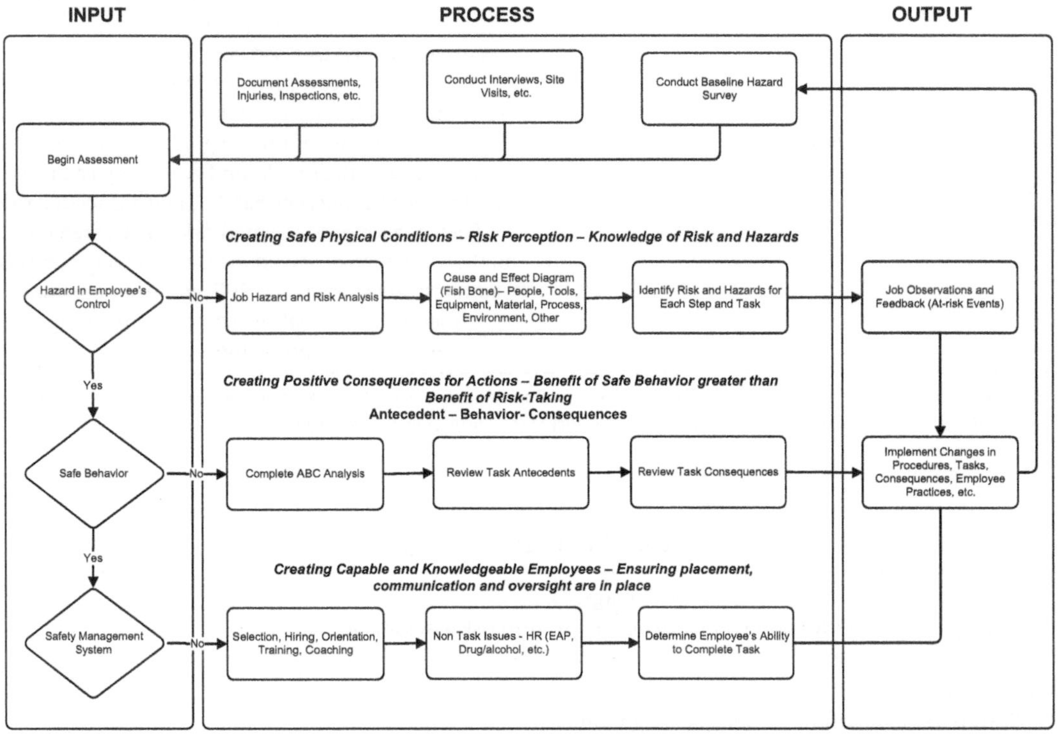

Figure 12.3

Risk Improvement Model. *Source: Adapted from Roughton, J., Crutchfield, N. (2008). Job hazard analysis: a guide for voluntary compliance and beyond. Chemical, petrochemical & process. Elsevier/ Butterworth-Heinemann.*

Summary

The JHA is the foundation for any successful safety management system. A safety culture can only exist when a full understanding of ongoing jobs, steps, and tasks are defined and the various hazards and associated risk are managed and controlled.

The objective of a JHA process is to reduce the potential of finding hazards only after a job has begun and a loss-producing event has happened. The JHA process builds an increased resilience into the organizations as hazard and associated risk are controlled before the job is begun.

The JHA process begins with a full assessment of the organization. This assessment develops an initial inventory and lists all the jobs being performed to accomplish the goals of the organization. The inventory is expanded to identify the potential inherent or built-in hazards and associated risk.

A properly designed JHA is used as a comprehensive training aid that ensures all employees know the primary requirements of each assignment and how to make safe choices when performing specific tasks.

When the JHA process is considered a primary component of the safety management system, it provides a method to bring reality to discussions about the nature of work being performed by employees. The completion of JHAs provides all levels of the organization not just a better understanding but a greater appreciation of the various tasks that are included in the job as a whole.

From our experience, even the simplest of jobs may require the coordination of many components when its steps and tasks, tools/equipment/materials, skills and abilities, general environment, and required administrative guidelines are considered. A simple job thought to be "an easy point of beginning" can be surprisingly complex.

A member of the central safety committee should be designated to guide and administer the JHA process. This person should use cross-functional ad hoc teams to develop and maintain diversity and expertise for development of a portfolio of JHAs.

Clear and concise expectations must be defined before training an Ad Hoc JHA team. As each committee and subcommittee must have a charter, the JHA process charter must be developed to clearly define and outline the team's roles and responsibilities in the development of a comprehensive portfolio of JHAs.

If you have a JHA process, it becomes part of the network of communications that spreads job information and influences safe behaviors throughout the organization. It covers both the job design for potential latent, built-in errors that are hidden and for potential active errors by the employee.

Chapter Review Questions

1. Discuss the process for developing a job inventory and its importance.
2. Discuss the importance and benefits of the JHA to the safety culture.
3. Compare and contrast the benefits and drawbacks of a JHA process.
4. Discuss why it is important to get employees involved in the JHA process.
5. Discuss how to select a JHA Ad Hoc team and its benefits.
6. Discuss how to build a case for having a JHA process.
7. Discuss how to select a job for analysis.
8. Discuss and define steps and tasks of a job.

Bibliography

Army System Safety Management Guide. (November 2008). Department of the Army, Pamphlet 385–16, Public Domain. Retrieved from http://bit.ly/11f0gib.

Conducting a Job Hazard Analysis (JHA). (n.d.). Oregon Occupational Safety and Health Division (Oregon OSHA), Public Domain, Permission to Reprint, Modify, and/or Adapt as necessary. Retrieved from http://bit.ly/WXGJhK.

How to Identify Job Hazards. (n.d.). Saskatchewan Labour, Occupational Health& Safety, Partners in Safety. Retrieved from http://bit.ly/Z3mCS9.

Job Safety Analysis: A Fundamental Tool for Safety. (n.d.). Commonwealth of Virginia, Workers' Compensation Services. Retrieved from http://bit.ly/U1YXPd.

Managing Worker Safety and Health. (n.d.). Illinois OSHA Onsite Safety & Health Consultation Program, Public Domain, Adapted for Use. Retrieved from http://bit.ly/WTsneh.

Managing Worker Safety and Health, Appendix 9-4 Hazard Analysis Flow Charts. (n.d.). Missouri Safety & Health Consultation Program (OSHA), Public Domain, Adapted for Use. Retrieved from http://on.mo.gov/XVZ6BD.

Roughton, J. (1995, April). How to develop a written job hazard analysis. In *Presentation, professional development course (PDC)*. Orlando, Florida: National Environmental Training Association Conference.

Roughton, J., & Crutchfield, N. (2008). Job hazard analysis: a guide for voluntary compliance and beyond. In *Chemical, petrochemical & process*. Elsevier/Butterworth-Heinemann. Retrieved from http://amzn.to/VrSAq5.

Roughton, J., & Florczak, C. (1999, January). *Job safety analysis: A better method*. Safety and Health, National Safety Council. pp. 72–75.

Volume 1: Concepts and Principles, Human Performance Improvement Handbook, Public Domain. (2009). U.S. Department of Energy. Retrieved from http://1.usa.gov/WdIqoP.

PART 4

Tools to Enhance Your Safety Management System

CHAPTER 13

Education and Training—Assessing Safety Training Needs

A picture in the head is worth more than a word in the ear.
—Richard Gandy

No one learns as much about a subject as one who is forced to teach it.
—Peter Drucker

Share your knowledge. It is a way to achieve immortality.
—Dalai Lama XIV

Introduction

The safety culture and safety management system are supported by effective, comprehensive education and training. The culture of an organization has a powerful influence over employee norms, habits, and behaviors as they complete their daily assignments and tasks. The strength of the underlying unconscious perceptions, thoughts, beliefs, etc. of the organization can override the best safety-related training designed to influence behavior.

Training may espouse clear beliefs about the value and importance of safety, yet on leaving the classroom employees revert back to the behavior actually driving the organization. The impact of training can only be sustained in the organization if the nature of the forces driving the culture is reviewed and the training is structured to overcome or reduce those forces (System Safety Training, 2000, chap. 14).

We have found, through experience, that training is more complicated than telling employees how to perform their job or just to "watch out" or "be safe". Information and training transfers the essential knowledge from the designated "trainer" to a "learner" (employees). The education and training must be provided in a manner that the employee can accept and immediately use in their assigned tasks (Roughton & Crutchfield, 2010; Roughton & Lyons, 1999, pp. 31–33; Roughton & Whiting, 2000).

> *Telling ain't training.*
>
> **Harold Stolovitch**

As part of your network, a strong positive link must be developed with the trainers and/or training department of your organization. They may or may not be able to provide specific safety training and may need to rely on your expertise. This your relationship with that department is strong, it should be able to assist in the assessment, structuring, design, and delivery of the required safety-related training.

For a strong safety culture, the organization must have developed emphasis on the positive values of the safety management system. The culture should have deeply embedded beliefs in the value of continuous education and training, as it ensures that all employees have a working knowledge of how the system operates and their responsibilities for continuous learning.

On completion of this chapter you will be able to:

- Discuss and contrast the differences between education and training.
- Define methods used to conduct a training needs assessment.
- Discuss how the specific, measurable, attainable, realistic, and timely (S.M.A.R.T.) method should be used to set training goals and objectives.
- Discuss and list the types of training methods.
- Discuss why an understanding of employee experience is needed in developing training content.
- Discuss why communication is important to the safety culture.

Education and Safety Performance

An effective safety culture is a product of a learning organization. As hazards and associated risks are identified, the ability to use the information becomes the key to establishing and implementing the necessary controls. Leadership and employees must be knowledgeable and understand core concepts of the safety management system. As organizations evolve and change, an ongoing education and training process keeps the information flowing through the organization that is timely and designed to meet the changes in hazards and associated risk.

> *Learning in organizations means the continuous testing of experience, and transformation of that experience into knowledge accessible to the whole organization, and relevant to its core purpose.*
>
> **(Senge, Kleiner, Roberts, Ross, & Smith, 1994)**

The objective of education and training is to change or improve employee performance and/or alter behavior. Similar to setting goals and objectives, in order to improve performance, training must be based on specific, measurable, attainable, realistic, and timely objectives (S.M.A.R.T.) (Best Practices for the Development, Delivery, and Evaluation of Susan Harwood Training Grants, 2010). It is not what the employee will know after completion of training but how well the employee will be able to use their acquired skills and knowledge to perform a given task after the training has been completed. Refer to Appendix R for Characteristics of Good Training Programs.

> *Remember, training is not what is ultimately important. . . performance is.*
>
> **Marc Rosenberg**

For example, many safety training sessions consist of classroom lectures, reviewing videos, or reading materials. While good information may be provided, unless a blend of hands-on, physical demonstration and follow-up within a short time frame is used, the potential of retaining information is dramatically reduced. Figure 13.1 provides Average Retention Rates.

> *Too often, training programs are pulled "off of the shelf" with limited or no effort made to adapt the training to the specific application. This generic training approach may be easy to implement but fails to ensure that employees understand why they must follow a set of established rules. Just training employees, for the sake of it, will not fix a perceived problem. Therefore, the appropriate training must be designed with specific guidelines and supplemental information that will enhance your training objectives.*
>
> **(Roughton & Lyons, 1999, pp. 31–33; Roughton & Mercurio, 2002)**

> *Though it seems hard to believe, instructors are frequently asked to develop courses intended to teach people what they already know, or to use instruction to solve problems that can't be solved by instruction.*
>
> **Robert Mager**

In order for safety training to be successful, it must have the support of all levels of the organization. If a safety management system has been selected and established, safety training becomes part of a formal structure accepted by the leadership team. For example, in ANSI Z10-2012, an organization should develop processes that define safety competencies, provide appropriate education and training, and ensure employees have access to training in a language they understand. This training should be timely and instructors should be competent

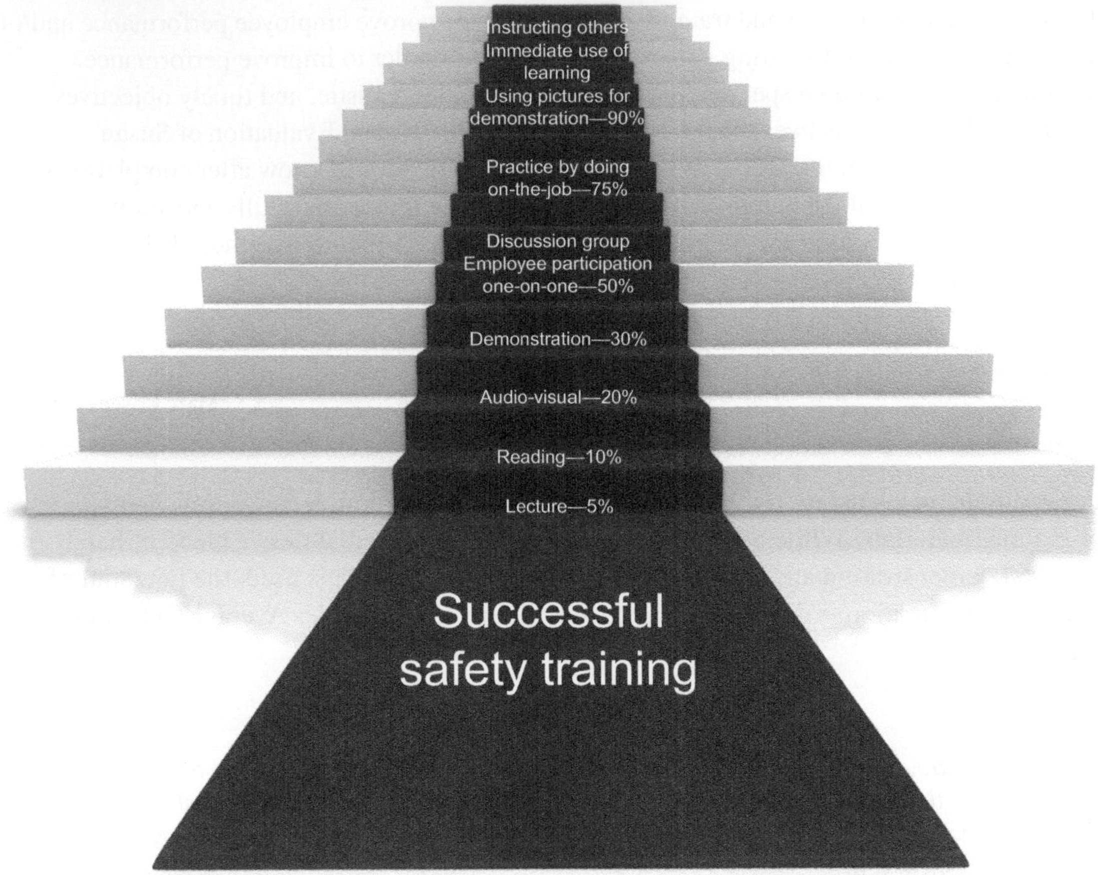

Figure 13.1
Average Retention Rates. *Source: Adapted from Roughton, J. & Mercurio, J. (2002).* Developing an effective safety culture: A leadership approach. *Butterworth-Heinemann.*

(Occupational Health and Safety Systems, 2012). Refer to Chapter 5, "Overview of Basic Safety Management Systems", and Table 13.1 for an "Overview of Education, Training, Awareness, and Competency".

If the activity-based safety system (ABSS) is used, supervisors/team leaders become mentors/coaches who will reinforce safety-related information after training sessions are completed. In the weekly and monthly meetings, employees could demonstrate what they have learned and share feedback on the effectiveness of the training materials and content. Refer to Chapter 11, "Developing an Activity-Based Safety System".

Training elements suggested in the Occupational Safety and Health Administration (OSHA) basic guidelines provide an outline for the design and the implementation of an effective training program. Refer to Table 13.2 for a list of "OSHA Training Development Guidelines".

Table 13.1: Overview of Education, Training, Awareness, and Competency

The organization shall establish a process to: • Define and assess the Occupational Health and Safety Management System (OHSMS) competence needed for employees and contractors; • Ensure through appropriate education, training, or other methods that employees and contractors are aware of applicable OHSMS requirements and their importance in the company to carry out the responsibilities as defined in the OHSMS; • Ensure effective access to, and remove barriers to participate in education and training as defined in the organization's OHSMS; • Ensure training is provided in the language trainees understand; • Ensure training is ongoing and is provided in a timely fashion; and • Ensure trainers are competent to train employees.

Source: Occupational Health and Safety Systems (2012).

Table 13.2: OSHA Training Development Guidelines

The OSHA model has seven elements: 1. Determine if training is needed, 2. Identify training needs, 3. Identify goals and objectives, 4. Develop content and activities, 5. Develop evaluation methods, 6. Develop training documents, and 7. Develop improvement strategies. The model is designed to enhance your training program. Using this model, you can develop and administer safety training programs that address site-specific issues, fulfill the learning needs of employees, and strengthen the overall safety program.

Source: Training Requirements in OSHA Standards and Training Guidelines, Publication 2254, 1998 (Training Requirements in OSHA Standards and Training Guidelines, 1998).

As an organization is constantly changing, potential gaps can develop in training needs as new hazards and associated risks may develop. This gap may create situations where out-of-date instruction does not match the new requirements for control of the new hazards and associated risks. A link between the ongoing findings of the risk guidance card, hazard assessments, ABSS activities, and safety management system reviews and audits is critical to ensuring information is current. The training needs assessment is part of the ongoing analysis of your organization (Roughton & Crutchfield, 2010).

Conducting Education and Training Needs Assessment

The education and training needs assessment is the process of identifying employee performance requirements and the "gap" between safety-related performance required and what is actually being completed.

To reemphasize the need for following a structured format when designing safety training requires more than just selecting a video or creating a handout. The selection of training objectives cannot

> *A needs assessment is the process of identifying the 'gap' between performance required and current performance. When a difference exists, it explores the causes and reasons for the gap and methods for closing or eliminating the gap. A complete needs assessment also considers the consequences for ignoring the gaps.*
> **(Training and Development, Planning & Evaluating, n.d.)**

be emphasized strongly enough. The training must be designed to be of value to the leadership and employees with clear links to what needs to be accomplished.

Table 13.3 outlines Steps in Course Development based on ANSI/ASSE Z490.1-2001, "Criteria for Accepted Practices in Safety, and Environmental Training".

Table 13.4 provides Questions to be Asked When Assessing Training Needs from the U.S. Federal Aviation Administration designed to drill down into the specific needs of employees with training.

The National Institute for Occupational Safety and Health (NIOSH) developed a training intervention effectiveness research model. For further research on the NIOSH training model, see "A Model for Research on Training Effectiveness" (A Model for Research on Training Effectiveness, DHHS (NIOSH) Publication No. 99–142, 1999).

Understand the Direction of Training

You may not be directly involved in providing safety-related training depending on how the role of safety is defined at your organization. All instructors/trainers must be made aware that each time they are presenting to a group of employees or even to one person, they are influencing a network of employees who will spread word about the value or lack of value that the training provided.

Table 13.3: Steps in Course Development

- Determine training requirements
- Select training objectives
- Translate training objectives into performance terms
- Construct appropriate criteria measures
- Select and sequence the course content
- Select instructional strategies and methods
- Determine equipment requirements
- Determine number and type of trainers required
- Establish course prerequisites
- Identify and procure training aids
- Develop instructional materials
- Establish time allocations
- Evaluate course in trial operation
- Analyze test results and take actions indicated
- Follow up on course graduates

Source: Criteria for Accepted Practices in Safety, Health, and Environmental Training, Z490.1, 2001, Annex B, (Criteria for Accepted Practices in Safety, Health, and Environmental Training, 2001).

Table 13.4: Questions to Be Asked When Assessing Training Needs

The trainer must assess the needs in which he/she is going to provide training with the following questions in mind (all of which are important):
• What is the extent of safety knowledge of the leadership team and employees within the organization?
• What are the leadership team and employees tasks that involve safety knowledge?
• What are the background, experience, and education of the leadership team and employees?
• What training has been provided in the past?
• What is the leadership team's attitude toward safety and training?
• Is training being provided to leadership and employees?
• Will the leadership team and employees be trained in hazard and associated risk analysis?

Source: Adapted from US FAA System Safety Training, Chapter 14, Public Domain (System Safety Training, 2000, chap. 14).

At no time should training be performed in a nonprofessional manner or given the appearance of being given just because it's a regulatory requirement that needs to be "checked off". When thinking through the development of a safety-related training session, keep in mind that the session is a "node" that has many connections in the overall network of communications within the organization. If done poorly, it immediately degrades the safety culture.

When you understand the many ingredients that make up a good education and training program, the task of blending the required elements can be challenging. What you say and the way you say it can go a long way toward effective communication, increasing the influence of the safety function and improving relations with employees.

> A successful trainer listens . . . *to the ideas and opinions of others and ask for clarity when they are confronted with information or situations they do not understand.*
>
> **(Effective Communications 4.A.01, n.d.)**

The Concepts of Education and Safety Training

To understand the concepts of education and safety training, we must first define what education and training really mean. Do not assume that the terms "Education" and "Training" are universally understood. Oregon OSHA, in the introduction to effective training, provides several definitions:

> What is Education?
>
> - That which leads out of ignorance,
> - Anything that affects our knowledge, skills, and attitudes/abilities (KSA's),
> - The "why" in safety educates about the natural (hurt and health) and system consequences (discipline, reward) of behavior, and
> - Primarily increases knowledge and attitudes
>
> **(An introduction to effective safety training, n.d.)**

What is Training?

- One method of education,
- The "how to" in safety, and
- Primarily increases knowledge and skills.

A specialized form of education that focuses on developing or improving skills, and the focus on performance.

(An introduction to effective safety training, n.d.)

A cautionary note from the US Department of Energy, Human Performance Handbook warns that training is not the sole solution to safety-related issues. For example, we have experienced training used as the end-all for many investigations and used as a short cut to complete a form, not completing a full analysis of what happened in a loss-producing event:

. . . retraining workers to do work that is already basically memorized and automatic, performed with little conscious thought because of the nature of the work, is a waste of time and is an insult to the worker. It is very hard to train a worker not to repeat something he or she did not intend to do in the first place. Training is not the solution in these instances.

(Volume 1: Concepts and principles, human performance improvement handbook, 2009)

Course Development Process

To be effective, safety training should be provided and refresher sessions developed as warranted. Refer to Table 13.5 for "Examples of Types of Safety Training Needed".

A lesson plan is an instructional prescription, a blueprint describing the activities the instructor and student may engage in to reach the objectives of the course. The main purpose is to prescribe the key events that should occur during the module.

Robert Mager

Table 13.5: Examples of Types of Safety Training Needed

- Orientation training for new, relocated, transferred, and contractor employees,
- Standard Operating Procedures (SOPs), i.e., Job Hazard Analysis (JHAs),
- Hazard recognition and control,
- Training required by regulatory agencies,
- Training for emergency response and emergency drills, and
- Injury prevention and loss-producing events.

Source: Training Requirements in OSHA Standards and Training Guidelines, Publication 2254 (Training Requirements in OSHA Standards and Training Guidelines, 1998).

Different training methodologies may be necessary depending on the skills, expertise, and experience of employees. All training materials must be written and/or adapted to meet specific needs as to what is the best media for use with each group. The method and media will vary based on what works best—computer-based training, hardcopy workbooks, brief handouts, one-on-one or group training sessions, audio or video, etc. (An introduction to effective safety training, n.d.).

The objective is to make the training interesting, of value, and presented in an environment where the employee can learn.

Refer to Table 13.6 for examples of "Common Types of Training Methods".

Developing Learning Activities

An understanding of adult learning theory is necessary to complete a successful education and training assessment. As discussed in Chapter 1, "The Perception of Safety", we are faced with a workforce that has been raised on entertainment delivered by very creative media sources. The era we are in can use media delivery systems that range from smart-phones, electronic tablets, computer gaming, Internet resources, reality television, movies, and other social media resources. In addition, we can go well beyond delivering information to a centralized group session and can use electronic media to bring together employees who are widely scattered geographically.

Table 13.6: Common Types of Training Methods

Method	Discussion
Case study	Actual or hypothetical situation.
Lecture	Oral presentation of material, usually from prepared notes and visual aids.
Role play	Employees improvise behavior of assigned fictitious roles.
Demonstration	Live illustration of desired performance.
Games	Simulations of real-life situations.
Stories	Actual or mythical examples of course content in action.
Brainstorming/Discussion	Facilitated opportunity for employees to comment.
Question	Employees question the trainer and receive answers to questions.
Small group	Employees divide into sub groups for discussion or exercise.
Exercises	Various tasks related to specific course content.
Instruments/Job aids	Tools, equipment, and materials used on the job.
Ice-breakers	This can include multiple things, such as asking each employee to discuss what they want to see from the class, show a video, or discuss a picture that relates to your subject.
Reading	Employees read material prior to, during, and/or after the session.
Manuals	Handbooks or workbooks distributed to employees.
Handouts	Selected materials, usually not part of a manual.

Source: Adapted from OR-OSHA's "An Introduction to Safety Training" (An introduction to effective safety training, n.d.).

As a result, many of the individuals entering the workforce will have years of experience with interactive and dynamic learning methods indirectly developed. They have experiences not found in the past generations and may be extremely well versed in use of electronic media. Education and training must be creative and take into account this skill level of the audience. Be imaginative in your choice of methods and materials, and ensure that you use all internal and available external resources.

Lesson Learned # 1

In developing formal presentations, the historical reason for putting large amounts of information on each slide has been based on the economics of developing a presentation. It is quicker to put more information on each slide and the same slides are used as a hardcopy handout. The more information per slide, the fewer pages that have to be printed out.

A better approach is to have two versions of the materials. One is designed as a handout, which can have multiple slides on each printed page. The second version can be greatly expanded to keep each slide limited to only one key thought. This keeps the focus on the thought being presented. It also has the benefit of using larger fonts that can be read from a distance!

A presentation on the Internet provides a look at "How not to use PowerPoint", which has interesting insights (McMillan, 2009).

Organizations such as the American Society of Training and Development (Training Development Website, n.d.) can provide up-to-date methods on organizational learning and adult education.

No two training groups are exactly alike. The training needs, level of motivation, educational background, and many other factors can affect the training environment. We have experienced a wide variation in how groups respond even when giving the same information to groups within the same organization. One group would be totally involved and engaged while the next group, an hour later, might just as well have been rocks!

Performance Deficiency

Safety-related training is used to close gaps in knowledge about hazards and associated risk, as well as the necessary controls that are to be consistently used. A safety-related problem may be a precursor to general performance issues impacting productivity or other organizational effectiveness or drivers. For example, an employee may not have an understanding of the required specific sequence of tasks, actions, or the decision rules to follow in taking corrective actions. The job may be one that requires problem-solving skills and not specific rules to follow or skills directly associated with the task (Roughton & Crutchfield, 2008; Volume 1: Concepts and principles, human performance improvement handbook, 2009).

> Skill-based performance *involves highly practiced, largely physical actions in very familiar situations in which there is little conscious monitoring. Such actions are usually executed from memory without significant conscious thought or attention.*
>
> Rule-based performance *is called the rule-based level because people apply memorized or written rules. These rules may have been learned as a result of interaction with the facility, through formal training, or by working with experienced workers.*
>
> The level of conscious control is between that of the knowledge- and skill-based modes. The rule-based level follows an IF (symptom X), THEN (situation Y) logic.
>
> Knowledge-based performance *is a response to a totally unfamiliar situation (no skill or rule is recognizable to the individual). The person must rely on his or her prior understanding and knowledge, their perceptions of present circumstances, similarities of the present situation, and similarities to circumstances encountered before, and the scientific principles, and fundamental theory related to the perceived situation at hand.*
>
> **(Volume 1: Concepts and principles, human performance improvement handbook, 2009)**

Understanding the type of job (skill, rule, and/or knowledge-based performance) aids in identifying and determining the appropriate training solution. By analyzing the skill, rule, or knowledge gap at hand you can determine the role, if any, education and training will play in resolving performance needs.

Establishing Learning Objectives

In establishing learning objectives, an effective safety culture should view the provision of hazard and associated risk control information as part of the obligatory orientation and training should be considered essential. This initial training sets the stage for future desired behaviors and decision-making. Ongoing training ensures leadership, and employees are kept aware and knowledgeable of the hazards and associated risk over time. With loss-producing events being rare occurrences, the potential for "scope drift" always exist.

A safety-related training learning objective is a brief, clear statement of what the leadership team and employees will be able to do as a result of what they gain through the learning process. The groundwork for a learning objectives begins with the training assessment. The learning objectives should include a focus on the specific job skills with the steps and tasks that have higher potential for a loss-producing event.

To be effective, training assessment should list the behaviors and actions that the risk guidance card and ongoing hazard identification have determined to be essential in the regard to control or prevention of injury or damage. Learning objectives are important because instructional strategies and evaluation techniques are an outgrowth of the learning objectives (System Safety Training, 2000, chap. 14).

> *Now it's time to describe the instructional outcomes (the need to do's); it's time to construct a verbal word picture that will help guide you in developing the instruction and help guide students in focusing their efforts.*
>
> **Robert Mager**

Unless we know what we want our training program to do, how can we measure the success in terms of upstream or pre-loss measures? Too often safety-related training is measured by the number of people in the classroom and not by the ability to understand and use the materials.

Guidelines for Writing Learning Objectives

The learning objective is a statement that describes the learning outcome, rather than a learning process or procedure. It describes results, rather than the means of achieving those results. It is important that all learning objectives be well-written and in clear measurable terms so both the trainer and the employee will understand what is expected as a result of the training session. You need to write objectives to ensure that the design and selected instructional content and procedures are organized to take advantage of the learner's own efforts and activities (An introduction to effective safety training, n.d.).

Target Audience

The general background of the intended trainee is assessed along with assigned job duties, previous training history, and overall experience. A part of the organizational culture assessment should have also included reviewing the "general emotional climate, behavioral norms, and attitudes toward training" (System Safety Training, 2000, chap. 14). Is training considered a necessary evil or is it considered an embedded positive activity and part of the organization's DNA? An indicator of the status of training is whether resistance is present in the form of excuses as to why employees did not have the time to take the training or if leadership asks immediately how long the training will take without considering the intent or need for the training.

Lesson Learned # 2

Nathan was once on a construction site before a shift began and was able to observe a subcontractor orientation session. The construction workers were milling around in front of the construction office trailer waiting for the trainer. At the designated time, a person came out of the office, placed a tape recorder on top of a wooden fence post, turned it on, and went back into the office. The recorder loudly played the safety message while the workers continued to mill around, smoke, and chat. At the end of the tape, the person returned, got the tape recorder, and went back into the office.

What do you think of that organization's opinion of training?

It is important to take time "up front" to pinpoint the expectations of employees by talking with a sample target audience to determine if they have mastered any prerequisite skills required needed for the training to be developed. A tour of a facility or operation provides an opportunity to ask simple and direct questions of employees. The key is to listen to employees for their expectations, as employees will often volunteer information to a sincere and skilled listener about what they see is needed to improve their performance (System Safety Training, 2000, chap. 14).

Lesson Learned # 3

The previously stated split between listening (80%) and talking (20%) applies to training. This basically says that when someone is talking, do not interrupt until they have completed their statement or they have finished. Steven Covey had as his fifth of seven habits for highly effective people the need to "Seek first to understand, then to be understood" (Covey, 1990, p. 151).

One thing that is important is to ensure that all courses and activities are designed with the adult learner in mind. The following Table 13.7 provides "Examples of Learning Objectives for Adult Learning".

The importance of gaining as much insight as possible on the organizational culture is reflected in how training is perceived, developed, and presented. A well-designed training curriculum will not have much impact if the general climate of the organization is detrimental to learning. As with the role of safety, how the training role is viewed and perceived is an indicator of where the organization places its values.

> *. . . employees are mostly influenced by the behavioral norms of an organization. Behavioral norms refer to the peer pressures that result from the attitudes and actions of the employees/management as a group. Behavioral norms are the behaviors a group expects its members to display. Before attempting to make changes in an organization, it is important to identify existing norms and their effects on employees.*
> **(System Safety Training, 2000, chap. 14)**

Table 13.7: Examples of Learning Objectives for Adult Learning

- Always include goals or objectives that let employees know what is coming.
- Adult learners must be given time to reflect or think about each point of learning. Focus on one item at a time. They should not have to take a lot of notes when you want them to listen.
- Design time into the process to let the employee reflect or think about each point of learning.
- Include samples, stories, and scenarios that apply the learning directly to something to which the employee can relate.
- Include a list of acronyms. We use a lot of safety acronyms and assume that everyone understands them. It is important that you clarify any acronym that may be used during the training session.
- Flag important information. Use a "parking lot" list (this can be simple notes listed on a flipchart) to keep ideas fresh in the trainer's and employees', minds.
- Adults do not effectively learn by simply being told. They must have a chance to digest the material and, whenever possible, apply the learning to something they can relate to. Ask questions and involve the employees as much as you can. This could include open-end questions and exercises that must be performed in class before you move to the next level. The key is to design active participation in the learning process whenever possible.
- Information more easily enters the long-term memory when it is linked to old memories or can be related to something the learner has experienced.
- The short-term memory is linear, works best through lists, and is the only conscious part of the brain.
- Giving adult learners an advanced organizer, such as workshop goals or objectives, helps them to retain information.
- Let them know what is important: what to focus on every time there is a change in points or a new topic to discuss.
- The mind pays more attention to what is novel than what is ordinary.

Source: Adapted from Dan Petersen, "An introduction to effective safety training", OR-OSHA 105, Public Domain, Permission to Reprint, Modify, and/or Adapt as necessary, (An introduction to effective safety training, n.d.; Petersen, 1993).

Conducting Site-Specific Education and Training

Unrealistic expectations are usually a result of a failure to understand what constitutes an effective training program.

■ *Lesson Learned # 4*

James was involved in training sessions when dealing with hazardous waste operations. The training had to be provided given the specific requirement for annual eight-hour refresher training and yet meet the production needs of the operation. A flexible schedule was set with training that began at 4 am and was conducted over several days to meet the requirements of the specific regulatory standard. The session took into consideration how to best keep the employees attention and involvement at this early hour. This was done by using a blend of classroom and hands-on activities.

Although no generalizations apply to every adult learner, it is helpful when planning training sessions to keep the following characteristics of adult learners in mind: (System Safety Training, 2000, chap. 14).

- Rote memorization may take more time as the student gets older, but purposeful learning can be assimilated as fast or faster by an older adult or by high school student. Despite the cliché that "old dogs can't learn new tricks", most healthy adults are capable of lifelong learning.
- Adults want to know the "big picture" and just want answers to their questions: "Why is it training important?" and "How can I apply the training to real life?"
- Adults function in roles, which mean that they are capable of and desirous of participating in decision-making about learning.
- Adults do not like to be treated "like children" and especially do not appreciate being reprimanded in front of others. Adults have specific objectives for learning and generally know how they learn best. James remembers his military days in the Air Force, where one instructor sat at his desk, lectured, and then asked questions. If the student did not get the correct answer for the question, he would then throw a piece of chalk at the student.
- Adults have experienced learning situations before and have positive and/or negative consequences about learning.
- Adults have had a wealth of unique experiences to invest in learning and can transfer knowledge when new learning is related to old learning. Adults recognize good training and bad training when they see it.

(An introduction to effective safety training, n.d.; Geigle, n.d.; Roughton & Crutchfield, 2008)

The trainer should try to determine the causes of the performance deficiencies and tailor the training to the needs of the organization. For example, when individuals are motivated to perform well but lack skills or knowledge, an ideal training opportunity exists (System Safety Training, 2000, chap. 14).

Communication

Effective communication is the most important tool used by the safety management system. It is assumed that employees know how to communicate information and concerns about immediate hazards and associated risks. In most safety-related training we have experienced, the employee is simply told to report unsafe conditions without further explanation. In addition, the quality of the communication network must be reviewed to eliminate obstacles and resistance to the message being sent.

A part of the learning process is the need for training in the basics of communication. Information is essentially defined as a "meme", as we discussed in Chapter 3, "Analyzing and Using Your Network", as a "packet of information" moving from one mind (a sender) to another mind (the receiver) in a manner that the information sticks and is incorporated into the thought process of the receiver. Communication is defined as "a process of transferring thoughts, ideas, messages and information from one individual, or group, to another." This perspective has effective communication being more than just transmitting a message. The message must be clear, accurate, and above all, understood by the person or persons with whom you are communicating (Effective Communications 4.A.01, n.d.).

> *Oral communication possesses a greater risk of misunderstanding as compared to the written form of communication. Misunderstandings are most likely to occur when the employee involved in the conversation may have a different understanding or mental models of the current situation or use terms that are confusing. Therefore, confirmation of verbal exchanges of operational information between individuals must occur to promote understanding and reliability of the communication.*
>
> **(Volume 2: Human performance tools for individuals, work teams, and management, human performance improvement handbook, 2009)**

We stressed one-on-one discussions in detail in Chapter 11, "Developing an Activity-Based Safety System", concerning the need to verify information presented in a safety meeting.

Effective training should follow the method of a three-way communication. The sender is the trainer who communicates the message. The employee is the receiver who gets the message, decodes the message based on current perceptions, background, and/or past history of similar messages. The employee may or may not interpret the meaning of the message as was intended by the sender/trainer.

> *In learning you will teach, and in teaching you will learn.*
>
> **Phil Collins**

Only when the employee has shown that the message was understood does true communication exist. The language of the message must mean the same thing to the both the receiver and the sender or miscommunication will occur (Effective Communications 4.A.01, n.d.). If the message contains hazard information and is not understood, the probability of a loss-producing event increases.

Lesson Learned # 5

An example we have both experienced involves a combination of miscommunication and failure of training to convey its desired message. A maintenance technician was working on a defective machine and had asked a co worker to stand by the power source to make sure that no one turned on the equipment. As he placed his hands into the machine, he called to his co worker to "Make sure it's not on". The co-worker heard "Make sure it's on" and flipped the switch, activating the machine and causing a severe injury. A verbal message was misinterpreted but more important was that a critical safety procedure was ignored—reenergizing or locking out equipment.

In a three-way communication, the receiver of a message or training should be able to repeat the message in a paraphrased form to ensure the sender knows the receiver understood the message or training. The overall process is dependent on the receiver knowing that if the message or training is not understood, he or she can, without fear, request a clarification, confirmation, or repetition of the message or training (Volume 2: Human performance tools for individuals, work teams, and management, human performance improvement handbook, 2009).

Not only is the communication of the message of importance, but the source and its verification as being true must be considered. The following advertisement is an example of always assessing and confirming the source of training materials and its author.

> An insurance commercial has a scenario where a man is describing how he could diagram an accident with his smart phone app to a young lady. She tells him, "I did not think that State Farm® had all of these apps!" The man then asks the question, "Where did you hear that?" The young woman relies, "The Internet". To which he replies, "And you believed it?" She answers "YES! They cannot put anything on the Internet that is not true". He asks, "Where did you hear that?" Together they say at the same time, "The Internet!" The young woman then says, "Here comes my date, I met him on the internet. He is a French model". Which he obviously is not, but she had believed what she saw on the Internet was always true. Her date gives a "Bonjour" and they walk off together to the amazement of the man.
>
> **(State Farm Insurance, State of Disbelief, TV advertisement, 2012)**

The source of information used must be verified before you conduct any type of training. Once the information is released, it moves as meme though the communications network, spreading and influencing the organization. Both good and poor messages can go viral through a network with mixed results for the improvement of the safety culture.

Summary

The safety culture and safety management system are supported by effective, comprehensive education and training. The culture of an organization has a powerful influence over employee norms, habits, and behaviors as they complete their daily assignments and tasks.

For a strong safety culture, the organization must have developed strong emphasis on the positive values of the safety management system. The culture should have deeply embedded beliefs in the value of continuous education and training of its employees.

An effective safety culture is a product of a learning organization. As hazards and associated risk are identified, the ability to use the information becomes the key to the establishment and the implementation of the necessary controls.

The objective of education and training is to change or improve employee performance and/or alter behavior.

The education and training assessment is the process of identifying employee performance requirements and the "gap" between safety-related performances required and what is actually being completed.

When thinking through the development of a safety-related training session, keep in mind that the session is a "node" that has many connections in the overall network of communications within the organization. If done poorly, it immediately degrades the safety culture.

Different training methodologies may be necessary depending on the skills, expertise, and experience of employees. All training materials must be written and/or adapted to meet specific needs as to what is the best media for use with each group.

An understanding of adult learning theory is necessary to complete a successful education and training assessment.

Safety-related training is used to close gaps in knowledge about hazards and associated risks as well as the necessary controls that are to be consistently used. A safety-related problem may be a precursor to general performance issues impacting productivity or other organizational effectiveness or organizational drivers.

In establishing learning objectives, an effective safety culture should view the provision of hazard and associated risk control information as part of the obligatory orientation and training should be considered essential. This initial training sets the stage for future desired behaviors and decision-making.

The learning objective is a statement that describes the learning outcome, rather than a learning process or procedure. It describes results, rather than the means of achieving those results.

The importance of gaining as much insight as possible on the organizational culture is reflected in how training is perceived, developed, and presented. A well-designed training curriculum will not have much impact if the general climate of the organization is detrimental to learning.

Effective communication is the most important defense in the prevention of loss-producing events.

In a three-way communication, the receiver of a message or training should be able to repeat the message in a paraphrased form to ensure the sender knows the receiver understood the message or training.

Chapter Review Questions

1. Compare and contrast the difference between education and training.
2. Define methods used to conduct a training needs assessment.
3. Discuss how the S.M.A.R.T. method can be used to set training goals and objectives.
4. Discuss and list the types of training methods.
5. Discuss why an understanding of employee experience is needed in developing training content.
6. Discuss why communication is important to the safety culture.

Bibliography

A Model for Research on Training Effectiveness, DHHS (NIOSH) Publication No. 99–142. (October 1999). National Institute for Occupational Safety and Health (NIOSH), Public Domain. Retrieved from http://1.usa.gov/VaMLLn.

An introduction to effective safety training. (n.d.). Oregon Occupational Safety and Health Division (Oregon OSHA), OR-OSHA 105 Public Domain, Permission to Reprint, Modify, and/or Adapt as necessary. Retrieved from http://bit.ly/WF1Yor.

Best Practices for the Development, Delivery, and Evaluation of Susan Harwood Training Grants. (September 2010). Occupational Safety and Health Administration (OSHA), Public Domain, Permission to Reprint, Modify, and/or Adapt as necessary. Retrieved from http://1.usa.gov/Z5nKoy.

Covey, S. R. (1990). *The 7 habits of highly effective people: Powerful lessons in personal change*. A Fireside Book, Simon and Schuster, Inc., Retrieved from http://amzn.to/WsCA5g.

Criteria for Accepted Practices in Safety, Health, and Environmental Training. (July 2001). American National Standards Institute (ANSI), Secretariat American Society of Safety Engineers (ASSE), Z490.1.

Effective Communications 4.A.01. (n.d.). Elisted Professional Military Education, Public Domain. Retrieved from http://bit.ly/U5JGwJ.

Geigle, S. J. (n.d.). Developing OSH Training, Train the Trainer Series. OSHAcademy™ Course 721 Study Guide, Permission to Reprint, Modify, and/or Adapt as necessary. Retrieved from http://bit.ly/WESN7n.

McMillan, D. (April 2009). *How not to use Powerpoint*. Retrieved from http://bit.ly/WZvCnz.

Occupational Health and Safety Systems. (2012). The American Industrial Hygiene Association. ANSI/AIHA Z10.

Petersen, D. (1993). *The challenge of change: Creating a new safety culture*. Safety Training Systems.

Roughton, J. E., & Crutchfield, N. (2008). *Job hazard analysis: A guide for voluntary compliance and beyond. Chemical, petrochemical & process*. Elsevier/Butterworth-Heinemann. Retrieved from http://amzn.to/VrSAq5.

Roughton, J., & Crutchfield, N. (November 2010). Safety culture – six basic safety program elements. *Ezine Articles*. Retrieved from http://bit.ly/Z5oolV.

Roughton, J., & Lyons, J. (April 1999). *Training program design, delivery, and evaluating effectiveness: An overview*. American Industrial Association, The Synergist.

Roughton, J. E., & Mercurio, J. J. (2002). *Developing an effective safety culture: A leadership approach, Adapted for Use*. Butterworth-Heinemann. Retrieved from http://amzn.to/X8Gaz8.

Roughton, J. E., & Whiting, N. E. (2000). *Safety training basics: A Handbook for safety training program development*. Government Institutes. Retrieved from http://amzn.to/111dlWB.

Senge, P. M., Kleiner, A., Roberts, C., Ross, R., & Smith, B. (1994). *The fifth discipline field book: Strategies and tools for building a learning organization*. N. Brealy.

State_Farm_Insurance. (2012, June). State Farm® - State of Disbelief (French Model). Retrieved from http://www.youtube.com/watch?v=rmx4twCK3_I.

System Safety Training. (December 2000). Federal Aviation Administration (FAA), Public Domain. Retrieved from http://1.usa.gov/VPo9dw.

Training and Development, Planning & Evaluating. (n.d.). Office of Personnel Management – Training and Development, Planning, and Evaluating, Public Domain. Retrieved from http://1.usa.gov/WUkEzi.

Training Development Website. (n.d.). Retrieved from http://bit.ly/XQHOUZ.

Training Requirements in OSHA Standards and Training Guidelines. (1998). Occupational Safety and Health Administration (OSHA), Publication 2254, Public Domain, Permission to Reprint, Modify, and/or Adapt as necessary. Retrieved from http://1.usa.gov/XFQkYt.

Volume 1: Concepts and Principles, Human Performance Improvement Handbook Public Domain, (2009).

Volume 2: Human Performance Tools for individuals, Work Teams, And Management, Human Performance Improvement Handbook Public Domain, (2009).

CHAPTER 14

Assessing Your Safety Management System

If you want to reach a goal, you must "see the reaching" in your own mind before you actually arrive at your goal.

—Zig Ziglar

Introduction

A safety management system assessment provides a comprehensive assessment of the current state of the safety culture. The effectiveness of the safety management system should reflect the real values and beliefs of the organization. A structured and detailed assessment should provide specific details about potential gaps and opportunities for further improvement, not just in the safety management system but in your organization as well.

> The safety management system is akin to standing the organization in front of a mirror and seeing the reality that it reflects. A good mirror does not lie! Not taking an assessment seriously or not conducting assessments is like looking into a distorted carnival mirror that does not give a true picture of reality.

The full implementation of a safety management system provides the structure that guides decisions and activities that are extremely important for the long-term success of your safety culture (Geigle, 2011). ANSI Z10-2012 calls for a process to "gather and review information to identify Occupational Health and Safety Management System (OHSMS) issues, which includes the processes necessary to establish or improve its management system" (Occupational Health and Safety Systems, 2012). Review or assessment is found in all quality, management, and safety systems and follows the "Plan–Do–Check–Act" (Shewhart, 1931) concept as we discussed in Chapter 6, "Selecting Your Process".

The problem faced by organizations is not in the finding safety-related information, which is readily available. It is in the implementation and daily use of the information at the different decision-making levels of the organization. It is important that safety-related information is considered to add value to individuals' daily work activity. It must be readily understood and used by leadership and employees.

Many resources are readily and economically available that provide the content for a safety process, its programs, and criteria. The issue is in the communication and feedback required to ensure that the safety management system is being utilized and is effectively meeting its goals and objectives. An assessment determines how effective the process is working and if it is incorporated into the daily activities of the organization.

When you complete this chapter, you will be able to:

- Discuss the coordination needed for a safety management system assessment.
- Discuss how to avoid the mentality of "blaming the system".
- Discuss types of safety management system assessments.
- Discuss the selection and development of an assessment team.

- Discuss preassessment activities.
- Discuss the elements of a safety management system assessment.
- Discuss the importance of the final report and action planning.
- Discuss why it is important to communicate the results of the assessment.

Planning for Your Safety Management System Assessment

Safety management systems use six basic core principles as we discussed in Chapter 5, "Overview of Basic Safety Management Systems". When improving, building, or sustaining a safety culture, you immediately increase your probability for effectiveness if the assessment keeps its primary focus on these core principles. The assessment will be valuable only if it leads to improved performance in meeting clearly identified safety goals and objectives. Refer to Table 14.1, "Six Basic Core Principles of a Typical Management System".

Within the six core principles, each element may require a number of subcategories that give guidance to implementing the selected safety management system. System elements must be viewed from the vantage point of the organization and how a system improves the safety culture. If you are just beginning to implement a safety management system, a phased approach should be considered to not overwhelm the leadership and employees with system elements that they cannot immediately lead or manage.

Trying to immediately implement elements just because you think that they will be of benefit can be detrimental to the success of the safety management system. You are not trying to implement a program. You are trying to implement a process that changes the overall safety culture and changes the habits that may be preventing improvement (Duhigg, 2012). As discussed in several other chapters, you are trying to change the perception of safety, extend your influence within the organization, and expand your internal network for the benefit of the organization. A personal assessment should verify how effectively you are accomplishing these goals and objectives as well as implemention of a safety management system.

Table 14.1: Six Basic Core Principles of a Typical Management System

• Management leadership • Employee involvement • Risk and hazard identification and assessment • Hazard prevention and control • Information and training • Evaluation of program effectiveness

Lesson Learned # 1

In addition to assessing the safety management system. Look into your own mirror and assess:

- What is the current perception held by the organization about safety-related activities?
- What is your level of influence within the organization? How are you perceived in the organization?
- What is the span and depth of your network in your organization?

If you have been tasked with developing or revitalizing a safety management system, the questions in Lesson Learned #1 may have identified areas that can provide quick-fix solutions to begin changing the current perception held by the organization about you as well as current safety-related activities. You will identify the span, influence, control and depth of your network. If your level of influence is found to be low, your potential for creating change will be limited. The overriding question is, what is your reach into the overall organization when advising on the desired safety management system?

> *If you want to reach a goal, you must 'see the reaching' in your own mind before you actually arrive at your goal.*
>
> **Zig Ziglar**

Avoiding a "Blame the System" Mentality

In many cases we tend to blame the safety management system as the problem when the result we get is not what was either intended or because it did not perform as well as we expected. This is a "shoot the messenger" mindset that leads to an ongoing search for a more effective system, even when it is known that most systems have the same basic core principles. You can lose sight of what is trying to be accomplished and waste time and resources as new structures and formats are developed and efforts restarted. Leadership and employees lose interest in the current safety management system, knowing that it may change after all of their effort. You may desire to evolve from one basic system to a more formally recognized format, but this can be normally accomplished using a gradual process as we briefly discussed in Chapter 1, "The Perception of Safety".

To combat a cycle of "Blame the System" mentality, a constant active search for defects in the system, as Six Sigma techniques stress, is necessary. The objective is not to change the system but to find and fix the defects or errors in the current system (How to Effectively Assess and Improve your Safety and Health Program through Safety and Health Evaluation, n.d.; Six Sigma Measure Phase, Data Collection Plan and Data Collection, n.d.).

> Six Sigma Learning about defeats in the System
>
> - A defect is defined as anything outside of customer specifications.
> - An opportunity is the total quantity of chances for a defect.
>
> **(Six Sigma Measure Phase, Data Collection Plan and Data Collection, n.d.)**

A safety management system assessment goes beyond just audits of safety program elements where the intent is to find only errors or problems and ignore what is being done correctly. A continuous negative approach makes the system appear to be just a "police state" function. Individual activities such as safety walk through, incident investigations, etc., and their use or lack of use are "artifacts" that give indications about the value placed on the safety management system. These activities are program elements that are sub-parts that are integrated into the overall safety management system. Refer to Chapter 2, "Assessing and Analyzing your Organizational Culture".

Types of Safety Management Assessments

As with any process, a safety management system consists of inputs, process, and outputs. These three categories provide another dimension of the safety management system (How to Effectively Assess and Improve your Safety and Health Program through Safety and Health Evaluation, n.d.).

To be effective and timely, an assessment should be part of an overall strategy scheduled into your annual planning process. The assessment can be spread over time to make use of available windows of opportunity given all the other activities that must be completed by the leadership and employees who should be involved.

An assessment should include members drawn from the leadership team, safety committee members, and employees with specific knowledge of the topic under discussion. The best evaluators will be employees who can provide an unbiased and fresh look at the organization. While many employees will be candid in their comments, care must be taken to not shut down critical comments or discussions as this results in not getting the solid honest feedback essential to any assessment.

> *Before various measurement systems are devised, their purpose and limitations need to be thoroughly understood.*
>
> **(Spear, 2010)**

Refer to Figure 14.1, "Basic Safety Management System Assessment Process".

Two types of assessments are used:

- An internal self-assessment.
- An external or third-party assessment.

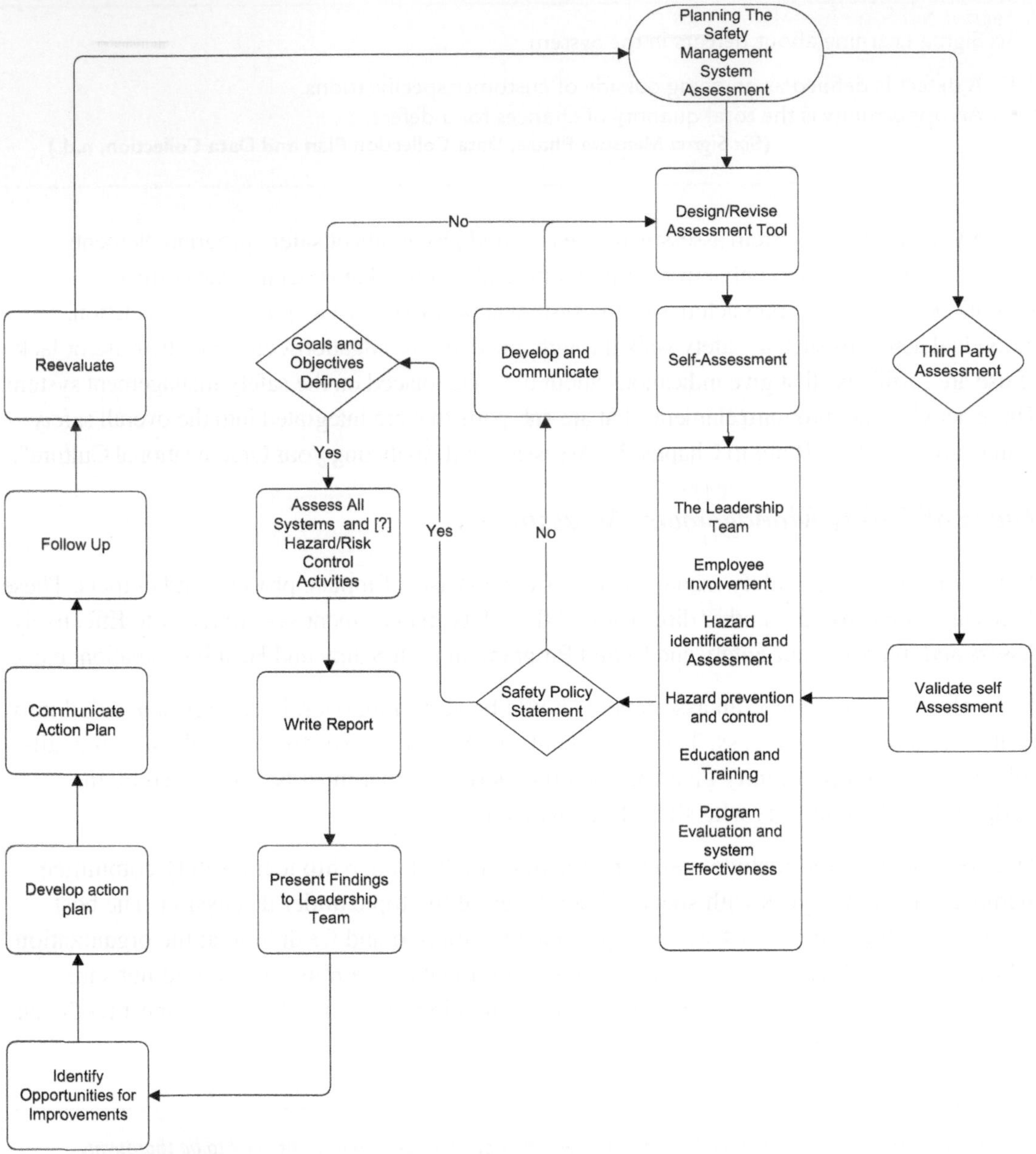

Figure 14.1
Basic Safety Management Assessment Process.

Internal Self-Assessments

This type of assessment uses your leadership team and employees to review, analyze, document, and communicate findings regarding the status of the system to all leadership and employees in the organization.

> *Self-assessment is the formal or informal process of identifying one's own opportunities for improvement by comparing present practices and results with desired practices, results, and standards. Because no one knows better how things are done in the facility than those working in the performing organizations, self-assessments can be the most effective means of identifying latent weaknesses in the organization and in the facility.*
>
> **(Volume 2: Human performance tools for individuals, work teams, and management, human performance improvement handbook, 2009)**

Lesson Learned # 2

James, as a regional safety manager, set up a system review method that had each plant take on one element of the organization's safety management system under review each month. As the system had 10 sections, a full review would be completed in 10 months. Each review was placed on the formal agenda schedule and discussed over a lunch or staff meeting with designated employees participating. A list of questions was provided to guide the internal assessment. The benefit was that time was spread throughout the year and the safety management system was kept visible as it was formally placed on agendas throughout the year.

To be consistent with your assessment, a customized self-assessment worksheet or checklist is used that covers all of the core principles and subelements that are to be considered under each principle. The key is to consistently use a regularly scheduled self-assessment to evaluate current performance and compare the results against the ideal or expected. Refer to Appendix S "Safety Management System Assessment Worksheet" and Appendix T, Attributes of a Safety Management System.

> *An organization may choose to use a qualitative or a quantitative evaluation system based on its size, operations, services, or culture.*
>
> **(Occupational Health and Safety Systems, 2012)**

Third-Party Assessments

Third-party assessments are conducted by an outside resource selected or designed to provide an unbiased view of the system. While internal self-assessments are an essential part of the assessment process, an outside point of view or perspective can uncover or identify additional gaps that have been unconsciously overlooked by the internal assessment. The day-to-day familiarity of "how things get done around here" may have allowed the process to gradually drift away over the course of time from the initial guidelines and standards established. If the organization has completed an internal self-assessment, the independent review will serve as verification (calibration) of management commitment to safety.

> *Reviews of facility activities by outside organizations or agencies provide an opportunity to reveal 'blind spots' to facility management and personnel that could otherwise remained hidden. Independent oversight can identify conditions, processes, and practices that fall short of industry best practices and those that could lead to degraded facility performance if uncorrected.*
>
> **(Volume 2: Human performance tools for individuals, work teams, and management, human performance improvement handbook, 2009)**

Refer to Figure 14.1, "Basic Safety Management Assessment Process".

Selecting the Assessment Team

To ensure effective employee involvement, a cross-functional, diverse team is used to administer the assessment. By maintaining a diverse perspective, the different experiences, skills, and expertise can bring additional insights. The team membership should be rotated using a percentage of new members for each assessment.

Refer to Table 14.2 for a "Review Process Checklist and Structure, Suggested Review Structure".

The team members should have been trained in the concepts and elements of the safety management system and have the technical expertise to review process and its supporting elements and programs.

> *Gain buy-in from plant floor through crosses functional collaboration. In order to drive the vision of the executive, collaboration amongst the functional groups is critical: EH&S, quality, safety, manufacturing, engineering, production, and even the supply chain. Without regular sharing of vision and effort, it becomes far more difficult to achieve the performance benefits of the Leaders.*
>
> **(Ismail, 2012)**

Preassessment Activities

An initial coordination is made with the location having a safety management system review. The preassessment is to begin the coordination with local leadership on the requirements of the assessment, its planning, and communication.

Table 14.2: Review Process Checklist and Structure, Suggested Review Structure

Suggested team members:
• The safety manager or coordinator or a person knowledgeable about the safety management process to act as assessment coordinator. • A member of leadership or safety management from a different company location. A safety consultant with an understanding of the organization might be considered. • One member of the location leadership team, for example, production/plant manager. • Hourly employees on a selected basis, particularly from another department or plant.

Source: Roughton & Mercurio (2002).

- Establish the dates and times of the assessment that are mutually acceptable to the local leadership and schedules of both the location and the assessment teams.
- Coordinate the assessment team for the available dates and times.
- Request the required documents and information to be available at the beginning of the assessment:
  - Background information about the location's current activities and overall status.
  - Previous assessments, five years' worth if possible, including other internal and external surveys or assessments by insurance loss control, other consultants, etc.
  - Current policies, procedures, programs, job hazard analyses, inspection reports (machine/equipment checklist, property/facility specific), safety meetings, and safety committee meetings, etc.
  - Leadership and employees' perception survey results.
  - Past history of loss-producing events—injuries, damage reports, near misses, insurance claims reports (workers' compensation, vehicle, liability, property, incident investigations).

Safety perception surveys are an important tool used to ask questions that employees can answer anonymously and in confidence. They provide the ability to measure the answers to questions and summarize responses. They can provide a statistical validation of the perceptions of employees about the safety culture. Surveys must be carefully selected and structured to prevent personal bias and a skewing of the responses.

Refer to Appendix U, "Safety Management Perception Questionnaire" for an example of a perception survey, an instrument provided by Bob Lapidus and Dr Mike Waite that provides a validated format that has been used to gain insights from employees (Lapidus & Waite, 2013).

The US Department of Energy, Human Performance Improvement Handbook, Volume II, has several surveys and tools that can be used to aid in defining the organizational safety climate, human performance gap analysis tool, and a job site conditions self-assessment (Volume 2: Human performance tools for individuals, work teams, and management, human performance improvement handbook, 2009).

Opening Meeting

The opening meeting is used to refresh with the local leadership team and employees the purpose, scope, intent, and activities that will be needed for the assessment:

- Discuss the objectives and scope of the assessment and its intent.
- Discuss the assessment process, checklists/forms, and the resources needed (schedule of interviews, employee group meetings, focus groups, etc.).
- Introduce assessment team members and share expertise and skills.
- Discuss personal safety considerations, for example, personal protective equipment, visitor guidelines, escorts areas, security clearances, etc.
- Establish communication, media to use (Internet, Skype™, emails, etc.), develop schedule based on activities, timing.
- Develop schedule and agenda for the closing meeting.

The preassessment meeting should be brief and, if all the preassessment activities have been completed, should take approximately 30 minutes–one hour (Safety and Health Achievement Recognition Program, n.d.).

Safety Management Safety System Assessment Activities

Activities involved in the assessment of a safety management system include:

- An initial tour to become familiar with the nature of operations, worksites, and facilities.
- Review of requested documents and materials provided.
- Begin the process using more in-depth observations of activities, specific high-hazard jobs, and the general environment, etc.
- Interview employees while doing observations.
- Communications and feedback to the leadership team and employees.

As we have stressed throughout this book, any activity that is undertaken must be considered as having the potential for either improving or degrading the safety culture of the organization. The methods used for coordination and the completion of the assessment must be conducted in a manner that is seen as professional, well organized, respectful, and considered of value to the location's leadership and employees. An assessment has the potential to be seen as critical of the leadership and, if not communicated and properly presented, may create issues within the organization.

Initial Location Tour

When the opening meeting has concluded, the assessment team completes a tour of the location to familiarize itself with the nature of the equipment, tools, materials, processes in use, and general activities. This tour provides the team with an initial perspective of conditions, hazards, and associated risks.

Document Reviews

The review of the location's documentation determines the scope and quality of each of the specific activities required by the safety management system that is in use. The documentation should provide a comprehensive and clear picture of how the location has been administering and managing the safety management system.

> *There are many ways to categorize safety performance measures. They are often classified as trailing or leading indicators, outcome or process oriented, results or activity-based measures, downstream factors or upstream factors, and/or qualitative or quantitative metrics.*
>
> **(Spear, 2010)**

> *Without measurement, accountability becomes meaningless.*
>
> **(Spear, 2010)**

> Problems with Current Measure(s):
>
> - Relying on any single metric is problematic.
> - Occupational Safety and Health Administration (OSHA) rates do not drive superior safety and health performance.
>   - Overly inclusive.
>   - Not very accurate.
> - The more pressure you put on them, the less accurate they get.
> - Safety and Health (S&H) measurement mindset is one of tracking failure or showing loss avoidance, not positive contribution to the business.
> - S&H metrics often undermine management creditability.
>
> **(Newell, 2001)**

Leadership and Employee Interviews

The primary goal is to collect information about employee involvement in the safety management system using three methods:

- Formal interviews with the leadership team.
- Formal interviews with small groups of employees.
- Informal one-on-one interviews with employees in their work environment.

The process starts with formal interviews with the senior leader/general manager. After a discussion that acquires the senior leader's insights and concerns, interviews would be conducted with selected individuals of the leadership team (managers, supervision). The interviews will be made to include all workshifts and at various times throughout each workshift.

Meetings with small groups of employees (optimum size is six employees) will be held with a representative number of employees to discuss their understanding and use of the safety management systems. The meetings should last approximately one hour based on the size of the group and its engagement.

Informal one-on-one interviews will be conducted with employees to verify feedback received from the small group meetings. It should be stressed that all interviews are to be considered confidential. The number of the small groups and individuals to be is interviewed dependent upon the size of the facility.

During walk-through of work areas for additional observations, the assessment team should interview employees one-on-one in their work areas about their training and perceptions of the safety management system. Any topic is acceptable if it leads to a better understanding of problems or concerns regarding the workplace. If safety-related training is effective, employees will be able to describe the hazards inherent in their work and the effectiveness of the controls that are in place. In this discussion, an indicator of the depth of the safety culture is how well employees can describe what is expected of them in any safety-related activities.

As examples, employees' perceptions and/or opinions about the ease or difficulty in reporting hazards and the time it takes to get a response will validate the real effectiveness of any hazard reporting system. If employees indicate that the system for enforcing safety rules and work practices is inconsistent or confusing, a gap in this system element is identified.

Communication and Feedback

The assessment team leader is the focal point of all information from the team and provides location leadership with updates on the assessment process. If any potentially high severity hazards and associated risks are identified, the assessment is stopped and the problem/issue is immediately relayed to leadership for corrective actions. At the end of each day or workshift, a meeting should be held with the leadership team and employees to update findings, additional documentation needs, and immediate concerns about the assessment.

Before the final assessment closing meeting, the assessment team provides a verbal review of the findings and recommendations with the leadership team. The assessment is to be a mutual effort and should not "blindside" leadership with its findings in any final written report.

Review of Site Conditions

The review of site conditions can provide a very strong indicator about the true safety culture of the organization. If the documentation shows that inspections are routinely completed, yet the conditions indicate uncontrolled hazards and that associated risks are clearly present, then a gap exists between what the organization espouses and what beliefs and norms are actually in place. Conditions are "artifacts", tangible visible signs about what is allowed to exist. A well-written inspection program may be available but obviously is not properly used. Refer to Chapter 2, "Analyzing the Organizational Culture".

Presenting Results of the Safety Management System Assessment

The methods for presenting the finding of the assessment should be determined in the initial planning meeting with a clearly defined format, style, and structure. The structure should follow the outline of the safety management system's content for consistency over time. A wide range of methods can be used that can involve either quantitative and/or qualitative information. A scoring method is usually followed to allow a measurement of the finding along a predetermined scale.

Lesson Learned # 3

James recalls an assessment that he participated in, when at the initial meeting of leadership team members and employees, everyone was given a paper showing the past scores received in previous assessments. At the final assessment review, unfortunately, the scores were lower than a previous assessment. An immediate problem arose as the assessment team was challenged to prove why it had rated the location lower. Resistance to assessment results must be considered in the initial planning and throughout the assessment by documenting in detail why and how any gaps are identified and backup information that can be logically discussed.

The use of interactive spreadsheets and custom-designed software is available and has been successfully used. By using a spreadsheet or similar format, scoring can be tracked over time to show the improvement or lack of improvement of the safety management system.

> *Measuring safety is difficult because it is difficult to predict the impact that new safety metrics will have on individual behavior, attitudes, and the overall safety climate.*
>
> **(Spear, 2010)**

Developing the Action Plan

After the leadership team has reviewed gaps identified by the assessment, opportunities and obstacles for improvement are discussed with the assessment team. The location would be responsible for establishing an action plan with goals and objectives needed to close the gaps identified in the final report. The action plan should describe the specific actions that will be taken, responsibilities, and the estimated completion dates for corrective actions.

Refer to Chapter 4, "Setting the Direction for the Safety Culture" for an overview of the S.M.A.R.T. method for developing an action plan's goals and objectives.

Communicating the Assessment and the Action Plans

Throughout the assessment, members of leadership and employees will have been involved in the meetings, discussions, reviews, and observations. It is important that all employees be provided with the results of the assessment, its findings, recommendations, and the action plan that has been developed. In our experience, too often employee involvement is not recognized by a sharing of information.

After the action plan has been developed, the findings should provide:

- A summary of the final assessment report.
- Information on how the assessment was completed and who was involved.
- A list the assessment findings and opportunities for improvement.
- The action plan that may have resulted from the assessment.

Communicating results in a timely manner fulfills the commitments made to employees and demonstrates the organization's support for the safety management system.

Example Assessment and Action Plan

Refer to Appendix V, "Example Safety Management System Assessment and Action Plan" for an example outline that provides the various activities that might be included in the assessment of a safety management system. It follows a strategy, measurement, treatment outline to establish how the assessment would be organized, the metrics required, and how hazards and associated risk would be treated.

This example is based on the ANSI Z10-2012 content (Occupational Health and Safety Systems, 2012) to show how an assessment could follow any selected safety management system to ensure that all of its components are reviewed. With this example, you will note the shift from reviewing elements to implementing specific actions. This was done to emphasize that the assessment may identify areas that must receive immediate action, and planning should consider that this may occur.

The various activities can be placed into a Gantt chart or in a project management form to show when the activities would be scheduled. Activities may vary as to when they might be reviewed or actions taken and they do not have to be done in a specific order.

Summary

A safety management system assessment provides a comprehensive review of the current state of the safety culture. The effectiveness of the safety management system should reflect the real values and beliefs of the organization.

The full implementation of a safety management system provides the structure that guides decisions and activities that are extremely important for the long-term success of your safety culture.

As discussed in several other chapters, you are trying to change the perception of safety, extend your influence within the organization, and expand your internal network. A personal assessment should verify how effectively you are in accomplishing goals as well as implementing the safety management system.

A safety management system assessment goes beyond just audits of safety program elements where the intent is to find only errors or problems and ignore what is being done correctly. A continuous negative approach makes the system appear to be just a "police state" function.

To be effective and timely, an assessment should be part of an overall strategy scheduled into your annual planning process.

An assessment should include members drawn from the leadership team, safety committee members, and employees with specific knowledge of the topic under discussion.

By maintaining a diverse perspective, different experiences, skills, and expertise can bring additional insights. The team membership should be rotated using a percentage of new members for each assessment.

An initial coordination is made with the location having a safety management system review. The preassessment begins the coordination with local leadership on the requirements of the assessment, its planning, and communication.

The opening meeting is used to refresh the local leadership team and employees with the purpose, scope, intent, and activities that will be needed for the assessment.

The review of the location's documentation determines the scope and quality of each of the specific activities required by the safety management system that is in use. The documentation should provide a comprehensive and clear picture of how the location has been administering and managing the safety system.

An important goal is to collect information about employee involvement in the safety management system.

The assessment team leader is the focal point of all information from the team and provides location leadership with updates on the assessment process.

The review of site conditions can provide a very strong indicator about the true safety culture of the organization.

After the leadership team has reviewed gaps identified by the assessment, opportunities and obstacles for improvement are discussed with the assessment team.

Chapter Review Questions

1. Discuss and develop the coordination needed for a safety management system assessment.
2. Discuss how to avoid the mentality of "blaming the system".
3. Discuss the types of safety management systems assessments.
4. Discuss the selection and development of an assessment team.
5. Discuss some of the preassessment activities before conducting a safety management system assessment.
6. Discuss the elements of a safety management system assessment.
7. Discuss the importance of the final report and action planning.
8. Discuss why it is important to communicate the results of the assessment to all employees.

Bibliography

Duhigg, C. (2012). *The power of habit: Why we do what we do in life and business*. Canada: Doubleday.

Geigle, S. J. (2011). Safety Management System Evaluation, Course 716 Study Guide. OSHAcademy™ Course 703 Study Guide, Permission to Reprint, Modify, and/or Adapt as necessary. Retrieved from http://bit.ly/YKv7Br.

How to Effectively Assess and Improve your Safety and Health Program through Safety and Health Evaluation. (n.d.). Oregon Occupational Safety and Health Division (Oregon OSHA), OR-OSHA 116, Public Domain, Permission to Reprint, Modify, and/or Adapt as necessary. Retrieved from http://bit.ly/106AhmR.

Ismail, N. (April 2012). Environmental, Health and Safety, Going Beyond Compliance, White Paper. Aberdeen Group, A Harte-Hanks Company. Retrieved from http://bit.ly/12XJnnk.

Lapidus, R. A., & Waite, M. J. (2013). *Safety management perception questionnaire*. Lapidus & Waite. Permission to reproduce.

Managing Worker Safety and Health. (n.d.). Illinois OSHA Onsite Safety & Health Consultation Program, Public Domain, Adapted for Use. Retrieved from http://bit.ly/WTsneh.

Newell, S. A. (2001). *A new paradigm for safety and health metrics: Safety and health metrics: Framework, tools, applications, framework, tools, applications, and opportunities*. Reprinted with Permission, ORC.

Occupational Health and Safety Systems. (2012). The American Industrial Hygiene Association. ANSI/AIHA Z10.

Roughton, J. E., & Mercurio, J. J. (2002). *Developing an effective safety culture: A leadership approach, Adapted for Use*. Butterworth-Heinemann. Retrieved from http://amzn.to/X8Gaz8.

Safety and Health Achievement Recognition Program. (n.d.). Oregon Occupational Safety and Health Division (Oregon OSHA), SHARP Program, Public Domain, Permission to Reprint, Modify, and/or Adapt as necessary. Retrieved from http://bit.ly/XHf9nd.

Schein, E. H. (2004). *Organizational culture and leadership. The Jossey-Bass business & management series*, John Wiley & Sons.

Shewhart, W. A. (1931). *Economic control of quality of manufactured product.*

Six Sigma Measure Phase, Data Collection Plan and Data Collection. (n.d.). tutorialspoint, Permission to Reprint/Modify/Adapt for Use. Retrieved from http://www.tutorialspoint.com/six_sigma/index.htm.

Spear, J. E. (2010). Measuring Safety and Health Performance, A Review of Commonly-Used Performance Indicators. Spear Consulting, LP. Retrieved from http://bit.ly/11DELYw.

Volume 2: Human Performance Tools For Individuals, Work Teams, And Management, Human Performance Improvement Handbook, Public Domain, 2009.

CHAPTER 15

Becoming a Curator for the Safety Management System

Knowledge is power.
—Sir Francis Bacon

Introduction

Throughout this book, a central theme has been the emphasis on communication and the flow of information through an organization. The safety culture is dependent on leadership and employees receiving and understanding the importance of maintaining open lines of communication, working toward reducing barriers and resistance to the message being sent.

As the focal point for safety-related information, you are the organization's memory regarding the documentation of hazards, associated risk, safety management system performance, regulatory compliance, job hazard analysis, etc. As the organizational source for valid and timely safety-related knowledge, you must be able to track and locate information as quickly as possible to get it into the hands of the appropriate decision-makers for further action.

From day one, you must adopt a library and filing methodology and structure that can organize all the media you must use to accomplish your assigned responsibilities. This would include hard copies and e-copies of materials, websites, reports, loss data, risks assessments, books, and much more. Your objective is to have a methodology that can access and retrieve the specific information for a timely and quick response to risk and hazardous situations or requests for assistance.

The task of converting data into useful information of value is a universal requirement. To be successful, converting your knowledge from theory into usable tools for the organization requires application of the concepts we have covered in other chapters.

The objective of this chapter is to outline techniques and concepts that allow you to organize and efficiently access research material and information needed by the safety management system. Whether you are writing and developing an article, a book, a safety program, training session, or even working on an advanced degree, you must do in-depth research to find related material for your project.

When you complete this chapter, you will able to:

- Discuss the importance of becoming an information curator.
- Define and discuss curation.
- Discuss various concepts to organize information.
- Discuss several tools that might be used in organizing information.

The Importance of Becoming an Information Curator

We live in an era where we must deal with a fire hose stream of information and data. Organizations live on information and data that is received from many sources, both internal and external. As a result, your information and message must be clear, well designed, and written in terms that can break through the background constant clutter of useful and useless information that goes through an organization.

> *Curation—The act of curating, of organizing and maintaining of a collection of artworks or artifacts.*
>
> **(Curation, n.d.)**
>
> The term curation is now used to cover the gathering, organizing, and maintaining of all types of information

Influencing the leadership team and employees is based on you becoming a knowledgeable person who knows where to find answers and provides the documentation for inquiries and issues. Again, the more the organization comes to you for answers, the greater the potential for increasing the positive perception about the safety management system and, in turn, improving the safety culture.

An article by Edward Curry, Andre Freitas, and Sean O'Riain provides insight into data curation—"Data curation is a process that can ensure the quality of data and its fitness for use".

> *Making decision based on incomplete, inaccurate, or wrong information can have disastrous consequences. Decision making knowledge workers need to have access to the right information and need to have confidence in that information. Data curation can be a vital tool to ensure knowledge workers have access to accurate, high-quality, and trusted information that can be reliably tracked to the original source, in order to ensure its credibility.*
>
> **(Curry, Freitas, & O'Riain, n.d.)**

They suggest that:

The following data quality dimensions are highly relevant within the context of enterprise data and business users:

Discoverability & Accessibility–Addresses if users can find the data and then access it in a simple manner.
Completeness—Is all the requisite information available?
Interpretation—Is the meaning of data ambiguous?
Accuracy—Is the data correct?
Consistency—Does the data contradict itself?
Provenance & Reputation—Is the data legitimate?"
Timeliness—Is the information up-to-date?

(Curry et al, n.d.)

If you follow a curation method, your information resources can be better used in establishing more effective plans with goals and objectives for improvement.

ANSI Z10-2012 requires that a process be established to create and maintain documents and records.

The organization shall establish a process to create and maintain documents and records... in order to:

1. Implement an effective OHSMS.
2. Demonstrate or assess conformance with the requirements of this standard.

(Occupational Health and Safety Systems, 2012)

Function of a Safety Management System

As we have discussed before, we have defined safety as an "emerging property" that develops from the interaction of the many components of a complex adaptive system. As such you will, by necessity, be required to have a working knowledge of just about anything and everything associated with the organization.

> *...Providing visibility to key decision makers to use the tools such as dashboards and analytics will ensure that decisions are made intelligently and at the right time to prevent any adverse events that can result from a catastrophic failure.*
>
> **(Ismail, 2012)**

For a safety culture to be maintained, it must have the best possible information to preserve the level of knowledge required to keep risk and hazards clearly identified. As people come and go, reorganizations occur, and new endeavors begin, information can be lost without a comprehensive method for maintaining documents or records. Resources must be expended unnecessarily in redoing materials or searching for information that had already been identified, developed, or is now unavailable because it is not categorized correctly. In addition, information may be on file that is not required or pertinent, is outdated, or duplicated.

Researching and Curating Information

We have found a direct parallel exists between writing a book and managing a safety management system. A book requires a central theme that is supported by expertise, solid references, correct in-text citations, and research. During the course of writing this book, we accumulated several thousand documents as well as many other reference materials that we had to read, categorize, and sort through to find safety culture-related information. Finding the appropriate documents is relatively easy given the scope and depth of the Internet. To retrieve, organize, and be able to provide references and proper citations is the challenge. You must be able to cite your sources to support your conclusions and recommendations.

Lesson Learned # 1

When writing our book *Job Hazard Analysis: A Guide For Voluntary Compliance And Beyond*, we found organizing data an enormous task. It was a real challenge to keep up with citations, references, figures, appendices, tables, etc. Keeping track of the research and changes in chapters was a constant battle. We quickly realized there was a major need to organize materials in binders, composition notebooks, hardcopy, electronic, and Internet resources that were used to support our opinion, experience, concepts, and research. The lesson we learned was to establish a plan in the beginning of a project on how all types of media and information will be gathered, organized, and maintained. This was different from our prior experience in file management.

> As quoted in the New York Times by Alex Williams, *in the case of curate, which the Oxford Dictionary simply defines as 'to look after and preserve,' its standard 'museum' meaning dominated until the mid-90s, when references to curating hotel libraries and cd-of-the-month clubs started to pop up in periodicals, said Jesse Sheidlower, a lexicographer with the Oxford English Dictionary.*
>
> **(Williams, 2009)**

In doing research on any topic, you will amass numerous documents that you must wade through to find, mark, or annotate the specific information that you need. This may be for a recommendation, to build a needed program or process, or to simply answer a question. On many occasions, an article may be of interest to you or applicable to the organization but not needed until a future date.

This process requires continuous effort as the fire hose of information flows your way. No one has found the perfect answer to fully manage and control this flood of information. However, as part of any process, an ongoing effort is needed to keep your citations, references, and library of information organized. To remain effective and current, a knowledge and information management process is part of the core of any true safety management system.

Information found using an Internet browser can be bookmarked. However, the problem with bookmarks is they can get out of hand and make the document harder to relocate. As part of the research process, related literature in pdf format can be organized by bookmarking, adding comments, and/or notes. This entails creation of proper citations and a bibliography as part of your library.

> It has been said, "if you cannot find it, it does not exist!" It may be the best of safety articles or risk-related information, but if you cannot put your hands or eyes on it, then it is only a figment of your imagination.

New Concepts for Organizing Information

The Internet has created easy access to a wealth of information and this access to information is no longer restricted to designated individuals as the use of smartphones, tablets, and other devices expands. In addition, quantity does not mean quality or that the material is true or correct, and you may have to respond to information brought to you from leadership and employees.

Refer to Figure 15.1, "Multiple Sources of Communications and Information".

As a note of caution, question everything you find on the Internet! Just because it exists, does not mean it is correct, accurate, or true. While you must keep track of trends and what others are advocating or have accomplished in your profession, a skeptical approach is always required. You have to become your own professional critic by questioning and interpreting

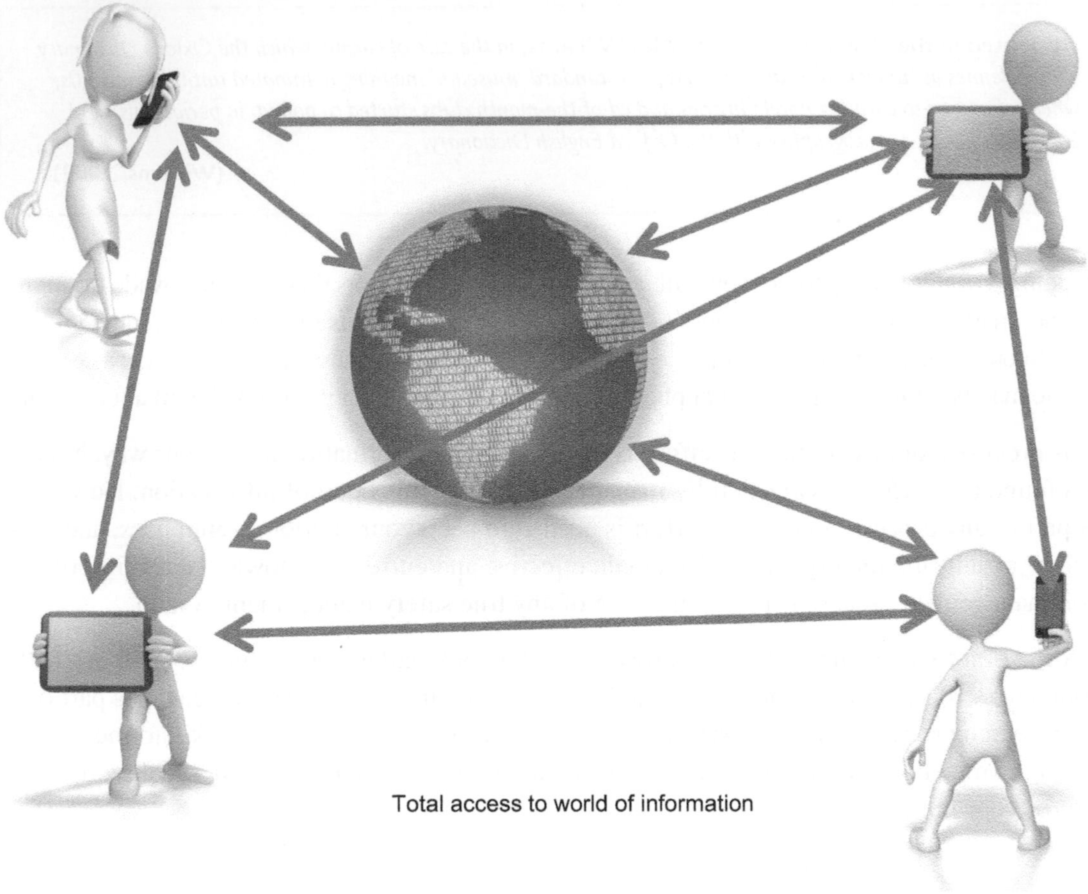

Figure 15.1
Multiple Sources of Communication and Information.

what you read. Keep your mind open to as many perspectives as possible. As the old saying goes, "think outside the box". The key is always to question your source until validated by your experience, expertise, or further research.

Managing Safety Management System Data

Safety management system documents, materials, and information are living documents that must be maintained and kept up-to-date.

The following checklists have been adapted from a data management planning checklist from the University of Edinburgh Research Data Management Guidance. They provide questions that can be used to address and share data (Research Data Management Guidance, 2011).

Step 1—Evaluate Your Data Needs

Questions before beginning to establish a data management process:

- What type of documents, materials, and information will be used?
  - Will they be reproducible? (How will copies be managed?)
  - What would happen if they got lost or became unusable later?
- Who will be the audience for your documents, materials, and information and how will they use them now and in the long run?
- Who controls the documents, materials, and information?
- Will there be any sharing requirements?
- Is any budget needed to gather and maintain the documents, materials, and information?
- How many documents, materials, and information will be generated and how often will they change?
- How long will these items be retained? For example, 5 years, up to 10 years, or permanently?
- Are tools or software needed to create/process/visualize the documents, materials, and information?
- What directory and file naming convention will be used?
- What file formats will be used? Are they long-lived?
- What project and identifiers will be assigned?
- Is there good project and data/information documentation? (How will they be inventoried or cataloged?)
- Are there any special privacy or security requirements? For example, personal data, high-security data, or proprietary?
- What will be the storage and backup strategy?
- When and where will the documents, materials, data, and information be published?
- Is there an organizational guideline or standard for sharing/integration?

(Research Data Management Guidance, 2011).

Step 2—Establish a Plan

A data management plan describes the creating, organizing, documenting, storing, and sharing of data. The plan should cover, protection, confidentiality, preservation, and curation. It provides a framework that supports the safety management system documents, materials, and information.

- What documents, materials, and information will be created or collected for the safety management system?
- What organizational policies and procedures will apply to the documents, materials, and information?
- What management practices (backups, storage, access control, archiving) will be used?
- What facilities and equipment will be required (hard disk space, backup server, and repository)?
- Who will own and have access to the documents, materials, and information?
- Who will be responsible for each aspect of the safety management data?
- How will use of documents, materials, and information be enabled and long-term preservation ensured to maintain the history of the safety management system?

(Research Data Management Guidance, 2011).

Technology and the Safety Management System

Technology has changed the landscape of how the safety management system is managed. Information can be moved easily and quickly well beyond the traditional boundaries of an organization. The safety management system and a strong safety culture can be the firewall that ensures employees know that pathways for communications will always be open and positive and that concerns will be addressed quickly and comprehensively through strong information and data management.

Summary

The safety culture is dependent on leadership and employees receiving and understanding the importance of maintaining open lines of communication and working toward reducing barriers and resistance to the message being sent.

Organizations live on information and data that is received from many sources both internal and external. As a result, your information and message must be clear, well designed and written, and in terms that can break through the background constant clutter of useful and useless information that goes through an organization.

For a safety culture to be maintained, it must have the best possible information to preserve the level of knowledge required to keep risk and hazards clearly identified. As people come and go, reorganizations occur, and new endeavors begin, information can be lost without a comprehensive method for maintaining documents or records.

Finding the appropriate documents is relatively easy given the scope and depth of the Internet. To retrieve, organize, and be able to provide references and proper citations is the challenge.

The Internet has created easy access to a wealth of information and this access to information is no longer restricted to designated individuals as the use of smartphones, tablets, and other devices expands.

Chapter Review Questions

1. Discuss the importance of becoming an information curator.
2. Define and discuss curation.
3. Discuss various concepts to organize information.
4. Discuss several tools that might be used in organizing information.

Bibliography

Curation. (n.d.). Wiktionary, a Wiki-Based Open Content Directory. Retrieved from http://bit.ly/VQ95KJ.

Curry, E. Freitas, A., & O'Riain, S. (n.d.). The role of community-driven data curation for enterprises. Digital Enterprise Research Institute, National University of Ireland, Galway, Ireland. Retrieved from http://bit.ly/W8TQ1O

Ismail, N. (April 2012). *Environmental, health and safety, going beyond compliance, white paper*. Aberdeen Group, A Harte-Hanks Company. Retrieved from http://bit.ly/12XJnnk.

Occupational health and safety systems. (2012). The American Industrial Hygiene Association. ANSI/AIHA Z10.

Research Data Management Guidance. (August 2011) Permission to reprint, modify, and adapt for use, Creative Commons Attribution 2.5 UK: Scotland License. Edinburgh University Data Library. Retrieved from http://www.ed.ac.uk/is/data-management.

Williams, A. (October 2009). *On the tip of creative tongues, curiate*. The New York Times. Retrieved from http://nyti.ms/ZDUdP4.

Final Words: Organizing and Sharing

Ignorance isn't what you don't know, it's what you know wrong.
—Yogi Berra

A safety management system is dependent on the information that it can provide to the organization. The strength of a safety culture is based on information flow and its value to the organization. Keeping your information organized is not an easy thing, yet it is one of the more important aspects necessary for a safety management system. The following discussion covers several of the ways we tried to manage the data and information for this book. We believe these resources would be useful in your own data management.

An Approach to Organizing Information

We found various resources and tools that were available to organize and maintain information, which were generally easy to use. We used the following software available on the Internet to manage this project. Refer to Figure 1, "Planning Your Document" and Figure 2, "Basic Coordination of Literature and Media".

- Docear
- JabRef
- Google Scholar
- Google Search
- Google Drive
- Dropbox

Docear

Docear is an "academic literature suite", which allowed us to increase our structuring and sharing of information. This mind map based tool greatly aided in managing our book content. We feel that if used properly, Docear can provide an excellent framework to tie the safety management system elements and their information together so that they can be tracked and accessed effectively and efficiently (Docear, n.d.).

Docear allowed us to bring together information into a single application in which we established a digit media library and mind mapping technology. It included an academic search engine, file manager, mind mapping and note-taking tool, and reference manager. Built in to Docear is Docear4word, an add-on for Microsoft Word that allows inserting citations into documents and create bibliographies (Docear4Word, n.d.; Mindmap, n.d.).

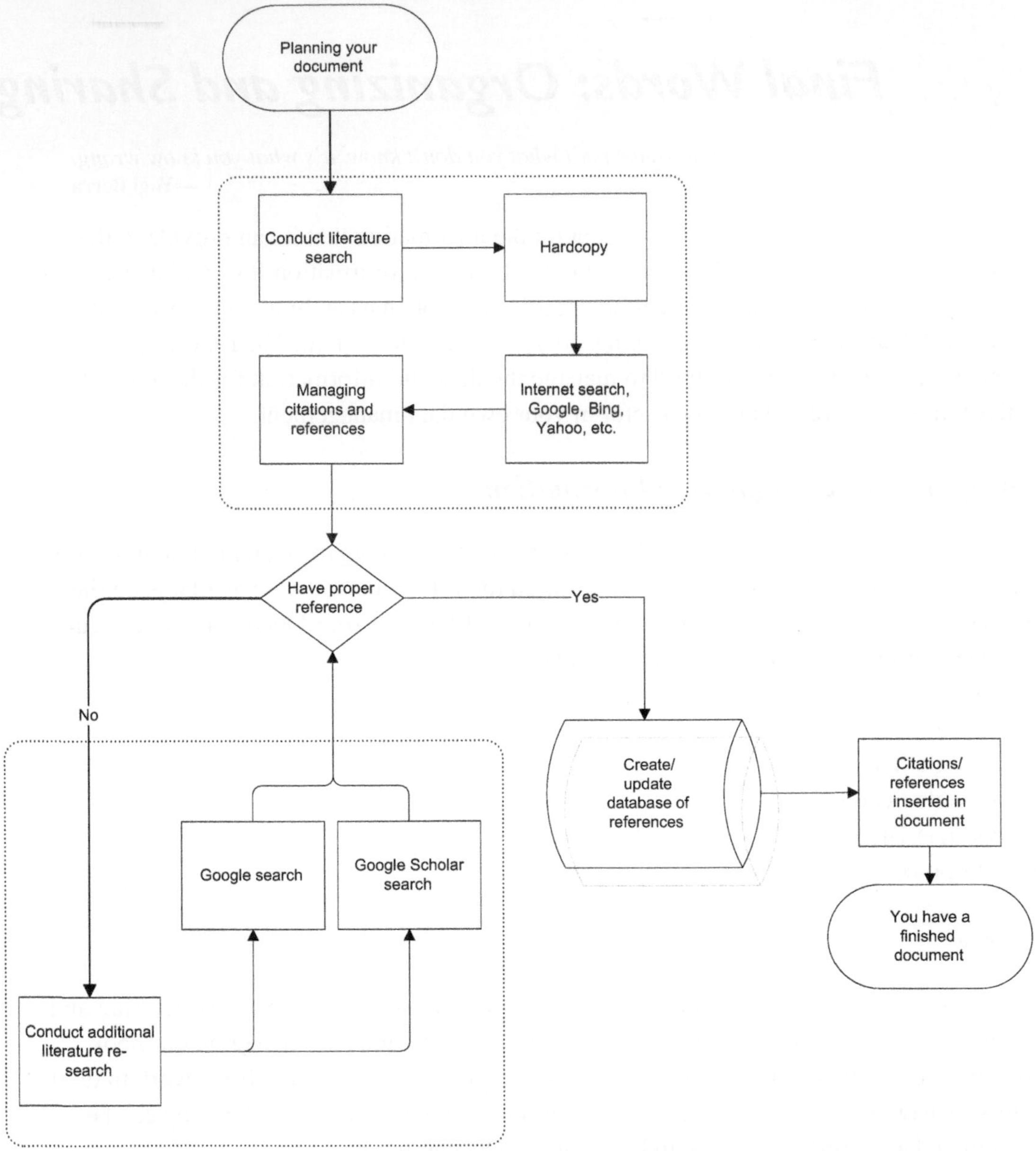

Figure 1 Planning Your Document.

> *Docear is free and open source, based on Freeplane (a mind mapping platform), funded by the German Federal Ministry of Technology and developed by scientists from around the world, among others from OVGU and the University Of California, Berkeley*
>
> **(Docear, n.d.)**

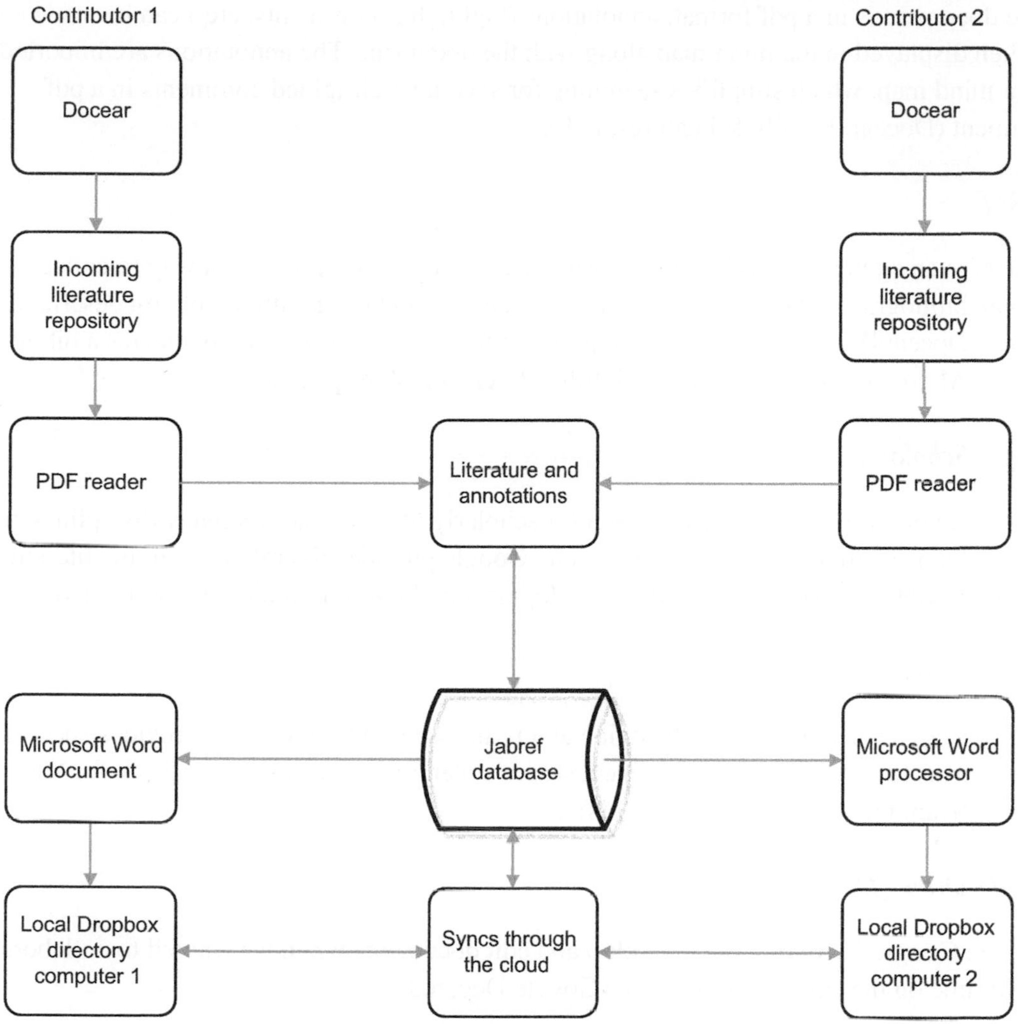

Figure 2 Basic Coordination of Literature and Media.

> "Mind Map is a diagram used to visualize and outline information. A Mind Map is often created around a single word or text, placed in the center, to which associated ideas, words, and concepts are added" (Mindmap, n.d.).
>
> A mind map is similar to network maps as described in Chapter 3, "Analyzing and Using Your Network".

Documents are saved into a "literature repository" subdirectory. Using Docear, documents are then imported into a mind map. The mind map provides a graphic picture showing all the documents in the literature repository for ease of organization. Any document can be directly opened by links embedded in the mind map.

If the document is in a pdf format, annotations (highlights, comments, etc.) can be made and are then displayed in the mind map along with the document. The annotations are imported into a mind map, which simplifies searching for specific highlighted comments in a pdf document (Docear, Details & Features, n.d.).

JabRef

"JabRef is an open source bibliography reference manager and provides a way to collect and manage bibliography data. It is incorporated in Docear and used with the plugin described earlier, 'Docear4Word'. It allows the import of citations and automatically creates a bibliography in a Microsoft Word document" (JabRef Reference Manager, n.d.).

Google Scholar

Google Scholar provides a way to search for scholarly literature across many disciplines and sources such as articles, books, abstracts, etc. Google provides the reference to the literature (called a "Bibtex") that can be used in developing a bibliography (Google Scholar, n.d.).

Google Search

Google Search is a popular search engine and is an essential tool. Google provides specific search terms to narrow down and refine a search statement. Refer to Appendix W for "Google Related Search Criteria (Google Insights)".

Google Drive (Doc)

We used Google Drive to jointly develop and edit documents when we wanted to collaborate in real time on the content of the book (Goggle Doc, n.d.).

Dropbox

We used Dropbox, which is a "cloud" file storage and retrieval application that allowed us to quickly and easily share files and documents. It essentially allowed us to have our own "private drive" to move information between computers in real time (Dropbox, n.d.).

With the wide variety of applications, software, and online tools constantly changing, it is beyond the scope of this discussion to try to identify the full array of current tools available beyond the few we mention at the time of this writing.

Bibliography

Docear. (n.d.). Retrieved from http://bit.ly/Ypx4Pv.
Docear, Details & Features. (n.d.). Docear, Details & Feature. Retrieved from http://bit.ly/WcROh2.
Docear4Word. (n.d.). Docear. Retrieved from http://bit.ly/13gar5F.
Dropbox. (n.d.). Retrieved from http://bit.ly/YpvKwe.
Google Doc. (n.d.). Google. Retrieved from http://bit.ly/YR8C8i.
Google Scholar. (n.d.). Retrieved from http://bit.ly/W4iu3S.
JabRef Reference Manager. (n.d.). Retrieved from http://bit.ly/13g6D4w.
Mindmap. (n.d.). Wikipedia, the free encyclopedia. Retrieved from http://bit.ly/Y2JJWE.

Appendices

Appendix A: Safety-Related Job Descriptions

Retrieved from the Dictionary of Occupational Titles (Dictionary of Occupational Titles, n.d.).

Example 1

Industrial Safety and Health Engineer

Tasks

1. Plan, implement, and coordinate safety programs, which require application of engineering principles and technology to prevent or correct unsafe environmental working conditions.
2. Devise and implement safety or industrial health programs to prevent, correct, or control unsafe environmental conditions.
3. Examine plans and specifications for new machinery or equipment to determine if all safety requirements have been included.
4. Conduct or coordinate training of workers concerning safety laws and regulations, use of safety equipment, devices, clothing, and first aid.
5. Inspect facilities, machinery, and safety equipment to identify and correct potential hazards and ensure compliance with safety regulations.
6. Conduct or direct testing of air quality, noise, temperature, or radiation to verify compliance with health and safety regulations.
7. Provide technical guidance to organizations regarding how to handle health-related problems, such as water and air pollution.
8. Compile, analyze, and interpret statistical data related to exposure factors concerning occupational illnesses and accidents.
9. Install or direct installation of safety devices on machinery.
10. Investigate causes of industrial accidents or injuries to develop solutions to minimize or prevent recurrence.
11. Conduct plant or area surveys to determine safety levels for exposure to materials and conditions.
12. Check floors of plant to ensure they are strong enough to support heavy machinery.
13. Design and build safety devices for machinery or safety clothing.

14. Prepare reports of findings from investigation of accidents, inspection of facilities, or testing of environment.
15. Maintain liaison with outside organizations, such as fire departments, mutual aid societies, and rescue teams.

Example 2

Safety Manager

Tasks

1. Plan, implement, and coordinate programs to reduce or eliminate occupational injuries, illnesses, deaths, and financial losses.
2. Identify and appraise conditions that could produce accidents and financial losses and evaluate potential extent of injuries resulting from accidents.
3. Conduct or direct research studies to identify hazards and evaluate loss-producing potential of given system, operation, or process.
4. Develop accident-prevention and loss-control systems and programs for incorporation into operational policies of organization.
5. Coordinate safety activities of unit managers to ensure implementation of safety activities throughout organization.
6. Compile, analyze, and interpret statistical data related to exposure factors concerning occupational illnesses and accidents and prepare reports for information of personnel concerned.
7. Maintain liaisons with outside organizations, such as fire departments, mutual aid societies, and rescue teams to assure information exchange and mutual assistance.
8. Devise methods to evaluate safety programs and conduct or direct evaluations.
9. Evaluate technical and scientific publications concerned with safety management and participate in activities of related professional organizations to update knowledge of safety program developments.
10. May store and retrieve statistical data using computer.

Example 3: Adapted from an Internet-Posted Safety Manager Job Description

This key leadership role provides safety, health, and environmental (SH&E) support to distribution and sales centers, encompassing a large fleet, warehouses, route sales, and a highly distributed workforce. It is a hands-on position, reporting directly to the Region General Manager. There is a functional reporting relationship to the Corporate Group Leader for SH&E. The successful candidate must be able to travel within their assigned region (up to 50%).

Tasks

1. Lead the assigned region in a definitive improvement effort to achieve world class safety, health, and environment (SH&E) performance.
2. Adopt zero tolerance for injuries and incidents and drive this mentality throughout the region.
3. Provide leadership and resource support to regional and area safety committees.
4. Partner with safety coordinators and safety system leaders to ensure that all risk and safety systems are fully developed and implemented with the assigned region.
5. Ensure assigned regions meet all DOT, OSHA, EPA, and other applicable federal, state, and local regulatory requirements, including all risk management, insurance, fire, and security requirements.
6. Conduct regular walk-throughs of all work areas to identify hazards, evaluate controls, recommend improvements, and communicate with team members.
7. Promote safety visibility.
8. Determine SH&E best practices and train individuals/teams. Accomplish all regulatory training and related activities.
9. Lead formal incident investigations to identify root causes. Drive corrective action to prevent reoccurrence.
10. Manage modified work programs to ensure timely rehabilitation of disabled employees. Serve as the liaison with medical providers/clinic.
11. Manage workers' compensation claims and partner with the insurance carrier to promote quality care and cost-effective outcomes.
12. Actively participate as a key member of the National SH&E Network.

Experience—External

- Bachelor's degree in Organizational Management, Engineering, or similar field.
- 5+ years of experience creating, executing, and managing a region-based safety and training program.
- Thorough knowledge of OHSA and state regulations and requirements.
- Demonstrated leadership abilities.
- Strong organization skills.
- Ability to multitask effectively and efficiently.
- Strong communication and time management skills.
- Ability to take initiative without external direction.

Appendix B: Assessing the Perception about How You Are Perceived in the Organization

Based on our experience, the following questions may provide insight about the perceptions held regarding the role of safety and give indications about the overall safety culture.

This questionnaire will help define your professional brand.

Assessing Yourself	Comments
Who do you report to in the organization and into what department?	
Where is your office? · Is it in an area associated with leadership? In a separate building? Cubicle? In operations? · Is the safety office equipped to the same quality as other offices? · Is it in an accessible area or does its location act as an obstacle to employees needing to discuss a safety issue? · Is it adequate for the activities of the safety office? · Is it well organized with adequate file and storage areas?	
How well organized are you? Is your office well organized with regard to files, resources, manuals, etc.?	
Do you have access to the latest technology and is it kept up to date?	
Is your budget adequate for the mandates and needs of the safety management system?	
Are you included in business communications and in leadership team meetings?	
Are you involved in business planning sessions?	
What is your communication style? Does it parallel the communication style of the organization and the leadership team?	
Are you providing timely information in a form that the leadership team understands?	
What is your personal image? Do you dress and look professional? When others see you, do they view you as a peer? Professional? Leader?	
Are you seen as a person who can solve problems and is responsive to questions and requests?	

Appendix C: Safety Culture Traits and Indicators

The final safety culture policy statement issued by the US Nuclear Regulatory Commission in January 2012 provides the following traits of a positive safety culture:

1. "Leadership Safety Values and Actions—leaders demonstrate a commitment to safety in their decisions and behaviors;

2. Problem Identification and Resolution—issues potentially impacting safety are promptly identified, fully evaluated, and promptly addressed and corrected commensurate with their significance;
3. Personal Accountability—all individuals take personal responsibility for safety;
4. Work Processes—the process of planning and controlling work activities is implemented so that safety is maintained;
5. Continuous Learning—opportunities to learn about ways to ensure safety are sought out and implemented;
6. Environment for Raising Concerns—a safety conscious work environment is maintained where personnel feel free to raise safety concerns without fear of retaliation, intimidation, harassment, or discrimination;
7. Effective Safety Communication—communications maintain a focus on safety;
8. Respectful Work Environment—trust and respect permeate the organization; and
9. Questioning Attitude—individuals avoid complacency and continuously challenge existing conditions and activities in order to identify discrepancies that might result in error or inappropriate action."

(The Safety Culture Policy Statement, 2012)

The NRC defines trait as a pattern of thinking, feeling, and behaving that emphasizes safety, particularly in goal conflict situations (e.g., production, scheduling, and the cost of the effort versus safety).

Appendix D 25 Ways to Know Your Safety Culture Is Awesome

25 Ways to Know Your Safety Culture Is Awesome

1. There is visible leadership commitment at all levels of the organization.
2. All employees throughout the organization exhibit a working knowledge of health and safety topics.
3. There is a clear definition of the desired culture the organization wishes to achieve.
4. There is a lack of competing priorities–safety comes in first every time!
5. There is visible evidence of a financial investment in health and safety.
6. Opportunities for improvement are identified and resolved **before** a problem occurs.
7. There is regular, facility-wide communication on health and safety topics.
8. A fair and just discipline system is in place for all employees.
9. There is meaningful involvement in health and safety from everyone in the organization.
10. Managers spend an adequate amount of time out on the shop floor, where the people are.
11. Participation rates are at an all-time high, indicating that employees are highly motivated, and your marketing of health and safety initiatives is effective.
12. Employees are actively engaged in health and safety initiatives, producing tangible results for your company.

Continued

13. Your employees report high job satisfaction due to the company's commitment to their health and well being.
14. Safety is the first item on the agenda of every meeting.
15. Employees feel comfortable reporting safety issues to their supervisors.
16. Regular, detailed audits of the company's health and safety program are conducted by an external auditor.
17. Rewards and recognition of good behaviors are regularly given and serve to motivate continued health and safety performance.
18. Safety is a condition of employment.
19. Managers and supervisors respond positively to safety issues that are raised.
20. Safety is viewed as an investment, not a cost.
21. A high standard exists for accurate and detailed reporting of injuries and illnesses—nothing is swept under the rug!
22. There is a concrete definition of what success looks like for your health and safety program.
23. The organization has the willpower to make major changes when necessary.
24. Safety issues are dealt with in a timely and efficient manner.
25. All employees throughout the organization are empowered with the necessary resources and authority to find and fix problems as they see them.

(Middlesworth, n.d.)

Appendix E NASA Culture as an Organizational Flaw

"The Columbia Accident Investigation Board determined that organizational failures were as much to blame as technical failures for the Columbia accident. They identified the NASA culture as an organizational flaw that led to unintentional blind spots, group think, and silent safety. NASA's organizational structure for the Space Shuttle Program (SSP) utilized matrixed work forces and complex, geographically separated operations that hindered effective communication. The SSP's pyramid leadership structure allowed unqualified SSP managers to waive any/all technical requirements. In particular, the organizational structure and hierarchy blocked effective communication of technical problems, and was not conducive to upchanneling concerns over foam/debris strike on launch. Signals were overlooked, people were silenced, and useful information and dissenting views on technical issues did not surface at higher levels. What was communicated to parts of the organization was that foam debris strikes were not a problem. Often key decisions were made based on abbreviated PowerPoint briefings, not on thorough, data-supported research."

"Organizational structure had similar impacts at Davis-Besse. At the Davis-Besse nuclear power station, management did not follow up to ensure that industry and NRC-mandated surveillances of vessel head integrity were conducted properly. The plant executive management team apparently relied too heavily on NRC's resident inspectors to identify issues rather than conduct their own in-depth follow up of operational data, work orders, and maintenance. The NRC inspectors did not communicate plant surveillance discrepancies to their management. The Davis-Besse independent oversight function did not identify the deteriorating condition of RPV heads as evidenced by the presence of boric acid deposits over a period of years. System engineers failed to assimilate the secondary effects that were indicative of a serious problem with leakage of primary coolant. Neither the Davis-Besse Quality Assurance organization nor the independent Davis-Besse Nuclear Safety Review Board was effective in detecting or identifying adverse trends that were indicative of a deteriorating situation." (Columbia Space Shuttle Accident and Davis-Besse Reactor Pressure-Vessel Head Corrosion Event, 2005).

Appendix F: Sample Safety Policy (Managing Worker Safety and Health, n.d.)

Insert Company Name is committed to the Safety Management System that protects our employees, contractors, and visitors.

Employees at all levels of the organization are responsible and accountable for their own personal safety. Active involvement by all employees, at all times, and in every job is necessary for a successful safety system.

The leadership team will set an example and provide the necessary leadership in safety by developing safety policies and procedures, the appropriate level of training, equipment, and adequate resources to safely perform the job.

All employees will follow all plant-wide safety rules, policies, and procedures and cooperate with the leadership team in working toward improved safety in the organization.

Our goal is an injury-free workplace for all employees. By working together, we can achieve this goal (Managing Worker Safety and Health, n.d.)

Appendix G: Numerical and Descriptive Goals

Goals can be either descriptive or numerical, which can be measured in the form of numbers.

Numerical Goals

Numerical goals have the advantage of being easy to measure. Numerical goals require a full assessment of the time, resources, and budget if they are to be both attainable and comprehensive enough to serve as a milestone. The difficulty is in removing the element of luck and understanding how probability is just a roll of the dice.

We routinely see a goal about reducing injuries to a specific number. A goal of a certain number of injuries and illnesses is not feasible. It ignores the latent human error-related hazards that have not yet resulted in an injury and the "near miss" incidents that could have resulted in greater severity or other loss-producing events that by circumstance did not involve human injury.

For example, if you set a goal of reducing your injury rate or TCIR by 10%, 20%, or 30%, you may think this is a good objective. But many variables must be considered to obtain the type of objective. The reason is that the TCIR measurements is based on the number of work hours and the number of injuries, and any fluctuation of these numbers could drastically change your goals.

If a goal of zero hazards at any time is established, again, depending on the work environment, it may be difficult to reach depending on the factors of hazard identification skills and the complexity of the operation. If incentive or recognition awards are being used based on unachievable goals, employees will become disillusioned long before they can be reached.

"Goals and quotas can be so arbitrary. Improve productivity by 10%. How? Most spout out numbers with no plans to reach them. Natural fluctuations in the right direction are interpreted as success. Fluctuations in the opposite direction result in a series of fire drills that create more problems and much more frustration" (Kerr & Goode, 2008; "Managing Worker Safety and Health, Public Domain", 1994; "Safety and Health Management Basics, Module One: Management Commitment, Public Domain, Published With Permission", n.d.).

"Deming doesn't like quotas and goals because they focus on the outcome rather than the process. He argues that half the workers will be above average and half will be below—no matter what you do. If you have a stable system, then there is no use specifying a goal. You'll get whatever the system will deliver. If a goal is set beyond the capacity of the system, it will not be reached. If you don't have a stable system there is no reason to set a goal, because you have absolutely no way of telling what the system will produce" (Kerr & Goode, 2008).

Descriptive Goals

Descriptive goals are those that involve something being implemented or deployed. These goals are not numerical but can also be sufficiently inclusive and still attainable. In the review process, a numerical score is used for descriptive goals to determine the depth of their implementation and or quality.

Descriptive goals are used in establishing the overall safety management system. ANSI Z10-2005 has 21 categories, each of which could be defined as a goal that defines and is intended to assure that each element of ANSI Z10 is fully implemented ("Occupational Health and Safety Systems, ANSI/AIHA Z10", 2012). These goals are not directly quantifiable. Scoring systems can be used to establish a scale that describes the scope and depth of implementation.

For example, a goal of "vehicle drivers will receive safe driver training by December 31" is easily measurable as a numerical goal. Its success can be determined by finding out how many drivers were trained and what percent of the total is considered success. A descriptive goal would be "a driver training program will be selected, and implemented by June 31." This descriptive goal will define objectives needed—evaluating various driver safety programs, selecting the program, defining training schedules, etc. It should not be difficult to evaluate objectives and program results against this goal (Roughton & Mercurio, 2002; "Safety and Health Management Basics, Module One: Management Commitment, Public Domain, Published With Permission", n.d.).

Descriptive goals are assessed using objectives that are numerical in order to make their attainment more tangible.

Nathan tells the story of providing a supervisory safety training class covering basic criteria (hazard identification, accident investigation, inspection, etc.). This was at the request of the risk manager as the plant had a serious injury history. The audience consisted of long-term supervisors and the goal was to have all supervisors training in core safety principles. After several sessions, a supervisor spoke out. "This is a waste of time. We already know this stuff! We know what the problem is. We identify hazards, do good inspections, and spend extra time on accident inspections. Nothing is ever done about our efforts and we never get any feedback." His comments led to a very candid discussion. It turned out that a key middle manager was not on board and simply filed their work and ignored their efforts. The lesson learned spoke to the need for a S.M.A.R.T. type format (Meyer, 2003).

Nathan was charged as well since he was under the impression that the need for the class had been fully assessed and he was providing one element in an overall plan, which turned out to be the case. Never assume an assessment has been made—always ask for it or assure that one has been completed.

Appendix H: Comparison of Safety Management System Process Elements

Selected Safety Management Systems

Proposed I2P2 (Injury and Illness Prevention Program (I2P2), Adapted for Use, n.d.)	**Voluntary Protection Program (VPP)** (Voluntary Protection Program (VPP), Adapted for Use, n.d.)	**Department of Energy Voluntary Protection Program (DOE-VPP)** (Voluntary Protection Program (DOE-VPP), n.d.)	**California Model Injury and Illness Prevention Programs** (Guide to Developing Your Workplace Injury and Illness Prevention Program with checklists for self-inspection Adapted for Use, n.d.)	**OSHA Performance Evaluation Profile (canceled)** (Program Evaluation Profile (PEP), Adapted for Use, 1996)	**Building an Effective Health and Safety Management System, Partnerships in Injury Reduction, Alberta** (Building an Effective Health and Safety Management System, n.d.)
Management leadership	Management leadership	Management commitment	Management commitment/Assignment of responsibilities	Management leadership and employee participation	Management leadership and organizational commitment
Employee participation	Employee participation	Employee involvement	Safety communications		
Hazard identification and assessment	Hazard identification and assessment	Worksite analysis	Hazard assessment and control	Workplace analysis	Hazard identification and assessment
Hazard prevention and control	Hazard prevention and control	Hazard prevention and control	Safety planning, rules, and work procedures	Hazard prevention and control	Hazard control
Information and training	Information and training	Safety and health training	Safety and health training	Safety and health training	Worker competency and training
Evaluation of program effectiveness	Evaluation of program effectiveness				
			Accident investigation	Accident and record analysis	Incident reporting and investigation
				Emergency response	Emergency response planning
					Worksite inspections
					Program administration

Appendix I: Sample Safety Responsibilities Worksheet (Managing Worker Safety and Health, Illinois, Public Domain, Adapted for Use, n.d.; Roughton & Mercurio, 2002)

Everyone must be responsible for supporting the safety culture and the safety management system and for carrying out the provisions of the safety policy, goals, and objectives that pertain to the organization. These responsibilities provided in the worksheet are minimum expectations.

Job title

General statement about leadership expectations

Levels/limit of roles, responsibility, delegation, authority, and accountability (include some of the following: resources, expenditures, reported hazards, authority to shut down equipment, etc.)

List of specific goals and objectives on how the employee will be responsible for the given task and held accountable.

Appendix J: Sample Responsibilities for the Leadership Team (Managing Worker Safety and Health, Appendix 5.2, Sample Assignment of Safety And Health Responsibilities, Public Domain, Adapted for Use, n.d.; Roughton & Mercurio, 2002)

Top Leadership Team Members

The leadership team establishes and provides the leadership, budget, and resources for carrying out the organization's safety policy, goals, and objectives by:

- Establishing a policy to ensure that the organization stays in compliance with all applicable organizational, best management practices, regulatory requirements, etc.
- Providing a safe work environment and working conditions for all employees.
- Providing the leadership and resources to accomplish safety policies and procedures.
- Resolving any conflicts of production priorities where safety is a concern.
- Establishing the vision/mission, goals, and objectives.
- Supporting the safety professional and employees in their requests for information, training, other professional services, facilities, tools, and equipment needed to develop an effective safety process.
- Assigning clear and understandable responsibilities for the various aspects of the safety process. Ensuring employees with assigned responsibilities have adequate resources and authority to perform their assigned duties.
- Holding managers, supervisors, and employees accountable for assigned responsibilities by checking to ensure that they are meeting their responsibilities and by recognizing them, as appropriate.
- Evaluating the effectiveness of managers and supervisors in regard to the safety process.
- Keeping in touch with all employees, contractors, etc. who perform safety activities.
- Assisting in providing direction and authority for specific activities and visibly demonstrating involvement.
- Setting an example by demonstrating commitment (following safety rules and safe work practices).
- Ensuring that all vendors, customers, contractors, and visitors comply with the company safety policy.
- Thoroughly understanding all hazards, potential hazards, and associated risk that employees may be exposed to. Ensuring that a comprehensive program of prevention and control is established and operating properly.
- Providing a reliable system for employees to report hazardous conditions and other situations that appear hazardous. Making sure that responses to employees' reports are appropriate and timely.
- Encouraging all employees to use established hazard reporting system(s).
- Establishing an inspection system, for example, self-inspections, and reviewing the results periodically to ensure that proper and timely hazard corrections are made.

- Developing a preventive maintenance program to ensure that there is proper care and functioning of equipment and facilities.
- Reviewing loss-producing event reports to keep informed of causes and trends.
- Providing medical and emergency response systems and first aid facilities adequate for the size and hazards at the worksite.
- Requiring periodic drills to make sure that each employee knows what to do in case of an emergency.
- Developing training programs to help provide improvement and knowledge of all employees, managers, and supervisors. This training should help to recognize and understand hazards and to protect themselves and others.

Appendix K: Sample Responsibilities for Plant/Site Superintendents/ Managers (Managing Worker Safety and Health, Illinois, Public Domain, Adapted for Use, n.d.; Roughton & Mercurio, 2002)

Managers help to maintain safe working conditions in their respective area of responsibilities by:

- Providing the leadership and appropriate direction essential to maintain the safety policy as the fundamental value in the organization.
- Holding all supervisors accountable for their assigned safety roles and responsibilities, including their responsibility to ensure that employees under their direction comply with all safety policies, procedures, and rules.
- Evaluating the safety performance of supervisors, taking into account indicators of good performance, for example: safety activities leading to reducing loss-producing events, housekeeping, involvement in safety activities, a positive approach to safety problems and solutions, and a willingness to implement recommendations.
- Insisting that a high level of housekeeping is maintained, safe working procedures are established, and employees follow procedures and apply good judgment to the hazardous aspects of all tasks.
- Participating in regular inspections to observe conditions and to communicate with employees. Providing positive reinforcement and instruction during inspections and requiring the correction of any hazards.
- Actively participating in and supporting employee involvement in safety activities.
- Providing timely and appropriate follow-up on recommendations made by employee or safety committees.
- Ensuring that all new facilities, equipment, materials, and processes are reviewed for existing and potential hazards and associated risk.
- Ensuring that job hazard analyses (JHAs) are conducted periodically for all jobs, with emphasis on specific tasks so hazards can be identified and minimized or controlled.

- Ensuring that employees know about the management system and are encouraged to use the systems for reporting hazards and making safety suggestions. Ensuring that employees are protected from reprisal, their input is considered, and that their ideas are adopted when helpful and feasible.
- Ensuring that corrective action plans are developed where hazards are identified or unsafe acts are observed.
- Ensuring that all hazardous tasks are covered by specific safe work procedures or rules to minimize injury.
- Providing safety equipment and protective devices, as appropriate, and making sure employees understand how to properly use the equipment.
- Ensuring that any injured employee regardless of how minor the injury, receives prompt and appropriate medical treatment.
- Ensuring that all incidents are promptly reported, thoroughly investigated, and documented, and that safety recognition programs do not discourage reporting of incidents.
- Staying on top of incident trends and taking proper corrective action, as appropriate, to reverse trends.
- Ensuring that all employees are physically qualified to perform their assigned tasks.
- Ensuring that all employees are trained and retrained, as applicable, to recognize and understand hazards and to follow safe work procedures.
- Ensuring that supervisors conduct periodic safety meetings to review and analyze the causes of incidents and to promote free discussion of hazardous conditions and possible solutions.
- Using the safety manager to help promote aggressive and effective safety programs.
- Helping to develop and implement emergency procedures.
- Participating in safety program evaluation.

Appendix L: Sample Responsibilities for Supervision ("Managing Worker Safety and Health, Illinois, Public Domain, Adapted for Use", n.d.; Roughton & Mercurio, 2002)

Supervisors help to maintain safe working conditions in their respective area of responsibilities by demonstrating their commitment to safety efforts by:

- Being thorough and conscientious in following the safe work procedures and safety rules that apply the rule "Practice what you preach".
- Evaluating employee performance, for example, safe behavior and work methods.
- Encouraging and actively supporting employee participation. Providing positive reinforcement and recognition for individual and/or group performance.

- Maintaining up-to-date knowledge and skills required to understand specific hazards, for example, improperly functioning tools and/or equipment.
- Maintaining good housekeeping.
- Ensuring that the preventive maintenance program is being followed and that any hazards found are tracked to completion.
- Conducting frequent inspections, using a checklist, to evaluate physical conditions.
- Investigating loss-producing events thoroughly to determine the root cause and how hazards can be minimized or controlled.
- Holding employees accountable for their safety responsibilities and actions.
- Consistently and fairly enforcing safety rules and procedures.
- Discouraging short cuts.
- Providing on the job training in safe work procedures and the use and maintenance of equipment.
- Ensuring that each employee knows what to do in case of an emergency.
- Referring to the leadership team any resource needs you cannot resolve.

Appendix M: Sample Responsibilities for Employees ("Managing Worker Safety and Health, Illinois, Public Domain, Adapted for Use", n.d.; Roughton & Mercurio, 2002)

Employees should exercise care in the course of their work to prevent injuries to themselves and their coworkers by:

- Learning all safety rules and avoiding short cuts.
- Reviewing all safety educational material posted and distributed. If you do not understand something, ask questions.
- Taking personal responsibility for keeping yourself, your coworkers, and equipment free from hazards.
- Ensuring that you understand work instructions before starting a new task.
- Seeking information on hazardous substances that you may work with so that you understand the associated dangers that may exist and how to protect yourself.
- If in any doubt about the safety of a task, stopping and getting instructions from your supervisor before continuing. Ensuring that you completely understand instructions before starting work.
- Offering suggestions for reducing safety risks and discussing suggestions with supervisor.
- Ensuring that you understand what your responsibilities are in emergency situations.
- Knowing how and where medical help can be obtained.
- Reporting all loss-producing events and unsafe conditions to your supervisor or using the management system to report safety conditions.

Appendix N: Example of a Safety Committee Team Charter ("Managing Worker Safety and Health, Illinois, Public Domain, Adapted for Use", n.d.; Roughton & Crutchfield, 2008)

Safety Committee Team Name: ______________________________

Date: ___________________________________

Problem Statement: (Provides a clear statement describing the improvement opportunity or problem.)

Safety Committee Mission Statement: (Provides a clear statement describing what the committee is established to do.)

Description: (Describes the process to be improved or problem to be solved, or identifies the steps in the process from beginning to end.)

Background: (Identifies what has been happening and the importance of the safety committee.)

Scope: (Identifies the limits on the project to include if the committee will be able to pilot improvements/solutions or just make recommendations.)

Time Frame of Charter Committee Membership:
Date/Time for Committee Launch:

Process Owner:

Sponsor:

Team Leader(s):

Facilitator: Time Keeper for Meetings: Note Taker:

Committee Members:

List of Resources:

Safety Committee Member Contract:
We have read and understand the safety committee's charter. We understand our roles and responsibilities and have come to an agreement with the leadership team on the opportunity be addressed, the actions to be taken, and the limitations of the safety committee. If at any time it becomes apparent that the safety committee charter needs to be modified, we will consult the leadership team and come to an agreement on the modifications.

(Signatures of Senior Leadership and Safety Committee Members)

Adapted from Managing Worker Safety and Health, Illinois, Public Domain, Adapted for Use (n.d.).

Appendix O: Sample Activities and Results Measurements

The objective of any organization is to continue to expand the focus on process metrics (leading indicators) as the primary means of measuring the safety management system performance (i.e., loss-producing events are not the only measurement).

You want to define and implement an approach that provides measurements that are a combination of leading indicators and trailing indicators.

For example, reducing the dependence on TCIR as the trailing indicator by using performance level metrics.

The following are examples of activity metrics that can be used to monitor and assess your safety management system in a positive manner. Modify and adjust for your organization.

The Leadership Team

This category includes support for safety, perception of the safety culture, goal setting, establishing objectives based on operational procedures, and the employee perception of trust.

- Number of times leadership participated in preshift safety meetings.
- Number of physical hazard inspections with leadership involvement.
- Number of times safety action plans were reviewed with employees.
- Percent of actions completed in the safety action plan.
- Percent of projects with predesign safety reviews.
- Percent of employees, supervisors, and manager groups involved in goal-setting process.
- Percent of loss-producing reports reviewed by members of the leadership team.
- Number of safety-related walk-throughs taken by members of the leadership team.
- Percent of managers with specific safety roles.
- Number of safety goals identified in performance reviews.
- Number of safety goals that are met on schedule.
- Percent of sampled employees who can state organization safety goals and objectives.
- Number of times communication on the status of corrective actions is completed.
- Ratio of safety suggestions implemented versus those submitted and/or based on number of employees.
- Average length of time feedback is provided on a suggestion, corrective action, etc.
- Percent of safety suggestions by employees that were acted upon.

Employee Involvement

Track each employee's "meaningful" safety-related activities. The intent is to identify the level of involvement of each employee in the safety management system. An ideal safety culture will allow employees to be involved in various safety-related activities. Employees are not expected to be involved in every safety-related activity but should have safety assignments based on their skills and experience. For example, if an employee is assigned or volunteering to do hazard surveys, how many were completed in the desired time frame? If not completed, what were the circumstances that prevented completion?

- Number of hazard surveys.
- Number of investigations reviewed.
- Number of standard operating procedures.
- Number of job hazard analyses completed.
- Number of safety committee meetings attended.
- Number of preshift reviews conducted.
- Number of times safety meetings were led or presentations made.
- Number of objectives on action plan controlled by employees.

Hazard Recognition and Control

Hazard recognition and control activities should track leadership and employee involvement using elements of the safety management system:

- Number of new products/processes analyzed using a hazard identification method.
- Percent of employees involved in hazard identification.
- Percent of employees trained in hazard recognition techniques.
- Percent of employees involved in hazard correction.
- Number of hazard inspections performed.
- Number of hazards identified from the inspections.
- Percent of hazards assessed for severity using the risk guidance card.
- Average time between hazard identification and abatement completion.
- Percent of hazards corrected in a timely manner (timeliness is defined).
- Percent of physical hazards that have been identified but have not been corrected.
- Percent of tasks that have a job hazard analysis completed and developed/reviewed/revised versus expected.

Education and Training

- Number/Percent of employees receiving new employee orientation training.
- Number/Percent of employees receiving annual refresher job safety training.
- Number/Percent of transferred employees receiving refresher/area-specific safety training.
- Percent of safety training performed that is properly documented.
- Number of employees that received incident investigation training.
- Number of employees trained divided by the total number required to be trained.
- Percent of employees receiving specific ergonomic training.

Appendix P

Sample

Operator General Observations and Machine/Equipment-Specific
Daily Inspection Checklist.
(Customize this checklist for each type of machine and equipment.)

Date	Time	Signature
Location—Machine/Equipment		

Description	Yes	No	Comments
Did you attend a shift review?			
Was proper communication on problems provided by the previous shift operator?			

Note: When conducting the daily inspection of your machine/equipment, you must perform a risk assessment for each item listed using the risk guidance card.

Description	1	2	3	Comments
Are you using the proper gloves for the operation?				
Are you using the proper approved knives for the operation?				
Are all machine guards in place?				
Are all machine guards functioning properly?				
Are machines safe from loose objects/tools that could fall into moving parts?				
Is everyone kept a safe distance away from rotating parts?				
Do you have the correct tools for the machine required task?				
Did you walk around your equipment at the beginning of your shift to check for safety and housekeeping issues?				

Appendix Q: Sample Manager/Supervisor Daily/Weekly/Monthly Safety Activity Report

Month	Date of report
Department	Manager/Supervisor

Method for Improving the Safety Culture	
What have I personally done to communicate and/or train employees on the key elements of the safety management system?	
Management leadership	
Employee involvement	
Risk and hazard identification and assessment	
Hazard prevention and control	
Education and training	
Evaluation of program effectiveness	

Daily Safety Topic/Shift Review
What percentage of the required daily shift reviews were accomplished this week?
Daily topic 1. 2. 3. 4. 5. 6. 7.
One-on-one communication
How many of my direct reports did I contact this week to discuss their safety-related efforts?
Employees (list employees met with for one-on-one communications):
Safety walk-through
Were safety action plans developed for hazards identified? Number determined this week? _______ How many safety action plans remain open? _________ Hazards identified and safety plans are developed for each. If hazards with severe associated risk were identified using the risk guidance card, has the activity/operation been stopped until controls are in place? Did you inform leadership of the issue? If safety action is not completed, why not?

Specific Machine/Equipment Review
What percentage of machine/equipment reviews were completed with employees this week?
Machine/equipment reviewed:

Safety Meetings
How many safety meetings were held? ___ How many managers/supervisors attended? _______ Week One: Week Two: Week Three Week Four - Monthly combined meeting

Safety Training
What safety training was provided and what topics covered this week?

Incidents/Near Miss/Loss-Producing Event Investigations
Were incidents, near misses, and any loss-producing events and investigation reports completed and submitted to the safety department within 24 hours? If no, why not? (Note: Immediately report any event while the investigation is being completed.)

Did I participate in any other loss-producing incident reviews? List loss-producing events in my department with a brief description of the injury or damage.			
Incident	**Incident description**	**Root cause conducted**	**Action plan developed**

Appendix R: Characteristics of Good Training Programs

According to the Susan Harwood Training Grant Program (Best Practices for the Development, Delivery, and Evaluation of Susan Harwood Training Grants, Public Domains, 2010), A general review of training "best practices" reveals four characteristics that good training programs have in order to be effective:

- Accurate,
- Credible,
- Clear, and
- Practical.

Accurate

All training materials should be prepared by a qualified trainer who understands training techniques. This training must be updated as needed and facilitated by a qualified and experienced trainer with the appropriate level of training techniques and methods.

Credible

Trainers should have a general safety background or be a subject matter expert in a safety-related field. They should also have the experience of being able to train adults. In addition, they should have practical knowledge and experience in the field of safety, as well as experience in training facilitation that can contribute to a higher degree of credibility.

Clear

The training program development must not only be accurate but be believable and based on many resources to ensure that all areas of the given subject is provided. In addition, the training must be clear, concise, and understandable to the employee. If the material is only understandable to everyone with a college education or someone who understands the jargon, then the program falls short of meeting the organizational needs (Best Practices for the Development, Delivery, and Evaluation of Susan Harwood Training Grants, Public Domains, 2010).

Training developers should ensure that readability and language choices match the intended audience. Therefore, the training materials must be written in the language and grammar of the everyday speech of the employees (Best Practices for the Development, Delivery, and Evaluation of Susan Harwood Training Grants, Public Domains, 2010).

Practical

The training programs must present information, ideas, and skills that employees see as beneficial and useful in their working environment. There must be a successful transfer of knowledge that occurs when the employee can see how the safety-related information presented in a training session can be applied directly to his or her work environment.

Appendix S: Instructions for Using the Safety Management System Assessment Worksheet

("58 Attributes of Excellence of a Safety, Health and Ergonomic Program", n.d.; "Managing Worker Safety and Health", n.d.; "Managing Worker Safety and Health, Appendix 5.2, Sample Assignment of Safety And Health Responsibilities", n.d.; "Safety and Health Program Assessment Worksheet Instructions", n.d.; "Safety and Health Program Assessment Worksheet – 33A", n.d.; "Safety and Health Program Assessment Worksheet", 2007)

The Occupational Safety and Health Administration (OSHA) has several assessment worksheets that can be used as a starting point for measuring a safety management system and should be expanded to meet your organization's needs.

The following is a brief worksheet from the OSHA SHARP program that is divided into the following six safety management system elements:

- Management Leadership
- Employee Involvement
- Risk and Hazard Identification and Assessment
- Hazard Prevention and Control
- Education and Training
- Performance and Measurement

Attributes or Subelements in Each Category

Use Appendix T, Attributes of a Safety Management System to act as a guideline for discussion as you complete the safety assessment.

Attributes

Each attribute or subelement is listed as a survey question. All attributes are positive statements that the evaluator agrees with in varying degrees of continuity. The attributes are considered building blocks to effective safety management systems.

Rating Instructions

Each survey question or attribute has five possible ratings: 0, 1, 2, 3, or NA. The value for each rating is described in the following table.

Rating	Description	Example Rating for Audit System	Descriptor
0	No discernible or meaningful indication that the item is even partially in place	No or mostly no	None
1	Some portion or aspect is present, although major improvement is needed		Some
2	Item is largely in place, with only minor improvements needed	Yes or mostly yes	Most
3	Item is completely in place		All
NA	Not applicable: Must have justification in the comments box why the item is not applicable		Not applicable

The key is to use your best professional judgment when you rate each attribute based on the information obtained during the assessment. To rate an attribute; place an "X" in the box under the desired rating indicator.

Rating Cues

The bits of information obtained in the assessment are rating cues. A rating cue is a fact or perception that prompts and supports the rating of a relevant topic. Each attribute is worded as a positive statement. All rating cues will either confirm (support) or negate (deny) the statement. The cues give weight to the rating for the individual attributes.

Cues confirm or deny the existence of the attribute, the extent of the attribute, the character of the attribute, and the effect of the attribute. Cues are found in observations and measurements, interviews, and reviews of documentation. There can be multiple cues: initial cues, corroborating cues, and conflicting cues. The following table illustrates how cues are used to rate attributes.

Rating value	
0	Eliminated by a single confirming cue
1	Requires a few confirming cues and one or more negating cues
2	Requires multiple confirming cues and a few negating cues
3	Eliminated by a single negating cue

Suggestions

In reality, there is a very small gap between the 0 (zero) and the 1 (one) rating and between the 2 (two) and the 3 (three) rating. There is a large gap between the 1 (one) and the 2 (two) rating. That gap is the difference between mostly no and mostly yes.

Look for things that are done well. Reinforce these things with the leadership team and employees for their good efforts. The more you can encourage small positive steps, the greater chance that significant positive change will follow and the greater the opportunity to return and provide comprehensive assistance.

Safety Management System Assessment Worksheet

The worksheet has been adapted, maintaining its content, to follow the outline of this book.

Leadership Team	**Yes**	**No**	**NA**
Top management policy establishes a clear priority for safety and health.			
Top management considers safety and health to be a line rather than a staff function.			
Top management provides competent safety and health staff support to line managers and supervisors.			
Managers personally follow safety and health rules.			
Managers delegate the authority necessary for personnel to effectively carry out their assigned safety and health responsibilities.			
Managers allocate the resources needed to properly support the organization's safety and health system.			
Managers support fair and effective policies that promote safety and health performance.			
Managers assure that appropriate safety and health training is provided.			
Top management is involved in the planning and evaluation of safety and health performance.			
Top management values employee involvement and participation in safety and health issues.			

Employee Involvement	**Yes**	**No**	**NA**
There is an effective process to involve employees in safety and health issues.			
Employees are involved in organizational decision-making in regard to safety and health policy.			
Employees are involved in organizational decision-making in regard to the allocation of safety and health resources.			
Employees are involved in organizational decision-making in regard to safety and health training.			
Employees participate in hazard detection activities.			
Employees participate in hazard prevention and control activities.			
Employees participate in the safety and health training of coworkers.			
Employees participate in safety and health planning activities.			
Employees participate in the evaluation of safety and health performance.			

Risk and Hazard Identification and Assessment	Yes	No	NA
A comprehensive baseline hazard survey has been conducted within the past five years.			
Effective safety and health self-inspections are regularly performed.			
Effective surveillance of established hazard controls is conducted.			
An effective hazard reporting system exists.			
Change analysis is performed whenever a change in facilities, equipment, materials, or processes occurs.			
Accidents are investigated for root causes.			
Material safety data sheets are used to reveal potential hazards associated with chemical products in the workplace.			
Effective job hazard analysis is performed.			
Expert hazard analysis is performed.			
Incidents are investigated for root causes.			

Hazard Prevention and Control	Yes	No	NA
Feasible engineering controls are in place.			
Effective safety and health rules and work practices are in place.			
Applicable OSHA-mandated programs are effectively in place.			
Personal protective equipment is effectively used.			
Housekeeping is properly maintained.			
The organization is properly prepared for emergencies.			
The organization has an effective plan for providing competent emergency medical care to employees and others present at the site.			
Effective preventive maintenance is performed.			
An effective procedure for tracking hazard correction is in place.			

Education and Training	Yes	No	NA
Employees receive appropriate safety and health training.			
New employee orientation includes applicable safety and health information.			
Supervisors receive appropriate safety and health training.			
Supervisors receive training that covers the supervisory aspects of their safety and health responsibilities.			

Education and Training	Yes	No	NA
Safety and health training is provided to managers.			
Relevant safety and health aspects are integrated into management training.			

Performance and Measurement	Yes	No	NA
Workplace injury/illness data are effectively analyzed.			
Hazard incidence data are effectively analyzed.			
A safety and health goal and supporting objectives exist.			
An action plan designed to accomplish the organization's safety and health objectives are in place.			
A review of in-place OSHA-mandated programs is conducted at least annually.			
A review of the overall safety and health management system is conducted at least annually.			
Safety and health program tasks are each specifically assigned to a person or position for performance or coordination.			
Each assignment of safety and health responsibility is clearly communicated.			
An accountability mechanism is included with each assignment of safety and health responsibility.			
Individuals with assigned safety and health responsibilities have the necessary knowledge, skills, and timely information to perform their duties.			
Individuals with assigned safety and health responsibilities have the authority to perform their duties.			
Individuals with assigned safety and health responsibilities have the resources to perform their duties.			
Organizational policies promote the performance of safety and health responsibilities.			
Organizational policies result in correction of nonperformance of safety and health responsibilities.			

No = Needs major improvement; Yes = Needs minor improvement; NA = Not Applicable.

Adapted from various OSHA-related safety and health program public domain assessment worksheets.

Appendix T: Attributes of a Safety Management System ("58 Attributes of Excellence of a Safety, Health and Ergonomic Program", n.d.; "Safety and Health Achievement Recognition Program", n.d.)

The following list was adapted from OSHA's 58 Attributes of Excellence of a Safety, Health and Ergonomic Program and can be used to develop a basic assessment of the safety management system.

- Leadership Commitment
  - The positive influence of the leadership team is evident in all elements of the safety management system.
  - All employees perceive that the leadership team is exercising positive leadership and can provide examples of positive leadership.
- Authority and Resources
  - Authority to meet assigned responsibilities exists for all employees.
  - Authority is granted in writing.
  - Authority is exclusively in the control of the employee holding the responsibility.
  - Each employee believes he or she deserves the authority granted to him or her.
  - Each employee understands how to exercise the authority granted to him or her.
  - Each employee uses his or her will to exercise the authority granted to him or her.
  - Responsibilities are being met appropriately and on time.
- Resources
  - Adequate resources (methods, equipment, and funds) are provided to meet responsibilities and are available to all employees.
  - The required and necessary resources are exclusively in the control of the each employee holding the responsibility.
  - Each employee is effectively applying resources to meet responsibilities.
- Accountability
  - Each employee is being held accountable for meeting his or her safety responsibilities.
  - Methods exist for monitoring performance of responsibilities.
  - Failure to meet assigned responsibilities is addressed and results in appropriate coaching and/or negative consequences.
  - Each employee meeting or exceeding his or her responsibilities is appropriately reinforced for his or her behavior with positive consequences.
  - Data related to key elements of safety performance are accumulated and displayed in the workplace to inform all employees of progress being made.
  - Each employee and team meets to revise goals and objectives so as to facilitate continuous improvement in safety use accountability data.

- Leadership Examples
  - The leadership team knows and understands all safety rules of the organization and the safe behaviors expected.
  - The leadership team throughout the organization consistently follows all rules and behavioral expectations.
  - Employees perceive the leadership team to be consistently setting positive examples and can illustrate why they hold these positive perceptions.
  - The leadership team at all levels consistently addresses the safe behavior of others by coaching and correcting poor behavior and positively reinforcing good behavior.
  - Employees credit the leadership team with establishing and maintaining positive safety values in the organization through their personal example and attention to the behavior of others.
- Employee Involvement
  - Employees accept personal responsibility to ensure that there is a safe workplace.
  - The leadership team provides opportunities and mechanism(s) for employees to influence the safety program design and operation.
  - There is evidence of leadership team support of employee safety interventions.
  - Employees have a substantial impact on the design and operation of the safety management system.
  - There are multiple avenues for employee involvement. These avenues are well known, understood, and utilized by all employees.
  - The avenues and mechanisms for involvement are effective in reducing loss-producing events and enhancing safe behaviors.
- Structured Safety Forum That Encourages Employee Involvement
  - A written charter or standard operation procedure (SOP) outlines the safety committee structure.
  - There is a structured safety forum in the goals. All employees throughout the company are aware of the forums.
  - Safety meetings are planned, using an agenda, and remain focused on safety.
  - Safety committees and/or employees of the committee hold regularly scheduled meetings.
  - Employees on the committee are actively involved in and contributing to discussion.
  - Minutes are kept and made available to all employees.
  - The leadership team actively participates in committee and employee meetings.
  - A method exists for systematic tracking of recommendations, progress reports, resolutions, and outcomes.
  - Employees are involved in selecting topics.
  - Involvement in the committee is respected and valued in the organization.
  - The safety committee is supplemented with other forums like employee and preshift safety meetings.

  - Clear roles and responsibilities are established for the committee and the leadership team.
  - There are open lines of communication between employees and forum meetings.
  - The safety committee analyzes safety hazards to identify deficiencies in the safety management system.
  - The safety committee assesses annual review of the safety management system.
  - Reviewed results are used to make positive changes in policy, procedures, and plans.
- Hazard Reporting System
  - A system for employees to report hazards is in place and is known to all employees.
  - The system allows for the reporting of physical and specific hazards and associated risk.
  - Supervisors and managers actively encourage the use of the system and employees feel comfortable using the system in all situations.
  - The system provides for self-correction through empowerment.
  - The system involves employees in correction planning as appropriate.
  - The system provides for rapid and regular feedback to employees on the status of evaluation and correction.
  - Employees are consistently reinforced for using the system.
  - Appropriate corrective action is promptly taken on all confirmed hazards.
  - Interim corrective action is immediately taken on all confirmed hazards where delay in final correction will put employees or others at risk.
  - The system provides for data collection and display as a means to measure the success of the system in resolving identified hazards.
- Hazard Identification (Third-Party Assessment)
  - Assessments are completed at appropriate intervals, with consideration of more frequent surveys in more hazardous, complex, and highly changing environments.
  - These assessments are performed by individuals competent in hazard identification and control, especially with hazards.
  - The assessments drive immediate corrective action on items found.
  - The assessments result in optimum controls for hazards found.
  - The assessments result in updated hazard inventories.
- Hazard Controls
  - Hazard controls are in place.
  - Hazard controls are selected in appropriate priority order, giving preference to engineering controls, safe work procedures, administrative controls, and personal protective equipment.
  - Once identified, hazards are promptly eliminated or controlled.
  - Employees get involved in developing and implementing methods for the elimination or control of hazards in their work areas.
  - Employees are fully trained in the use of controls and ways to protect themselves in their work area and utilize those controls.

- Hazard Identification (Change Analysis)
  - Operational changes in space, processes, materials, or equipment at the facility are planned.
  - Planned operational changes are known to responsible management and affected workers during the planning process.
  - A comprehensive hazard review process exists and is used for all operational changes.
  - The comprehensive hazard review process involves competent qualified specialists appropriate to the hazards anticipated and the operational changes being planned.
  - Employees actively participate in the comprehensive hazard review process.
  - The comprehensive hazard review process results in recommendation for enhancement or improvement in safety elements of the planned operational change that are accepted and implemented prior to operational start-up.
- Hazard Identification (Job Hazard Analysis)
  - Members of the leadership team and employees are aware that hazards can develop in existing job tasks, processes, and/or phases of activity.
  - One or more hazard analysis systems designed to address routine job, process, or phase hazards is in place at the facility.
  - All jobs, processes, or phases of activity are analyzed using the appropriate hazards analysis system.
  - All jobs, processes, or phases of activity are analyzed when there is a change, when a loss incident occurs, or on a schedule of no more than three years.
  - All hazard analyses identify corrective or preventive action to be taken to reduce or eliminate the risk of injury or loss, where applicable.
  - All corrective or preventive actions identified by the hazard analysis process have been implemented.
  - Upon implementation of the corrective or preventive actions identified by the hazard analysis process, the written hazard analysis is revised to reflect those actions.
  - All employees of the workforce have been trained on the use of appropriate hazard analysis systems.
  - A representative sample of employees is involved in the analysis of the job, process, or phase of activity that applies to their assigned work.
  - All employees of the workforce have ready access to, and can explain the key elements of, the hazard analysis that applies to their work.
- Hazard Identification (Routine Inspection)
  - Inspections of the workplace are conducted in all work areas to identify new, reoccurring, or previously missed safety hazards and/or failures in hazard control systems.

- Inspections are routinely conducted at an interval determined necessary based on previous findings or industry experience (at least quarterly at fixed work sites, weekly at rapidly changing sites such as construction, as frequently as daily or at each use where necessary).
- Employees at all levels of the organization are routinely involved in safety inspections.
- All employees involved in inspections have been trained in the inspection process and in hazard identification.
- Standards exist that outline minimum acceptable levels of safety and are consistent with federal or state requirements as applicable.
- Standards cover all work and workplaces at the facility and are readily available to all employees.
- All employees involved in inspections have been trained on the workplace safety standards and demonstrate competence and their application to the workplace.
- All inspections result in a written report of hazard findings, where applicable.
- All written reports of inspections are retained for a period required by law or sufficient to show a clear pattern of inspections.
- All hazard findings are corrected as soon as practically possible and are not repeated on subsequent inspections.
- Statistical summaries of all routine inspections are prepared, charted, and distributed to management and employees to show status and progress of hazard elimination.

• Safety Program Review
  - The safety management system is reviewed at least annually.
  - The criteria for the review are set against established guidelines or other recognized consensus criteria in addition to the internal goals and objectives and any other specific criteria.
  - The review samples evidence over the entire organization.
  - The review examines written materials, the status of goals and objectives, records of incidents, records of training and inspections, employee and management opinion, observable behavior, and physical conditions.
  - Review is conducted by an individual (or team) determined competent in all applicable areas by virtue of education, experience, and/or examination.
  - The results of the review are documented and drive appropriate changes or adjustments in the program.
  - Identified deficiencies do not appear on subsequent reviews as deficiencies.
  - A process exists that allows deficiencies in the program to become immediately apparent and corrected in addition to a periodic comprehensive review.
  - Evidence exists that demonstrates that program components actually result in the reduction or elimination of incidents.

Other Program Review Elements

- Written Safety Policy
  - There is a policy that promotes safety.
  - The policy is available in writing.
  - The policy is straightforward and clearly written.
  - The leadership supports the safety policy.
  - The policy can be easily explained or paraphrased by everyone in the organization.
  - The policy is expressed in the context of other organizational values.
  - The policy statement goes beyond compliance to address the safe behavior of everyone in the organization.
  - The policy guides all employees in making a decision in favor of safety when apparent conflicts arise with other values and priorities.
- Clear Goals Are Established and Communicated to All Employees
  - A set of safety goals exists in writing.
  - The goals directly relate to the safety policy and vision/mission.
  - The goals incorporate the essence of "a positive and supportive safety management system integrated into the workplace culture".
  - The goals are supported by the leadership team and can be easily explained or paraphrased by others in the organization.
- Clear Objectives Are Established and Communicated to All Employees
  - Objectives exist that are designed to achieve the goals.
  - The objectives relate to opportunities for improvements, are identified in a safety management assessment, and/or when using other comparable assessment tools.
  - The objectives are clearly assigned to responsible individual(s).
  - A measurement system exists that indicates progress on objectives toward the goal.
  - The measurement system is consistently used to manage work on objectives.
  - Others can easily explain the objectives in the organization.
  - All employees know measures used to track the objective progress.
  - Members of the workforce are active participants in the objective process.
- Company-Specific Work Rules
  - All rules are clearly written and relate to the safety policy.
  - All rules address potential hazards and associated risk.
  - Safe work rules are understood and followed as a result of training and accountability.
  - The leadership team supports work rules as a condition of employment.
  - Methods exist for monitoring performance.
  - All employees, including managers, are held accountable to follow stated rules.
  - Employees have significant input to the rules.

- Employees have authority to refuse unsafe work.
- Employees are allowed access to information needed to make informed decisions.
- Documented observations demonstrate that employees at all levels are adhering to safe work rules.

Adapted for use from 58 Attributes of Excellence of a Safety, Health and Ergonomic Program and Safety and Health Achievement Recognition Program, SHARP Program, Public Domain.

Appendix U

Safety Management Perception Questionnaire

Date completed: ______________________________

Department: ______________________________

Check (✓) One

—— Line Management—exempt (such as managers and supervisors in operations and production, or whatever the main function of the organization)

—— Staff Management—exempt (such as managers, supervisors, and other exempt-level internal advisors in departments such as human resources, finance, and purchasing)

—— Employee—nonexempt (includes first-line employees and employees who supervise but are not part of management)

Advisory Note:

The above are examples for sorting answers by different organizational personnel levels. If needed, create a tailored sort that would fit your own organization and then create a customized cover page.

If the above delineation works for your organization, delete or "white-out" this advisory note.

Safety Management Perception Questionnaire

Purpose: This questionnaire is designed to identify your perceptions about your organization's current occupational safety and health program.

Directions: Answer each question to the best of your knowledge by placing a check mark (✓) in the appropriate column.

Question						
Administration: Managers' and supervisors' performance of their safety duties.	**Do not know**	**Never**	**Rarely**	**Sometimes**	**Usually**	**Always**
1. Are achievable safety goals and objectives set?	☐	☐	☐	☐	☐	☐
2. Is employee safety performance evaluated?	☐	☐	☐	☐	☐	☐
3. Are managers' and supervisors' safety performance evaluated?	☐	☐	☐	☐	☐	☐
4. Do employees receive adequate recognition for desirable or safe behavior?	☐	☐	☐	☐	☐	☐
5. Is undesirable or unsafe behavior corrected?	☐	☐	☐	☐	☐	☐
6. Do managers and supervisors understand employees' safety needs, concerns, and problems?	☐	☐	☐	☐	☐	☐
7. Do managers and supervisors take action on an ongoing basis to deal with employees' safety needs, concerns, problems?	☐	☐	☐	☐	☐	☐
8. Are managers' and supervisors' safety expectations adequately communicated to employees?	☐	☐	☐	☐	☐	☐
9. Is there adequate two-way communication (dialogue) between employees and management?	☐	☐	☐	☐	☐	☐
10. Are employees properly assigned so they can do their jobs in a safe manner?	☐	☐	☐	☐	☐	☐
11. Are there enough people to do the work safely?	☐	☐	☐	☐	☐	☐
12. Are employees given authority to take action to prevent mishaps?	☐	☐	☐	☐	☐	☐
13. Is effective leadership demonstrated by example and attitude on the part of managers and supervisors?	☐	☐	☐	☐	☐	☐
Accountability: Consequences (recognition, reward, correction, discipline) are given for meeting or not meeting management's expectations.	**Do not know**	**Never**	**Rarely**	**Sometimes**	**Usually**	**Always**
14. Are managers and supervisors held accountable for taking action to prevent mishaps?	☐	☐	☐	☐	☐	☐
15. Are managers and supervisors held accountable for their safety results?	☐	☐	☐	☐	☐	☐
16. Are employees held accountable for working in a safe manner?	☐	☐	☐	☐	☐	☐

Question						
17. Are employees held accountable for maintaining workplace order?	☐	☐	☐	☐	☐	☐

Facilities/equipment/materials (Management): Buildings, grounds, equipment, work stations, materials, tools.	**Do not know**	**Never**	**Rarely**	**Sometimes**	**Usually**	**Always**
18. Are facilities, equipment, and work stations *designed* with safety in mind?	☐	☐	☐	☐	☐	☐
19. Are there *adequate* equipment, tools, and materials to do the job in a safe manner?	☐	☐	☐	☐	☐	☐
20. Is there proper *repair* of the facility and equipment to prevent safety problems?	☐	☐	☐	☐	☐	☐
21. Is housekeeping acceptable?	☐	☐	☐	☐	☐	☐

Procedures (Management): Rules, regulations, policies, protocols, standards, guidelines.	**Do not know**	**Never**	**Rarely**	**Sometimes**	**Usually**	**Always**
22. Are safety procedures *developed?*	☐	☐	☐	☐	☐	☐
23. Are safety procedures *satisfactory?*	☐	☐	☐	☐	☐	☐
24. Are employees *trained* in safety procedures?	☐	☐	☐	☐	☐	☐
25. Are employees *sufficiently trained* in safety procedures?	☐	☐	☐	☐	☐	☐
26. Is *follow-up or refresher* safety training provided?	☐	☐	☐	☐	☐	☐
27. Are safety procedures *enforced?*	☐	☐	☐	☐	☐	☐
28. Are safety procedures *followed* by managers and supervisors?	☐	☐	☐	☐	☐	☐

Problems: Unsafe practices, unsafe conditions, and management-related safety issues associated with the work environment.	**Do not know**	**Never**	**Rarely**	**Sometimes**	**Usually**	**Always**
29. Are safety problems regularly *identified?*	☐	☐	☐	☐	☐	☐
30. Are identified safety problems properly *analyzed* in a timely manner?	☐	☐	☐	☐	☐	☐
31. Are identified safety problems promptly *corrected?*	☐	☐	☐	☐	☐	☐

Question	Do not know	Never	Rarely	Sometimes	Usually	Always
Employees: All personnel (full-time, part-time, contract) including managers, supervisors, first-line employees.						
32. Are safety procedures *known* by employees?	☐	☐	☐	☐	☐	☐
33. Are safety procedures *complied with* by employees?	☐	☐	☐	☐	☐	☐
34. Do employees know *how to recognize* safety hazards and problems?	☐	☐	☐	☐	☐	☐
35. Are available tools, equipment, materials *used* by employees?	☐	☐	☐	☐	☐	☐
36. Do employees *report* hazards?	☐	☐	☐	☐	☐	☐
Add check marks in each column. Total number = 36:						
Insert percentages from next page:						

Percentages

36 = 100%	18 = 50%
35 = 97%	17 = 47%
34 = 94%	16 = 44%
33 = 92%	15 = 42%
32 = 89%	14 = 39%
31 = 86%	13 = 36%
30 = 83%	12 = 33%
29 = 81%	11 = 31%
28 = 78%	10 = 28%
27 = 75%	9 = 25%
26 = 72%	8 = 22%
25 = 69%	7 = 19%
24 = 67%	6 = 17%
23 = 64%	5 = 14%
22 = 61%	4 = 11%
21 = 58%	3 = 8%
20 = 56%	2 = 6%
19 = 53%	1 = 3%
	0 = 0%

Safety Management Perception Questionnaire, Permission to reproduce as is.

Appendix V: Example Safety Management System Assessment and Action Plan

Planning

1. Organizational Information and History
   - Develop a problem statement that defines the scope of the issues.
   - Develop understanding of organizational culture and structure, history of situation, past attempts to resolve, barriers or obstacles to past solutions, demographics, business environment, employee issues, regulatory issues/history.
   - Develop understanding of current activities, processes, operations, etc.
   - Develop understanding of problem-solving process in use by the organization.
   - Develop understanding of current safety management system and its administration.
   - Assess the current safety culture with a survey of leadership and employees.
2. Develop Overall Action Plan
   - Vision—desired future situation; develop the mission.
   - Establish team (leadership, employees)—skill sets required.
   - Assessment design and scheduling, sequencing actions.
   - Communication (media, contacts, formats, timing), time, budget.
   - Logistics – material, equipment, travel, etc.
3. Assess the Safety Management System
   - Review the Safety Management System and complete a gap analysis. Review each of its elements and assess the effectiveness of its implementation and success in reducing hazards and associated risk. See Chapter 14

Measurement and Analysis

- Risk Assessment and Analysis—severity, frequency, nature of exposures.
- Hazard Assessment and Analysis based on observations, inspections, surveys, etc.
- High-Hazard Activities from Risk and Loss Analysis review Job Hazard Analyses
- Loss-Producing Incidents—employee, facility/equipment/vehicles damage, third party.
- Organizational Change—technology, tasks, structure, human-social, environment, goals (French & Bell, 1984).
- High Severity Hazard Communications and Feedback.

Core Hazard and Risk Control Activities

- Risk and Occupational Health and Safety.
  - Review and observation of specific programs and processes to control high-risk situations and activities:
    - Fire/Explosion.
    - Electrical.
    - Motor vehicles, heavy equipment.
    - Confined space entry.
    - Lockout/Tag out.
    - Fall protection.
    - High pressure equipment/operations.
    - Hot work, welding.
    - Powered industrial trucks.
    - Housekeeping.
    - Hazard communication.
    - Personal protective equipment.
    - Respirator program.
    - Ergonomics.
    - Hearing conservation program.
    - Radiation/Lasers.
    - Bloodborne pathogens.
    - General life safety.
    - Others as identified.
  - Assist organization in developing or modifying control programs.
- Industry Specific—review specialty programs and process.

Tools for Use in Safety Management System

("Volume 2: Human performance tools for individuals, work teams, and management, human performance improvement handbook", 2009)

- Individuals.
  - Task preview.
  - Job-site review.
  - Questioning attitude—at the activity level.
  - Questioning attitude—work planning and preparation.
  - "Pause when unsure".
  - Procedure use and adherence.
  - Validate assumptions.

  - Signature (sign-off).
  - Effective communications.
  - Place keeping.
- Work Teams.
  - Prejob briefing.
  - Technical task prejob briefing.
  - Checking and verification practices.
    - Peer checking.
    - Concurrent verification.
    - Independent verification.
    - Peer review.
  - Flagging (marking/identifying).
  - Turnover—shift change information.
  - Post-job review—field review.
  - Technical Task Post-Job Review.
  - Project planning.
  - Problem solving.
  - Decision-making.
  - Project review meeting.
  - Vendor oversight.
- Management.
  - Benchmarking.
  - Observations.
  - Self-assessments.
  - Performance indicators.
  - Independent oversight.
  - Work product review.
  - Investigating events triggered by human error.
  - Operating experience.
  - Change management.
  - Reporting errors and near misses.
  - Culpability decision tree.
  - Employee surveys.
    - Organizational safety climate surveys.
    - Human performance gap analysis.
    - Job-site conditions self-assessment questionnaire.

Appendix W: Google Search—Narrowing the Search Terms

("Advance Search", n.d.; "Google, Inside Search", n.d.)

Explicit Phrase

- "Safety Culture".

Exclude Words

- "-" (dash) sign in front of the word—Safety Culture—training.

Site Specific Search

- Site: www.osha.gov.

Similar Words and Synonyms

- Use the "~" in front of the word.
- "Safety Culture" ~safety climate.

Specific Document Types

Looking for documents related to safety culture.

- "safety culture" filetype:ppt.
- "safety culture" filetype:doc.

This OR That

- The OR has to be capitalized.
- Safety Culture OR climate.

Word Definitions

Example:

- Define: Safety Culture.

Bibliography

Appendix A

Dictionary of Occupational Titles. (n.d.). Photius Coutsoukis and Information Technology Associates. Retrieved from http://bit.ly/XlUQcJ.

Appendix C

The Safety Culture Policy Statement. (2012, November). US Nuclear Regulatory Commission. Retrieved from http://1.usa.gov/ZuAPCN.

Appendix D

Middlesworth, M. (n.d.). 25 Signs You Have an Awesome Safety Culture, Permission to reprint. Ergonomics Plus. Retrieved from http://bit.ly/Uw9C2u.

Appendix E

Columbia Space Shuttle Accident and Davis-Besse Reactor Pressure-Vessel Head Corrosion Event. (2005, July). U.S. Department of Energy Action Plan Lessons Learned, Public Domain. Retrieved from http://1.usa.gov/YOWYft.

Appendix F

(Managing Worker Safety and Health, n.d.) Managing Worker Safety and Health. (n.d.). Illinois OSHA Onsite Safety & Health Consultation Program, Public Domain, Adapted for Use. Retrieved from http://bit.ly/WTsneh

Appendix G

Kerr, J., & Goode, W. B. (June 2008). Deming's Point 11, Eliminate Numerical Quotas/Goals. Joe and Wanda on Management. Retrieved from http://begoodventures.com/joeandwanda/?p=123.

Managing Worker Safety and Health, Public Domain. (November 1994). U.S. Department of Labor, Office of Cooperative Programs Occupational Safety and Administration (OSHA).

Meyer, P. J. (2003). What would you do if you knew you couldn't fail? Creating SMART Goals. Attitude Is Everything: If You Want to Succeed Above and Beyond. Meyer Resource Group, Incorporated.

Occupational Health and Safety Systems. (2012). ANSI/AIHA Z10 The American Industrial Hygiene Association.

Roughton, J. E., & Mercurio, J. J. (2002). *Developing an effective safety culture: A leadership approach*. Butterworth-Heinemann.

Safety and Health Management Basics, Module One: Management Commitment, Public Domain, Published With Permission. (n.d.). Oregon OSHA.

Appendix H

Building an Effective Health and Safety Management System. (n.d.). Partnerships in Health and Safety program, Partnerships in Injury Reduction (Partnerships). Retrieved from http://humanservices.alberta.ca/documents/whs-ps-building.pdf.

Guide to Developing Your Workplace Injury and Illness Prevention Program with checklists for self-inspection Adapted for Use. (n.d.). State of California, Department of and Industrial Relations. Retrieved from http://bit.ly/Xewi5m.

Injury and Illness Prevention Program (I2P2), Adapted for Use. (n.d.). Occupational Safety and Health Administration (OSHA). Retrieved from http://1.usa.gov/lIszWK.

Program Evaluation Profile (PEP), Adapted for Use. (1996). Occupational Safety and Health Administration (OSHA). Retrieved from http://1.usa.gov/VuKM1C.

Voluntary Protection Program (DOE-VPP). (n.d.). Department of Energy. Retrieved from http://hss.doe.gov/HealthSafety/wsha/vpp/index.html.

Voluntary Protection Program (VPP), Adapted for Use. (n.d.). Occupational Safety and Health Administration (OSHA). Retrieved from http://1.usa.gov/T0j6EW.

Appendix I

Managing Worker Safety and Health, Illinois, Public Domain, Adapted for Use. (n.d.). Illinois Onsite Safety & Health Consultation Program. Retrieved from http://on.mo.gov/13MAj83.

Roughton, J. E., & Mercurio, J. J. (2002). *Developing an effective safety culture: A leadership approach, Adapted for Use*. Butterworth-Heinemann.

Appendix J

Managing Worker Safety and Health, Appendix 5.2, Sample Assignment of Safety And Health Responsibilities, Public Domain, Adapted for Use. (n.d.). Missouri Department of Labor and Industrial Relations. Retrieved from http://on.mo.gov/VQwB9z.

Roughton, J. E., & Mercurio, J. J. (2002). *Developing an effective safety culture: A leadership approach, Adapted for Use*. Butterworth-Heinemann. Retrieved from http://amzn.to/X8Gaz82002.

Appendix K

Managing Worker Safety and Health, Illinois, Public Domain, Adapted for Use. (n.d.). Illinois Onsite Safety & Health Consultation Program.

Roughton, J. E., & Mercurio, J. J. (2002). *Developing an effective safety culture: A leadership approach, Adapted for Use*. Butterworth-Heinemann.

Appendix L

Managing Worker Safety and Health, Illinois, Public Domain, Adapted for Use. (n.d.). Illinois Onsite Safety & Health Consultation Program. Retrieved from http://www.illinoisosha.com/pdf/books/book-managing.pdf.

Roughton, J. E., & Mercurio, J. J. (2002). *Developing an effective safety culture: A leadership approach, Adapted for Use*. Butterworth-Heinemann.

Appendix M

Managing Worker Safety and Health, Illinois, Public Domain, Adapted for Use. (n.d.). Illinois Onsite Safety & Health Consultation Program. Retrieved from http://www.illinoisosha.com/pdf/books/book-managing.pdf.

Roughton, J. E., & Mercurio, J. J. (2002). *Developing an effective safety culture: A leadership approach, Adapted for Use*. Butterworth-Heinemann.

Appendix N

Managing Worker Safety and Health, Illinois, Public Domain, Adapted for Use. (n.d.). Illinois Onsite Safety & Health Consultation Program. Retrieved from http://www.illinoisosha.com/pdf/books/book-managing.pdf.

Roughton, J., & Crutchfield, N. (2008). *Job hazard analysis: A guide for voluntary compliance and beyond, chemical, petrochemical & process*. Elsevier/Butterworth-Heinemann.

Appendix R

Best Practices for the Development, Delivery, and Evaluation of Susan Harwood Training Grants. (2010, September). Occupational Safety and Health Administration (OSHA), Public Domain, Permission to Reprint, Modify, and/or Adapt as necessary. Retrieved from http://1.usa.gov/Z5nKoy.

Appendix S

58 Attributes of Excellence of a Safety, Health and Ergonomic Program. (n.d.). Oregon Occupational Safety and Health Division (Oregon OSHA), Public Domain, Permission to Reprint, Modify, and/or Adapt as necessary. Retrieved from http://bit.ly/W4ZY8A.

Managing Worker Safety and Health. (n.d.). Illinois OSHA Onsite Safety & Health Consultation Program, Public Domain, Adapted for Use. Retrieved from http://bit.ly/WTsneh.

Managing Worker Safety and Health, Appendix 5.2, Sample Assignment of Safety And Health Responsibilities. (n.d.). Missouri Safety & Health Consultation Program (OSHA), Public Domain, Adapted for Use. Retrieved from http://on.mo.gov/VQwB9z.

Safety and Health Program Assessment Worksheet. (2007). Oregon occupational safety and health Division (Oregon OSHA), Public Domain, Permission to Reprint, Modify, and/or Adapt as necessary. Retrieved from http://bit.ly/XLNphg.

Safety and Health Program Assessment Worksheet – 33A. (n.d.). Department of Labor and Industries WISHA OSHA, Public Domain, Adapted for Use. Retrieved from http://1.usa.gov/VRfgjK.

Safety and Health Program Assessment Worksheet Instructions. (n.d.). Occupational Safety and Health Administration (OSHA), Public Domain, Adapted for Use. Retrieved from http://1.usa.gov/Y95L9T.

Appendix T

58 Attributes of Excellence of a Safety, Health and Ergonomic Program. (n.d.). Oregon Occupational Safety and Health Division (Oregon OSHA), Public Domain, Permission to Reprint, Modify, and/or Adapt as necessary. Retrieved from http://bit.ly/W4ZY8A.

Safety and Health Achievement Recognition Program. (n.d.). Oregon Occupational Safety and Health Division (Oregon OSHA), SHARP Program, Public Domain, Permission to Reprint, Modify, and/or Adapt as necessary. Retrieved from http://bit.ly/XHf9nd.

Appendix U

Safety Management Perception Questionnaire, Permission to reproduce as is.

Notes from Authors of this questionnaire: This questionnaire may be reproduced without change in its entirety for non-commercial purposes without permission of Lapidus & Waite. Any manipulation or other utilization of this material must be authorized by Lapidus & Waite in writing. Copyright © 1995–2013, Robert A. Lapidus & Michael J. Waite. All Rights Reserved.

Appendix V

French, W., & Bell, C. (1984). *Organizational development: Behavioral science interventions for organizational movement.* Englewood Cliffs, NJ: Prentice-Hall.

Volume 2: *Human performance tools for individuals, work teams, and management human performance improvement Handbook*, 2009. U.S. Department of Energy. Retrieved from http://1.usa.gov/11Ex7vE.

Appendix W

Advance Search. (n.d.). Google. Retrieved from http://bit.ly/15lPe8Y.

Google, Inside Search. (n.d.). Retrieved from http://bit.ly/VQNT7j.

Glossary

ABC Analysis A root cause analysis tool for understanding why a person uses a specific behavior. ABC stands for Antecedent, Behavior, and Consequence.

ABSS (Activity-Based Safety System) Specific activities performed on a routine basis as part of the safety management system.

Administrative Controls Processes developed by the employer to control hazards not eliminated by engineering controls (e.g., safe work policies, practices and procedures, job scheduling or rotation, and training).

Antecedent Something that triggers behavior, including things that a person sees, hears, and thinks about prior to taking an action. Antecedents influence behavior to the extent that they consistently predict consequences.

Artifacts Visible organizational structures and processes. These are the tangible things you can see as you move about within facilities or review the environment of the organization. For example, general housekeeping that includes appearance of equipment and its maintenance, the visible work environment that includes the type and design of facilities, warning signage, written safety materials, policies, training rooms, and dress codes all indicate what the underlying assumptions and beliefs are in the organization.

Behavior An observable act. Safety behaviors are classified as safe or at-risk. Safe behavior is something that a person does to protect themselves from injury; at-risk behavior exposes a person to injury.

Charter The document that will provide guidance over the life of a committee and gives the reason for the establishment of the committee.

Cognitive Dissonance "In modern psychology, cognitive dissonance is the feeling of discomfort when simultaneously holding two or more conflicting cognitions: ideas, beliefs, values, or emotional reactions". (Cognitive dissonance, n.d.)

Consequence An outcome from behavior. Consequences control behavior; in other words, we generally behave the way we do because of what we expect the consequences to be.

Continuous Improvement Always striving to innovate, implement, and improve on current conditions. (Building an Effective Health and Safety Management System, 1989)

Culture A "pattern of shared basic assumptions that the group learned as it evolved its problems of external adaptation and internal integration. Over time this pattern of shared assumptions has worked well enough to be considered valid and, therefore, to be taught to new members as the correct way you perceive, think, and feel in relation to those problems". (Schein, 2004)

Curation The act of curating, organizing, and maintaining a collection of artworks or artifacts. (Curation, n.d.)

DMAIC (Define, Measure, Analyze, Improve, and Control) A more advanced approach than PDCA.

Dyad The connection between two nodes.

Emergent Property Any unique property that "emerges" when component objects are joined together in constraining relations to "construct" a higher-level aggregate object, a novel property that unpredictably comes from a combination of two simpler constituents. (Emergent Property, n.d.)

Employees This term is used to refer to front-line workers. Anyone who works in an organization (e.g. the leadership team, which includes managers, supervisors, and employees in the operation.) (Building an Effective Health and Safety Management System, 1989)

Engineering Controls Preferred method of hazard control if elimination is not possible; physical controls implemented at the design, installation, or engineering stages (e.g., guards, auto shutoff, etc.).

Espoused Values The content of the various strategies, goals, and core philosophies that are used by the leadership to guide the organization. What does the organization say about itself? We see these espoused values in the form of safety slogans, safety mission statements, and various commitments made with regard to safety.

Goals "General guidelines that explain what you want to achieve. They are usually long-term in nature and represent your overall vision". (Define Goals and Objectives, Step 2, 2003)

Habits An acquired behavior pattern regularly followed until it has become almost involuntary.

Hazard A situation, condition, or behavior that has the potential to cause an injury or loss. (Building an Effective Health and Safety Management System, 1989)

Hazard Assessment A process used to identify and evaluate the health and safety hazards associated with job tasks and provides a method to prioritize health and safety hazards. (Building an Effective Health and Safety Management System, 1989)

Hazard Control A method used to eliminate or control loss. (Building an Effective Health and Safety Management System, 1989)

Hub A node that links multiple nodes.

Incident A preventable, undesired, and unexpected event that results, or has the potential to result, in physical harm to a person or damage to property (loss or no loss). (Building an Effective Health and Safety Management System, 1989)

Job Hazard Analysis – aka Job Safety Analysis, Task Analysis Provides task-specific safety information regarding hazards present and preventive measures for avoiding potential incidents.

Leadership "The process of influencing people by providing purpose, direction, and motivation to accomplish the mission and improve the organization". (Army Leadership, ADP 6-22, 2012)

Leading Indicator The performance drivers that communicate how outcome measures are to be achieved; a measure that is upstream of incidents. Leading indicators are also used to measure activities and whether or not things are getting done. Leading indicators provide data to assess aspects of the safety system that need to be fixed. Examples include process element performance levels, safety perception survey results, number/percent of hazards identified/corrected, number of one-on one-contacts completed, number/percent of employees involved in meaningful incident prevention activities, contact rate, percent of those safe, comment quality, number of action plans completed, number/percent of near misses reported/investigated/preventive actions completed. Establish goals and measure performance for identified leading indicators.

Line Management Managers who have responsibility over production, including positions such as supervisor, superintendent, operations manager, and plant manager. At nontraditional workplaces, production employees with supervisory responsibilities (e.g., leads, lead hands) are also considered part of line management.

Loss-Producing Event Injury, illness, property damage, or any other loss.

Management Salaried/exempt employees including line managers (e.g. supervisors, superintendents, plant managers) and support departments such as Human Resources, Purchasing, Information Resources, safety staff, etc.

Meme "A unit of information in a mind whose existence influences events such that more copies of itself get created in other minds". (Brodie, 2009)

Mental Model The "images, assumptions, and stories which we carry in our minds of ourselves, other people, institutions, and every aspect of our World". (Senge, 2006)

Near Miss An undesired event that under slightly different circumstances could have resulted in personal harm, property damage, or loss. Also referred to as an incident. (Building an Effective Health and Safety Management System, 1989)

Networking The exchange of information or services among individuals, groups, or institutions; specifically, the cultivation of productive relationships for employment or business. (Networking, n.d.)

Node A specific group, department, or whatever you are trying to map a social network. You are a "node" within an interconnected group of nodes.

Objectives The specific paths you will follow to achieve a goal. They are statements of results or performance. They are short positive steps along the way to your organization's goals. (Safety Pays, n.d.)

PDCA (plan–do–check–act or plan–do–check–adjust) An iterative four-step management method used in business for the control and continuous improvement of processes and products.

Perception The process by which people select, organize, interpret, retrieve, and respond to information from the world around them. (Hunt, Osborn, & Schermerhorn, 1997)

Policy Statement A document that outlines the site's commitment to safety, health, and the environment and establishes basic expectations.

Politics Political affairs or business; especially competition between competing interest groups or individuals for power and leadership (as in a government). (Politics, n.d.)

Risk The chance or probability of injury, damage, or loss. It is the combination of severity and frequency/ exposure to hazards. (Building an Effective Health and Safety Management System, 1989)

Risk Perception the subjective judgment that people make about the characteristics and severity of a risk. (Risk perception, n.d.)

Root Causes The most basic causes that can be reasonably identified and fixed, and for which effective recommendations for recurrence can be generated. Tools such as 5-Whys, ABC analysis, and fishbone diagramming can be used to identify root causes for incidents.

Safety Action Plan A document that contains the site's safety and health goals and objectives, with responsibilities and target dates established for each objective. Safety action plans are commonly developed on an annual basis. Some sites also update their safety action plans as new goals and objectives are developed during the year.

Safety Management System A comprehensive business management system designed to manage safety elements in a workplace. (Safety Management Systems, n.d.)

Safety Perception Survey A tool that samples the perceptions of employees and managers, allowing for comparisons (e.g., between employees, supervisors, and top-level managers) and highlights areas where strengths and weaknesses in the safety management system

Scope Drift A rifle sighting term describing the condition where your telescope is not aligned with the rifle barrel. When aiming at a target through an unaligned rifle scope you will miss your target every time.

Six Sigma A methodology for pursuing continuous improvement in customer satisfaction and profit; a management philosophy attempting to improve effectiveness and efficiency. It is based on having not more than 3.4 defects per 1,000,000 opportunities. (Six Sigma Measure Phase, Data Collection Plan and Data Collection, n.d.)

Skills Assessment A process by which the skills necessary to safely perform roles and responsibilities for each position are assessed and documented. This also includes an assessment of skills needed by those involved in the safety activities (i.e., hazard inspections, one-on-one contacts).

S.M.A.R.T. Objectives Objectives defined in specific terms that are measurable, achievable, realistic, and timely.

Social Network Analysis (SNA) A tool that identifies and portrays the details of a network structure. It shows how a "…networked organization behaves and how that connectivity affects its behavior. SNA allows analysts to assess the network's design, how its member may or may not act autonomously, where the leadership resides or how it is distributed among members, and how hierarchical dynamics may mix or not mix with network dynamics" (Petraeus & Amos, 2006).

Supervisor A front-line manager with direct responsibility for employees. At nontraditional workplaces, production employees with supervisory responsibilities (e.g., leads, lead hands) are also considered supervisors.

Trailing Indicator A safety measure that is downstream of incidents. Trailing indicators are often measures of failure, such as total case incident rate (TCIR), lost workday case incident rate (LWCIR), and incurred workers' compensation costs. A trailing indicator is an outcome measure and they are not diagnostic; they do not tell you what part of the system is broken.

Upstream Accident Prevention Activities Proactive incident prevention activities that focus on those things that have the greatest impact on the prevention of loss producing incidents.

Bibliography

Army Leadership, ADP 6-22. (2012, August). Army Knowledge Online. Headquarter, Depart of the Army, Approved for Public Release; Distribution is Unlimited, Public Domain. Retrieved from http://bit.ly/Vs6YyK.

Brodie, R. (2009). *Virus of the Mind: The New Science of the Meme.* Hay House.

Building an Effective Health and Safety Management System. (1989). *Partnerships in Health and Safety program, Partnerships in Injury Reduction (Partnerships).* Reprinted/Modified and/or adapted with Permission. Retrieved from http://bit.ly/WsHteC.

Cognitive dissonance. (n.d.). Wikipedia, the free encyclopedia. Retrieved from http://bit.ly/105nAsv

Curation. (n.d.). Wiktionary, a Wiki-Based Open Content Directory. Retrieved from http://bit.ly/VQ95KJ

Define Goals and Objectives, Step 2. (2003). *Michigan State Government, public domain.* Michigan State Government, Public Domain. Retrieved from http://1.usa.gov/WlIGEa.

Emergent Property. (n.d.). Dictionary.com. Retrieved from http://bit.ly/YQRNKT

Hunt, J., Osborn, R., & Schermerhorn, J. (1997). *Basic Organizational Behavior.* New York: John Wiley and Sons.

Networking. (n.d.). Merriam-Webster, Incorporated. Retrieved from http://bit.ly/WYIbMF

Petraeus, D., & Amos, J. F. (2006, December). *The US Army/Marine Corps Counterinsurgency Field Manual, FM 3-24, MCWP 3-33.5., Approved for public release.* Headquarters Department Of The Army, Public Domain. Retrieved from http://bit.ly/105MvMj

Politics. (n.d.). Merriam Webster.com. Retrieved from http://bit.ly/15HqUzF

Risk perception. (n.d.). Wikipedia, the free encyclopedia. Retrieved from http://bit.ly/YCUcrI

Safety Management Systems. (n.d.). Wikimedia Foundation, Inc., Retrieved from http://bit.ly/15lZVIH

Safety Pays. (n.d.). *Oklahoma Department of Labor.* Oklahoma Safety and Health Management, Public Domain. Retrieved from http://bit.ly/14o9Duz

Schein, E. H. (2004). *Organizational Culture and Leadership.* The Jossey-Bass Business & Management Series. John Wiley & Sons.

Senge, P. M. (2006). *The Fifth Discipline: The Art & Practice of The Learning Organization. A Currency book.* Crown Publishing Group.

Six Sigma Measure Phase, Data Collection Plan and Data Collection. (n.d.). tutorialspoint, Permission to Reprint/Modify/Adapt for Use. Retrieved from http://www.tutorialspoint.com/six_sigma/index.htm

Index

Note: Page numbers with "f" denote figures; "t" tables; "b" boxes.

I

J

L

M

N

O

Printed and bound by CPI Group (UK) Ltd, Croydon, CR0 4YY
03/10/2024
01040319-0012